INORGANIC SOLVENTS AND REAGENTS

INORGANIC SOLVENTS AND REAGENTS

By
Raju Monga

2016

SBS Publishers & Distributors Pvt. Ltd.
New Delhi

ISBN 13 : 9789380090924

First Published in 2016

Published by:

SBS PUBLISHERS & DISTRIBUTORS PVT. LTD.

2/9, Ground Floor, Ansari Road, Darya Ganj,

New Delhi - 110002,

INDIA

Tel: 0091.11.23289119 / 41563911

Email: mail@sbspublishers.com

www.sbspublishers.com

Preface

Inorganic chemistry is the study of the synthesis and behavior of inorganic and organometallic compounds. An inorganic non-aqueous solvent is a solvent other than water that is not an organic compound. Solvents are a class of chemicals that are used for the purposes of dissolving, extracting, or suspending materials, usually without altering their chemical make-up. Solvents are usually liquids, though they can also be in the form of a gas or solid, and generally work on the principle that the solvent will have similar properties to the substance it is attempting to dissolve, extract, or suspend. A solvent is usually a liquid but can also be a solid or a gas. The quantity of solute that can dissolve in a specific volume of solvent varies with temperature. Common uses for organic solvents are in dry cleaning, as paint thinners, as nail polish removers and glue solvents in spot removers, in detergents and in perfumes. Solvents find various applications in chemical, pharmaceutical, oil and gas industries, including in chemical syntheses and purification processes. A reagent is a substance or compound that is added to a system in order to bring about a chemical reaction, or added to see if a reaction occurs. Although the terms reactant and reagent are often used interchangeably, a reactant is more specifically a substance that is consumed in the course of a chemical reaction. Solvents, although they are involved in the reaction, are usually not referred to as reactants. In organic chemistry, reagents are compounds or mixtures, usually composed of inorganic or small organic molecules, which are used to effect a transformation on an organic substrate. Examples of organic reagents include the Collins reagent, Fenton's reagent, and Grignard reagent. There are also analytical reagents which are used to confirm the presence of another substance.

Editor

Contents

Chapter 1

GREEN CHEMISTRY: NEW SYNTHESIS OF SUBSTITUTED CHROMENES AND BENZOCHROMENES VIA THREE-COMPONENT REACTION UTILIZING ROCHELLE SALT AS NOVEL GREEN CATALYST

Awatef Mohamed El-Maghraby

Department of Chemistry, Faculty of Science, South Valley University, Qena 83523, Egypt Correspondence should be addressed to Awatef Mohamed El-Maghraby; awatefelmaghraby@yahoo.com

ABSTRACT

Substituted 2-amino-4-aryl-7-hydroxy-4H-chromene-3-carbonitriles (6), 2-amino-4-aryl-4H-benzo[h]chromene-3-carbonitriles (7), and 3-amino-1-aryl-1H-benzo[f]chromenes-2-carbonitriles (8) were prepared, in good yields, via one-pot three-component reactions of aromatic aldehydes (1), malononitrile (2), and resorcinol (3) or α-naphthol (4) or β-naphthol (5) in refluxing ethanol or water in the presence of Rochelle salt as novel green heterogeneous and reusable catalyst.

INTRODUCTION

Aminochromenes represent an important class of organic compounds being the main components of many naturally occurring products. In addition, they are valuable precursors used for the synthesis of cosmetics, pigments [1], and potentially biodegradable agrochemicals [2]. Furthermore, fused chromenes are important constituents of pharmacologically active compounds, as these systems have displayed a broad spectrum of biological activities such as antimicrobial [3, 4], mutagenicity [5], antiviral [6], antiproliferative [7], sex pheromonal [8], antitumor [9], central nervous system (CNS) activities [10], and inhibitors of influenza virus sialidases [11, 12]. One-pot multicomponent reactions have received considerable attention in synthetic chemistry as they can produce target products from readily available starting materials in one reaction step without isolating the intermediates thus reducing reaction times, labor cost, and waste production [13].

In addition, water has emerged as a versatile solvent for organic reactions in the last two decades since it is readily available, inexpensive, environmentally benign, neutral, and a natural solvent [14, 15]. For these reasons, water has been used for MCRs as well [16, 17]. MCRs in water are of outstanding value in organic synthesis and green chemistry [16, 17]. Aminochromenes have been prepared by heating a mixture of malononitrile, aldehyde, and activated phenol or naphthols in refluxing DMF or acetonitrile in the presence of hazardous organic bases such as piperidine and triethylamine [18, 19].

Although different synthetic methods to prepare these heterocyclic systems have been reviewed [20–36], to the best of our knowledge, the use of clean solvents in combination with heterogeneous and reusable catalysts to synthesize these systems has not been largely reported [13, 25]. In continuation of our work concerning the synthesis and biological evaluation of new heterocycles [37–40] and aiming to explore the efficiency of Rochelle salt (R. S.) as a novel green heterogeneous and reusable catalyst in the one-pot reactions in the organic syntheses, we report herein our results on the utility of Rochelle salt (R. S.) as a green catalyst in the three-component condensations between aromatic aldehydes, active methylene reagents, and activated phenols.

RESULT AND DISCUSSION

Our synthesis began with the reaction of a mixture of aromatic aldehydes 1a–h, malononitrile (2), and resorcinol (3) in refluxing ethanol containing a catalytic amount of Rochelle salt to give 2-amino-4-aryl-7-hydroxy-4H-chromene-3-carbonitriles 6a–h (Scheme 1), (Table 1). The structures of the isolated products 6a-h were confirmed on the basis of their elemental analyses and spectral data. The IR spectrum of the reaction products showed the presence of both OH and NH_2 functions at 3496–3320 cm^{-1} and a cyano at ~2200 cm^{-1}. The ^1H NMR spectra displayed the presence of two singlets at δ = 6.37–6.87 and 9.39–9.61 ppm attributable to the amino (NH_2) and OH groups, respectively. Furthermore, the ^1H NMR gave strong evidence for the formation of compounds 6a–h. The data confirmed the presence of the H-4 proton at δ = 4.57–4.92 ppm, in addition to the signals of aromatic protons and other groups (see Table 3). Moreover, their structures were supported by both correct mass spectra and analytical data, which were compatible with the proposed structures for compounds 6a-h.

Table 1: Yields and melting points of the synthesized compounds 6–8

Compound	R	Yield %	Observed m.p.	Reported m.p.	References
6a	2,3-Dimethoxy	85	250	—	—
6b	2,5-Dimethoxy	85	170	—	—
6c	3-Benzyloxyphenyl	86	230	—	—
6d	4-MeO	86	112–114	111-112	[30, 32]
6e	4-Cl	90	184	164	[30]
6f	H	80	228	231	[32]
6g	3,4,5-Trimethoxy	85	210	205	[34]
6h	3-NO_2	84	170	169-170	[32]
7a	2,3-Dimethoxy	85	260	—	—
7b	2,5-Dimethoxy	75	240	—	—
7c	3,4,5-Trimethoxy	80	190	189	[35]
7d	4-Cl	90	235	232	[20]
7e	3-NO_2	82	210	212	[20]
7f	H	90	205	205	[20]
7g	4-MeO	84	204	205	[35]
8a	2,3-Dimethoxy	80	240	—	—
8b	2,4-Dimethoxy	80	205	—	—
8c	2,5-Dimethoxy	75	206	—	—
8d	3,4,5-Trimethoxy	85	209	—	—
8e	3-NO_2	81	195	190	[29]
8f	4-MeO	92	260	255	[26]
8g	H	90	280	278–280	[20]
8h	4-Cl	83	187	191	[29]

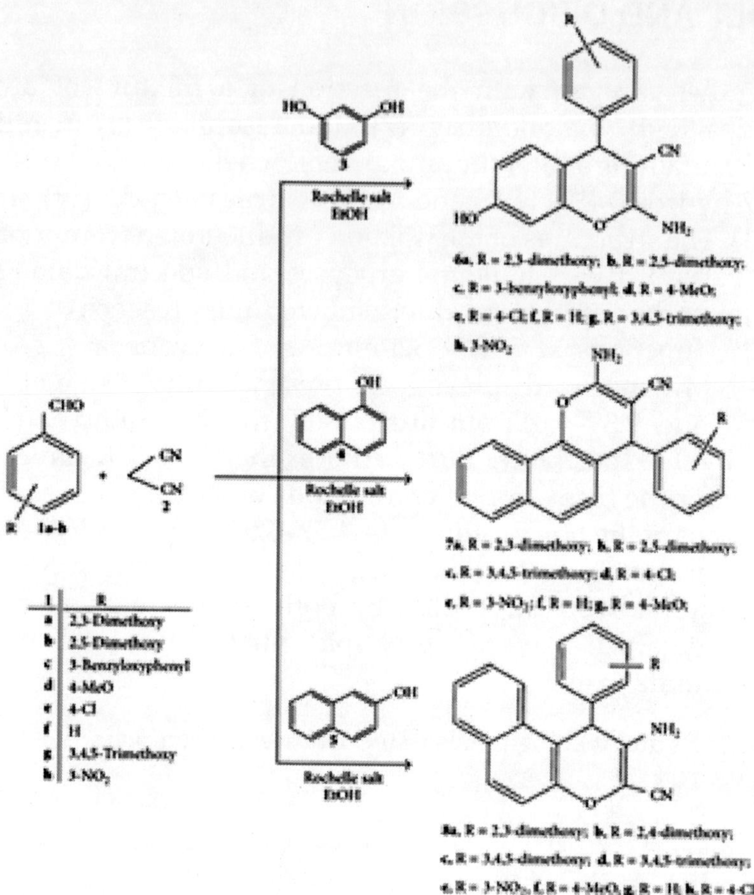

Scheme 1: Synthesis of compounds **6–8**.

Since our interest is in developing a synthetic approach with a view to synthesize new derivatives of the interesting aminochromenes, α-naphthol (4) and β-naphthol (5), good precursors for this purpose, were thus investigated. Reacting a mixture of aromatic aldehydes 1a–h, malononitrile (2) and α-naphthol (4) or β-naphthol (5), under the same reaction conditions, gave the 2-amino-4-aryl-4H-benzo[h]chromene-3-carbonitriles 7a-g and 3-amino-1-aryl-1H-benzo[f]-chromenes-2-carbonitriles 8a-h, respectively, in good yields. The structures of these products (7) and (8) were established by correct elemental analyses and spectral data, which were compatible with the assigned structures. (cf. Tables 2 and 3).

Table 2: Elemental analyses of the newly synthesized compounds 6–8.

Compound	Mol. formula/M.Wt.		Elemental analysis C%	H%	N%
6a	$C_{18}H_{16}N_2O_4$ (324.35)	Calc. Found	66.66 66.41	4.97 5.12	8.64 8.34
6b	$C_{18}H_{16}N_2O_4$ (324.35)	Calc. Found	66.66 67.01	4.97 5.13	8.64 9.01
6c	$C_{23}H_{18}N_2O_3$ (370.40)	Calc. Found	74.58 76.86	4.90 4.75	7.56 7.15
7a	$C_{22}H_{18}N_2O_3$ (358.39)	Calc. Found	73.73 74.11	5.06 5.20	7.82 8.07
7b	$C_{22}H_{18}N_2O_3$ (358.39)	Calc. Found	73.73 74.02	5.06 5.38	7.82 7.58
8a	$C_{22}H_{18}N_2O_3$ (358.39)	Calc. Found	73.73 74.11	5.06 5.74	7.82 8.21
8b	$C_{22}H_{18}N_2O_3$ (358.39)	Calc. Found	73.73 73.63	5.06 4.83	7.82 8.19
8c	$C_{22}H_{18}N_2O_3$ (358.39)	Calc. Found	73.73 73.52	5.06 5.24	7.82 7.63
8d	$C_{23}H_{20}N_2O_4$ (388.42)	Calc. Found	71.12 71.34	5.19 5.40	7.21 7.50

Table 3: Spectral data of the newly synthesized compounds 6–8.

Compound	IR (cm^{-1})	MS	^1H NMR (DMSO-d_6) (δ ppm)
6a	3419–3327 (OH and NH_2), 2190 (CN)	324 (M^+)	3.65 (s, 3H, OCH_3), 3.70 (s, 3H, OCH_3), 4.89 (s, 1H, H-4), 6.75 (s, 2H, NH_2), 6.45–6.74 (m, 6H, ArH), 9.61 (br s, 1H, OH).
6b	3430–3325 (OH and NH_2), 2210 (CN)	324 (M^+)	3.68 (s, 3H, OCH_3), 3.73 (s, 3H, OCH_3), 4.92 (s, 1H, H-4), 6.37 (s, 2H, NH_2), 6.53–6.75 (m, 6H, ArH), 9.51 (br s, 1H, OH).
6c	3430–3320 (OH and NH_2), 2215 (CN)	370 (M^+)	4.57 (s, 1H, H-4), 5.04 (s, 2H, $-CH_2-$), 6.75–6.82 (m, 5H, ArH), 6.87 (s, 2H, NH_2), 7.22–7.37 (m, 3H, ArH), 7.40–7.44 (m, 4H, ArH), 9.39 (br s, 1H, OH).
7a	3387, 3310 (NH_2), 2190 (CN)	358 (M^+)	3.64 (s, 3H, OCH_3), 3.79 (s, 3H, OCH_3), 5.14 (s, 1H, H-4), 6.69 (s, 2H, NH_2), 7.0–7.53 (m, 9H, ArH).
7b	3387, 3315 (NH_2), 2195 (CN)	358 (M^+)	3.71 (s, 3H, OCH_3), 3.80 (s, 3H, OCH_3), 4.54 (s, 1H, H-4), 6.72 (s, 2H, NH_2), 7.10–7.53 (m, 9H, ArH).
8a	3460, 3340 (NH_2), 2200 (CN)	358 (M^+)	3.85 (s, 3H, OCH_3), 4.01 (s, 3H, OCH_3), 4.62 (s, 1H, H-4), 6.66 (s, 2H, NH_2), 7.21–7.45 (m, 9H, ArH).
8b	3445, 3300 (NH_2), 2201 (CN)	358 (M^+)	3.82 (s, 3H, OCH_3), 3.90 (s, 3H, OCH_3), 4.54 (s, 1H, H-4), 6.72 (s, 2H, NH_2), 7.28–7.53 (m, 9H, ArH).
8c	3450, 3320 (NH_2), 2187 (CN)	358 (M^+)	3.73 (s, 3H, OCH_3), 3.82 (s, 3H, OCH_3), 4.55 (s, 1H, H-4), 6.69 (s, 2H, NH_2), 7.29–7.52 (m, 9H, ArH).
8d	3465, 3310 (NH_2), 2191 (CN)	388 (M^+)	3.70 (s, 3H, OCH_3), 3.81 (s, 3H, OCH_3), 3.89 (s, 3H, OCH_3), 4.50 (s, 1H, H-4), 6.64 (s, 2H, NH_2), 7.11–7.78 (m, 8H, ArH).

On the other hand, heating a mixture of aromatic aldehydes (1a–h), malononitrile (2), and resorcinol (3) in boiling water containing a catalytic amount of Rochelle salt gave 2-amino-4-aryl-7-hydroxy-4H-chromene-3-carbonitriles 6a–h, in excellent yields. In contrast, neither α-naphthol nor β-naphthol underwent the above one-pot three-component reactions in boiling water even upon heating for extended periods. When, a mixture of ethanol/water was used as a solvent in the previous reactions, the three phenols gave the desired products 6a–h, 7a–g, and 8a–h, in good yields. All known compounds were identical in all physical and spectroscopic aspects with the others which are reported in literatures.

CONCLUSIONS

We have discovered a green and efficient synthetic route to some new chromenes, namely, 2-amino chromenes, benzo[h]chromenes and benzo[f]chromenes, of expected biological interest, by utilizing Rochelle salt as novel green catalyst. To the best of our knowledge, this is the first time for utilizing Rochelle salt, as an efficient, green, and cheap catalyst in the one-pot three-component reactions.

EXPERIMENTAL

General

All melting points were measured on a Gallenkamp apparatus and are uncorrected. IR spectra were recorded with a Shimadzu FT-IR 8101 PC spectrophotometer in KBr disks. ^1H NMR spectra were recorded with Bruker AM 300 spectrometer at 300 MHz with DMSO-d_6 and $CDCl_3$ as solvents and TMS as an internal standards; chemical shifts (δ) are reported in ppm. Mass spectra were measured on a GCMS-QP1000 EX (EI, 70 eV) mass spectrometer. Analytical thin-layer chromatography (TLC) was performed on Merck silica gel 60 plates, 0.25 mm thick with F-254 indicator. Visualization was accomplished by UV light. Solvents for chromatography were reagent grad and used as received. Microanalyses were performed by the microanalytical Data Unit at Cairo University.

General Procedure for the Synthesis of 2-Amino-4-aryl-7-hydroxy-4H-chromene-3-carbonitriles 6a–h.

Method (A). To a mixture of equimolar amounts of aromatic aldehydes 1a–h, malonontirile (2), and resorcinol (3) (5 mmol) in ethanol or ethanol/water mixture (1:1) (10 mL), Rochelle salt (0.30 g) was added. Then, the reaction mixture was heated at reflux temperature for 2–4 h. After cooling to room temperature, the resulting solid products were collected by filtration, dried, and recrystallized from EtOH to give chromenes 6a–h.

Method (B). To a mixture of equimolar amounts of aromatic aldehydes 1a–h, malononirile (2), and resorcinol (3) (5 mmol) in H_2O (10 mL), Rochelle salt (0.3 g) was added. Then, the reaction mixture was worked up as described above to give chromenes 6a–h.

General Procedure for the Synthesis of 2-Amino-4-aryl-4H-benzo[h] chromene-3-carbonitriles 7a–g and 3-amino-1-aryl-1H-benzo[f]chromenes-2-carbonitriles 8a–h. To a mixture of equimolar amounts of aromatic aldehydes 1a–h, malononirile (2) and 1-naphthol (4) (or 2-naphthol (5)) (5 mmol) in ethanol or ethanol/water mixture (1:1) (10 mL), Rochelle salt (0.3 g) was added. The reaction mixture was refluxed for 4–8 h. After cooling to room temperature, the resulting solid products were collected by filtration, dried, and recrystallized from EtOH to give the products 7a–g and 8a–h, respectively.

Conflict of Interests

The author declares that there is no conflict of interests regarding the publication of this paper.

REFERENCES

1. G. P. Ellis, "Chromenes, chromanones, and chromones," in The Chemistry of Heterocyclic Compounds Chromenes, A. Weissberger and E. C. Taylor, Eds., chapter 2, pp. 11–139, John Wiley, New York, NY, USA, 1977.
2. E. A. A. Hafez, M. H. Elnagdi, A. G. A. Elagamey, and F. M. A. A. El-Taweel, "Nitriles in heterocyclic synthesis: novel synthesis of benzo[c]-coumarin and of benzo[c]pyrano[3,2-c]quinoline derivatives," Heterocycles, vol. 26, no. 4, pp. 903–907, 1987. View at

Scopus

3. M. M. Khafagy, A. H. F. A. El-Wahab, F. A. Eid, and A. M. El-Agrody, "Synthesis of halogen derivatives of benzo[h]chromene and benzo[a] anthracene with promising antimicrobial activities," Farmaco, vol. 57, no. 9, pp. 715–722, 2002. View at Publisher · View at Google Scholar · View at Scopus

4. A. H. Bedair, H. A. Emam, N. A. El-Hady, K. A. R. Ahmed, and A. M. El-Agrody, "Synthesis and antimicrobial activities of novel naphtho[2,1-b]pyran, pyrano[2,3-d]pyrimidine and pyrano[3,2-e] [1,2,4]triazolo[2,3-c]-pyrimidine derivatives," Farmaco, vol. 56, no. 12, pp. 965–973, 2001. View at Publisher · View at Google Scholar · View at Scopus

5. K. Hiramoto, A. Nasuhara, K. Michikoshi, T. Kato, and K. Kikugawa, "DNA strand-breaking activity and mutagenicity of 2,3-dihydro-3,5-dihydroxy-6-methyl-4H-pyran-4-one (DDMP), a Maillard reaction product of glucose and glycine," Mutation Research, vol. 395, no. 1, pp. 47–56, 1997. View at Publisher · View at Google Scholar · View at Scopus

6. A. Martínez-Grau and J. L. Marco, "Friedlander reaction on 2-amino-3-cyano-4H-pyrans: synthesis of derivates of 4H-pyran[2,3-b]quinoline, new tacrine analogues," Bioorganic and Medicinal Chemistry Letters, vol. 7, no. 24, pp. 3165–3170, 1997. View at Publisher · View at Google Scholar · View at Scopus

7. C. P. Dell and C. W. Smith, "Antiproliferative derivatives of 4H-naphtho[1,2-b]pyran and process for their preparation," European Patent Applications EP 537 949 21 Apr1. 993; Chemical Abstracts 119, 139102d, 1993.

8. G. Bianchi and A. Tava, "Synthesis of (2R)(+)-2,3-dihydro-2,6-dimethyl-4H-pyran-4-one, a homologue of pheromones of a species in the hepialidae family," Agricultural and Biological Chemistry, vol. 51, pp. 2001–2002, 1987.

9. S. J. Mohr, M. A. Chirigos, F. S. Fuhrman, and J. W. Pryor, "Pyran copolymer as an effective adjuvant to chemotherapy against a murine leukemia and solid tumor," Cancer Research, vol. 35, no. 12, pp. 3750–3754, 1975. View at Scopus

10. F. Eiden and F. Denk, "Synthesis and CNS-activity of pyran derivatives: 6,8-dioxabicyclo[3,2,1]octanes," Archiv der Pharmazie, vol. 324, no. 6, pp. 353–354, 1991. View at Scopus

11. P. W. Smith, S. L. Sollis, P. D. Howes et al., "Dihydropyrancarboxamides related to zanamivir: a new series of inhibitors of influenza virus

sialidases. 1. Discovery, synthesis, biological activity, and structure-activity relationships of 4-guanidino- and 4-amino- 4H-pyran-6-carboxamides,"Journal of Medicinal Chemistry, vol. 41, no. 6, pp. 787–797, 1998. View at Publisher · View at Google Scholar · View at Scopus

12. R. N. Taylor, A. Cleasby, O. Singh et al., "Dihydropyranocarboxamides related to zanamivir: a new series of inhibitors of influenza virus sialidases," Journal of Medicinal Chemistry, vol. 41, no. 6, pp. 798–807, 1998. View at Scopus
13. R. Maggi, R. Ballini, G. Sartori, and R. Sartorio, "Basic alumina catalysed synthesis of substituted 2-amino-2-chromenes via three-component reaction," Tetrahedron Letters, vol. 45, no. 11, pp. 2297–2299, 2004. View at Publisher · View at Google Scholar · View at Scopus
14. C. J. Li and T. H. Chan, Organic Reactions in Aqueous Media, Wiley, New York, NY, USA, 1997.
15. P. A. Grieco, Organic Synthesis in Water, Blackie Academic and Professional, 1998.
16. K. Kandhasamy and V. Gnanasambandam, "Multi-component reactions in water," Current Organic Chemistry, vol. 13, pp. 1820–11841, 2009.
17. C. K. Z. Andrade and L. M. Alves, "Environmentally benign solvents in organic synthesis,"Current Topics in Medicinal Chemistry, vol. 9, pp. 195–218, 2005.
18. A. G. A. Elagamay and F. M. A. A. El-Taweel, "Nitriles in heterocyclic synthesis: synthesis of condensed pyrans.," Indian Journal of Chemistry B, vol. 29, pp. 885–886, 1990.
19. M. M. Heravi, B. Baghernejad, and H. A. Oskooie, "A novel and efficient catalyst to one-pot synthesis of 2-amino-4H-chromenes by methanesulfonic acid," Journal of the Chinese Chemical Society, vol. 55, no. 3, pp. 659–662, 2008. View at Scopus
20. M. M. Heravi, K. Bakhtiari, V. Zadsirjan, F. F. Bamoharram, and O. M. Heravi, "Aqua mediated synthesis of substituted 2-amino-4H-chromenes catalyzed by green and reusable Preyssler heteropolyacid," Bioorganic and Medicinal Chemistry Letters, vol. 17, no. 15, pp. 4262–4265, 2007. View at Publisher · View at Google Scholar · View at Scopus
21. R. Ballini, G. Bosica, M. L. Conforti et al., "Three-component process for the synthesis of 2-amino-2-chromenes in aqueous media," Tetrahedron, vol. 57, no. 7, pp. 1395–1398, 2001. View at Publisher · View at Google Scholar · View at Scopus
22. L. Chen, X.-J. Huang, Y.-Q. Li, M.-Y. Zhou, and W.-J. Zheng, "A

one-pot multicomponent reaction for the synthesis of 2-amino-2-chromenes promoted by N,N-dimethylamino-functionalized basic ionic liquid catalysis under solvent-free condition," Monatshefte fur Chemie, vol. 140, no. 1, pp. 45–47, 2009. View at Publisher · View at Google Scholar · View at Scopus

23. J. Albadi, A. Mansournezhad, and M. Darvishi-Paduk, "Poly(4-vinylpyridine): as a green, efficient and commercial available basic catalyst for the synthesis of chromene derivatives," Chinese Chemical Letters, vol. 24, pp. 208–210, 2013.
24. M. Kidwai, S. Saxena, M. K. R. Khan, and S. S. Thukral, "Aqua mediated synthesis of substituted 2-amino-4H-chromenes and in vitro study as antibacterial agents," Bioorganic and Medicinal Chemistry Letters, vol. 15, no. 19, pp. 4295–4298, 2005. View at Publisher · View at Google Scholar · View at Scopus
25. H. M. Al-Matar, K. D. Khalil, H. Meier, H. Kolshorn, and M. H. Elnagdi, "Chitosan as heterogeneous catalyst in Michael additions: the reaction of cinnamonitriles with active methylene moieties and phenols," Arkivoc, vol. 2008, no. 16, pp. 288–301, 2008. View at Scopus
26. B. S. Kumar, N. Srinivasulu, R. H. Udupi et al., "Efficient synthesis of benzo[g]- and benzo[h]chromene derivatives by one-pot three-component condensation of aromatic aldehydes with active methylene compounds and naphthols," Russian Journal of Organic Chemistry, vol. 42, no. 12, pp. 1813–1815, 2006. View at Publisher · View at Google Scholar · View at Scopus
27. S. R. Kolla and Y. R. Lee, "$Ca(OH)_2$-mediated efficient synthesis of 2-amino-5-hydroxy-4H- chromene derivatives with various substituents," Tetrahedron, vol. 67, no. 43, pp. 8271–8275, 2011. View at Publisher · View at Google Scholar · View at Scopus
28. D. Kumar, V. B. Reddy, B. G. Mishra, R. K. Rana, M. N. Nadagoudac, and R. S. Varma, "Nanosized magnesium oxide as catalyst for the rapid and green synthesis of substituted 2-amino-2-chromenes.," Tetrahedron, vol. 63, pp. 3093–3097, 2007.
29. H. Sheibani, K. Saidi, M. Abbasnejad, A. Derakhshani, and I. Mohammadzadeh, "A convenient one-pot synthesis and anxietic activity of 3-cyano-2(1H)-iminopyridines and halogen derivatives of benzo[h]chromenes," Arabian Journal of Chemistry, 2011. View at Publisher · View at Google Scholar · View at Scopus
30. S. Makarem, A. A. Mohammadi, and A. R. Fakhari, "A multi-component electro-organic synthesis of 2-amino-4H-chromenes," Tetrahedron Letters, vol. 49, no. 50, pp. 7194–7196, 2008.View at Publisher · View at Google Scholar · View at Scopus

31. R. A. Mekheimer and K. U. Sadek, "Microwave-assisted reactions: three-component process for the synthesis of 2-amino-2-chromenes under microwave heating," Chinese Chemical Letters, vol. 20, no. 3, pp. 271–274, 2009. View at Publisher · View at Google Scholar · View at Scopus

32. S. Khaksar, A. Rouhollahpour, and S. M. Talesh, "A facile and efficient synthesis of 2-amino-3-cyano-4H-chromenes and tetrahydrobenzo[b] pyrans using 2,2,2-trifluoroethanol as a metal-free and reusable medium," Journal of Fluorine Chemistry, vol. 141, pp. 11–15, 2012.

33. M. P. Surpur, S. Kshirsagar, and S. D. Samant, "Exploitation of the catalytic efficacy of Mg/Al hydrotalcite for the rapid synthesis of 2-aminochromene derivatives via a multicomponent strategy in the presence of microwaves," Tetrahedron Letters, vol. 50, no. 6, pp. 719–722, 2009.View at Publisher · View at Google Scholar · View at Scopus

34. D. S. Raghuvanshi and K. N. Singh, "An expeditious synthesis of novel pyranopyridine derivatives involving chromenes under controlled microwave irradiatio," Arkivoc, pp. 305–317, 2010.

35. J. M. Khurana, B. Nand, and P. Saluja, "DBU: a highly efficient catalyst for one-pot synthesis of substituted 3, 4-dihydropyrano[3,2-c] chromenes, dihydropyrano[4,3-b]pyranes, 2-amino-4H-benzo[h] chromenes and 2-amino-4H-benzo[g]chromenes in aqueous medium," Tetrahedron, vol. 66, no. 30, pp. 5637–5641, 2010. View at Publisher · View at Google Scholar · View at Scopus

36. A. V. Borhade, B. K. Uphade, and D. R. Tope, "PbO as an efficient and reusable catalyst for one-pot synthesis of tetrahydro benzo pyrans and benzylidene malonitriles," Journal of Chemical Sciences, vol. 125, pp. 583–589, 2013.

37. A. A. Harb, A. M. El-Maghraby, and S. A. Metwally, "Nitriles in heterocyclic synthesis: a New synthesis of some 4H-naphthopyrans, 2H-benzothiopyrans and their fused derivatives,"Collection of Czechoslovak Chemical Communications, vol. 57, pp. 1570–1574, 1992.

38. A. M. El-Maghraby, "Synthesis of some new pyrazolo[4,3-b]pyridine-3-one and dihydro-pyrazolo[4,3-c][1,2]oxazine derivatives," Egyptian Journal of Chemistry, vol. 51, pp. 373–388, 2008.

39. O. S. Zaky, M. S. Moustafa, M. A. Selim, A. M. El- Maghraby, and M. H. Elnagdi, "Scope and limitations of a novel synthesis of 3-arylazonicotinates," Molecules, vol. 17, pp. 5924–5934, 2012.

40. A. M. Elmaghraby, I. A. Mousa, A. A. Harb, and M. Y. Mahgoub, "Three component reaction: an efficient synthesis and reactions of 3,4-dihydropyrimidin-2(1H)-ones and thiones using new natural catalyst," ISRN Organic Chemistry, vol. 2013, Article ID 706437, 13 pages, 2013. View at Publisher · View at Google Scholar

Chapter 2

ASYMMETRIC ORGANOCATALYSIS AT THE SERVICE OF MEDICINAL CHEMISTRY

Alfredo Ricci

Department of Industrial Chemistry "Toso Montanari", School of Science, University of Bologna, V. Risorgimento 4, 40136 Bologna, Italy

ABSTRACT

The application of the most representative and up-to-date examples of homogeneous asymmetric organocatalysis to the synthesis of molecules of interest in medicinal chemistry is reported. The use of different types of organocatalysts operative via noncovalent and covalent interactions is critically reviewed and the possibility of running some of these reactions on large or industrial scale is described. A comparison between the organo- and metal-catalysed methodologies is offered in several cases, thus highlighting the merits and drawbacks of these two complementary approaches to the obtainment of very popular on market drugs or of related key scaffolds.

INTRODUCTION

Over the past ten years, the field of enantioselective organocatalysis has had a significant impact on chemical synthesis [1, 2]. Currently, asymmetric organocatalysis is recognized [3] as an independent synthetic tool besides asymmetric metallic catalysis and enzymatic catalysis for the synthesis of chiral organic molecules. Multiple advantages compared with the other two catalytic domains are the reasons for the rapid growth and acceptance of organocatalysis. In general, organocatalysts are air- and moisture-stable and, thus, inert-equipments such as vacuum lines or glove boxes are not necessary. They are easy to handle even on large scale and relatively less toxic compared to transition metals. Moreover, frequently the reactions are conducted under mild conditions and high concentrations thus avoiding the use of large amounts of solvents and minimizing waste.

The organocatalysts can be classified by means of their interactions with the substrate or "mode of action" as covalent or noncovalent catalysts (Figure 1).

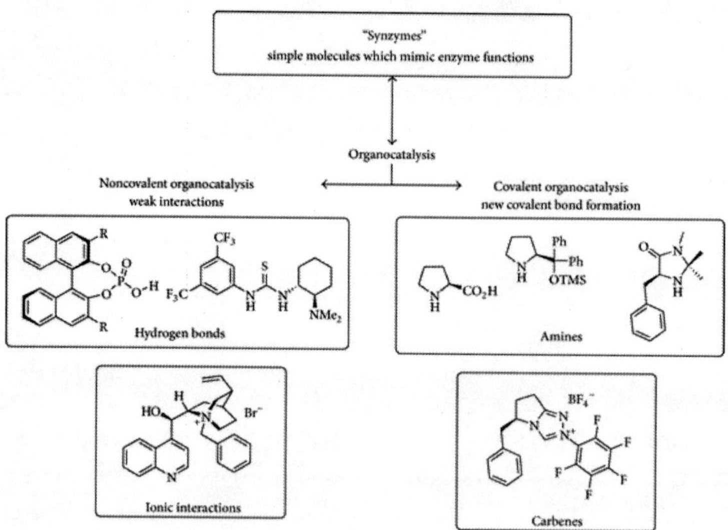

Figure 1: General classification of the activation mode of several representative classes of molecules in organocatalysis.

In covalent organocatalysis, a new covalent bond between the catalysts and the substrate is formed as in the case of aminocatalysis [4] and carbenes [5], leading to a strong interaction between the substrate and the reagent in the reaction. In the case of noncovalent interactions between the substrate and the catalyst, the activation of the substrate occurs via weak binding exemplified by hydrogen bonding [6] or ionic interaction as in the case of phase transfer catalysis [7].

The field of asymmetric organocatalysis has enjoyed phenomenal growth in the past 15 years [8–10] and during the "golden age" [11] of organocatalysis many researchers from academia and chemical industry were involved in this field, with most efforts focused on the development of novel organocatalysts, new reactivities and asymmetric methodologies. Moreover current developments in the field of the synthesis of structurally complex and/or polyfunctionalized molecules indicate that chemists have adopted the fundamental principles of biosynthesis as synthetic strategic key elements for their synthetic approaches [12]. Among these, cascade reactions [13, 14] employing a single catalyst capable of promoting each single step have gained in the recent years an important role in the efficient and rapid generation of molecules with complex architectures generally correlated with specificity of action and potentially useful biological properties [15]. Organocatalysts turn out to be particularly favourable when used in catalytic cascade reactions because they allow distinct modes of activation, which can often be combined [16, 17].

Despite their great development, the application of organocatalytic methodologies to the synthesis of active compounds in medicinal chemistry in the past years has rarely been reported. However, in most recent years organocatalytic methodologies for the synthesis of enantioenriched molecules for medicinal chemistry purposes have been gaining momentum being particularly attractive for the preparation of compounds that do not tolerate metal contamination. In academia, several groups have made a remarkable effort to show the great applicability of organocatalysts to the total synthesis of bioactive natural products [18] and of drugs [19] most of them currently available in the market such as oseltamivir, warfarin, paroxetine, baclofen, and maraviroc. These efforts mainly focused on the removal of barriers for scale-up by addressing issues such as catalyst loading, product inhibition, substrate scope, and bulk availability of designer catalysts which have drawn the attention of the companies [20, 21] that have begun to incorporate organocatalysis as a synthetic tool in some industrial scale processes [22, 23].

In this review some of the most recent representative applications of asymmetric organocatalysis to medicinal chemistry will be highlighted. Not only the access to market available drugs but also the access to drug candidates, to medicinal scaffolds, and to promising new compounds whose biological profile has not yet been fully explored will be reviewed. In some cases, a comparison between organo- and metal-catalysed methodologies aimed at the obtainment of the same medicinal targets will be reported and a few interesting industrial examples of the use of organocatalysis in medicinal chemistry taken from the literature will be discussed as well.

DISCUSSION

Noncovalent Organocatalysis

Hydrogen Bonding Catalysis

This ubiquitous interaction is one of the central forces in Nature. As an individual, hydrogen bonds are feeble and quite easy to break. However when acting together they become much stronger and lean each other. This phenomenon is called "cooperativity" (1 + 1 is more than 2). Some of the many vital functions that hydrogen bonds fulfil in biological systems are shown in Figure 2.

The simultaneous donation of two hydrogen bonds leads to a highly successful strategy for electrophilic activation, increased strength, and directionality relative to single hydrogen bonds. Therefore asymmetric catalysis via noncovalent bond interactions such as hydrogen bonding turns out to be a powerful synthetic strategy.

Original achievements regarding bis-hydrogen bond complexes have been reported through the years. It is worth noting that this two-point binding is a powerful strategy both in metal-centered catalysis and in organocatalysis as shown in Figure 3.

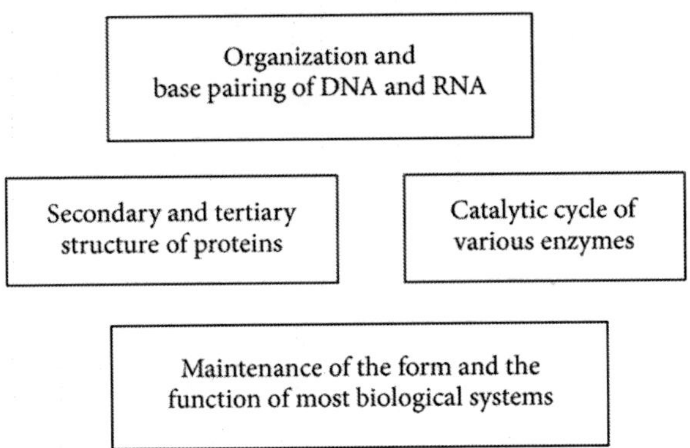

Figure 2: Some of the vital functions that hydrogen bonds fulfil in biological systems.

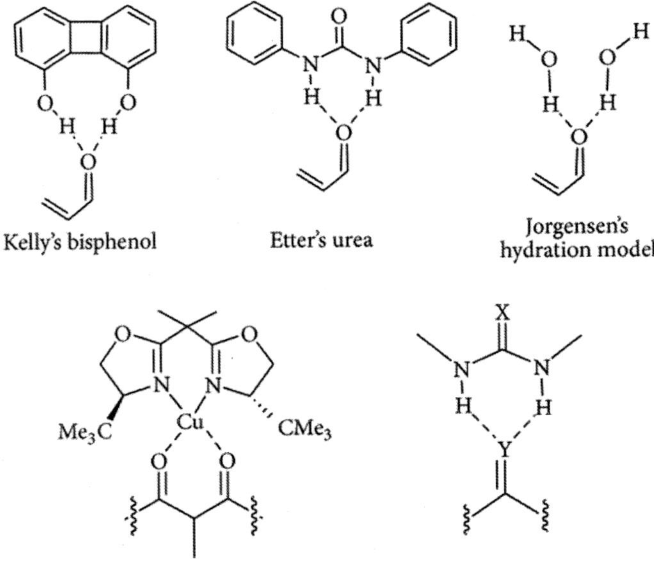

Figure 3: Original achievements regarding bis-hydrogen-bonded complexes.

However, whereas coordination to a chiral Lewis acid imposes limitations upon substrate structure, any Lewis base is in principle capable of engaging bifurcated hydrogen bonds so that this catalytic strategy has potential as a general paradigm for synthesis. Thiourea catalysts designed by Sigman and Jacobsen [24] and Corey and Grogan [25] as well as minimal peptides introduced by Davie et al. [26] appeared together with many others during the last decade; they all fall into this class of catalysts. These structures are highly modular and can be readily modified and finely tuned. Many organocatalysts have been recognized to be reminiscent of natural enzymes in their mode of action and substrate interaction/activation [27]. Their use in several bioinspired methodologies has led to envisaging efficient synthetic routes leading among the others to the obtainment of target compounds of interest in medicinal chemistry.

In the biosynthesis of fatty acids and polyketides the active site of polyketide synthase (PKS) clearly highlights the key role displayed by hydrogen bonds (Scheme 1) [28].

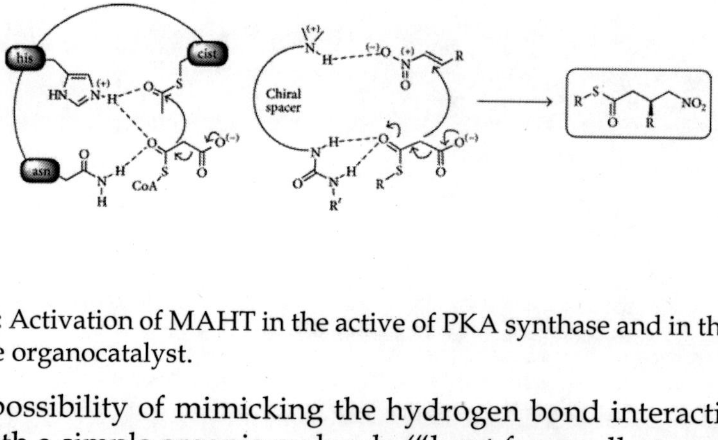

Scheme 1: Activation of MAHT in the active of PKA synthase and in the chiral core of the organocatalyst.

The possibility of mimicking the hydrogen bond interactions of PKS's with a simple organic molecule ("hunt for smallest enzymes" according to Schreiner) has been shown feasible by using double hydrogen bond-based organocatalysts. The potential of this strategy has been appreciated by synthetic chemists for many years and has been widely used in asymmetric catalysis [29].

The application of the enantioselective decarboxylative reaction of malonic half thioesters (MAHTs) to the synthesis of medicinal targets is exemplified by the synthesis of GABA receptor antagonists 3 and 4 using I and II as the organocatalysts (Scheme 2). The γ-nitrothioesters 1 and 2 easily achievable through these organocatalytic approaches occurring under mild conditions and tolerating both moisture and air are versatile building blocks for further modifications. Among them the formation of γ-butyrolactams by reduction of the nitro group followed by intramolecular cyclization leads to intermediates en route to the antidepressant (R)-Rolipram [30] 3 and to gram scale synthesis and transformation to (S)-baclofen ·HCl 4, a GABA receptor antagonist used in the treatment of spasticity [31].

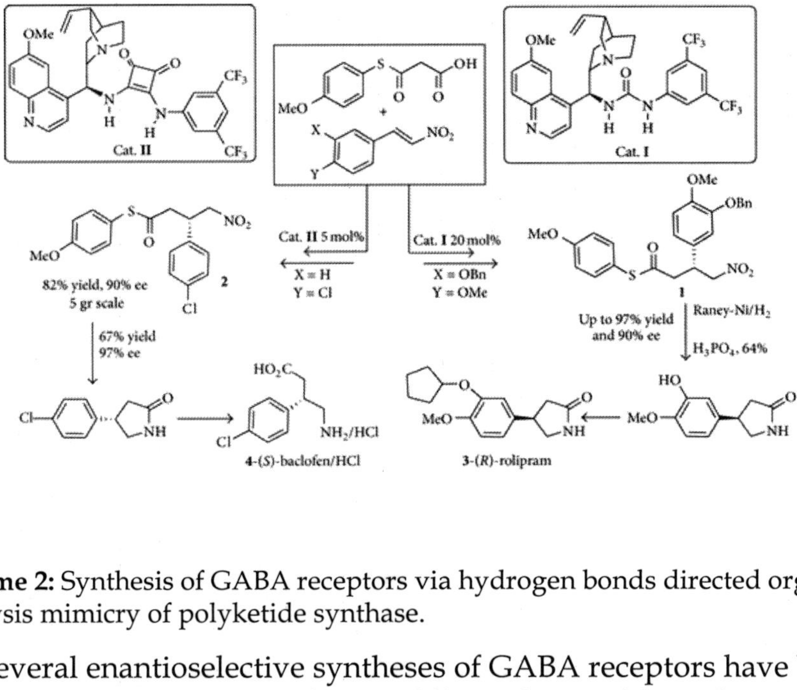

Scheme 2: Synthesis of GABA receptors via hydrogen bonds directed organocatalysis mimicry of polyketide synthase.

Several enantioselective syntheses of GABA receptors have been reported based on the use of a metal/ligand assembly as the catalyst system. So far (Scheme 3) the Rh(acac)-catalyzed asymmetric 1,4-additions of arylboronic acids to 4-aminobut-2,3-enoic acid derivatives led to (−)-(R)-baclofen ent -4 and to (−)-(R)-rolipram 3 in high yields and excellent enantioselectivities [32]. However the use of inert atmosphere (argon) and the to some extent difficult purification by flash chromatography prevent this methodology to be easily applied on large scale.

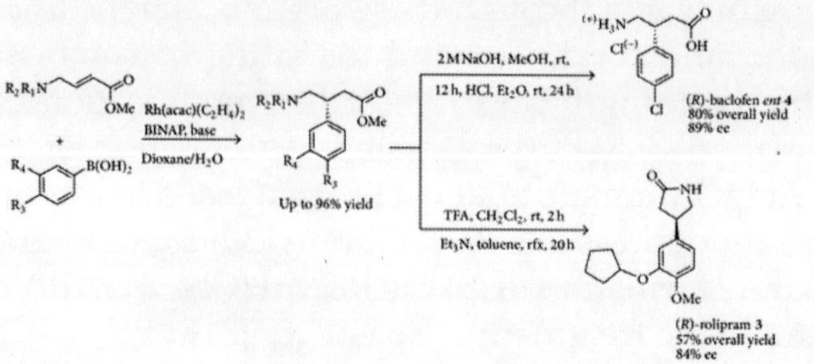

Scheme 3: Metal-catalysed synthesis of GABA receptors.

An alternative metal-catalysed system [33] in which the potential for scale-up is clear is shown in Scheme 4 and appears highly competitive with the organocatalysed approach. The chiral Lewis acid-catalyzed Michael addition of diethyl malonate to fully elaborated nitrostyrene 5 allows the nitroester 6 that upon reduction and saponification leads to the target compound 3. Both enantiomers of rolipram 3 can be accessed in a total of six steps and at 10 gram scale with excellent overall yields of 76% and without chromatography.

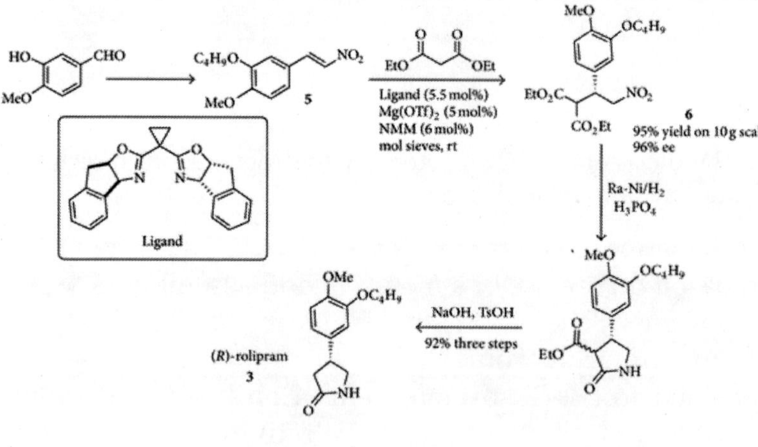

Scheme 4: Scalable metal-catalysed synthesis of both enantiomers of rolipram.

The use of magnesium is preferable to many other metal catalysts since toxicity issues are avoided. The dependence on solvents such as chloroform does, however, raise in this method toxicological and environmental issues.

The range of applications of bio inspired decarboxylative reactions is witnessed by the very recent [34] hydrogen-bond directed enantioselective decarboxylative Mannich reaction of keto acids with ketimines. Under the action of saccharide-derived amino thioureas as chiral catalysts (III), this reaction that can be run on a gram scale without any detriment on the reaction outcome leads to the expected trifluoromethylated 3,4-dihydro-quinazolin-2(1H)-one rings in very high yields and up to 99% ee. The potential application of this decarboxylative Mannich reaction in the domain of pharmaceutics is demonstrated in a new and efficient and shortcut synthesis of the anti-HIV drug DPC083 7 shown in Scheme 5. Herein the crucial role of the hydrogen bond interactions in building a complex rigid architecture responsible for the high stereoselectivity is highlighted.

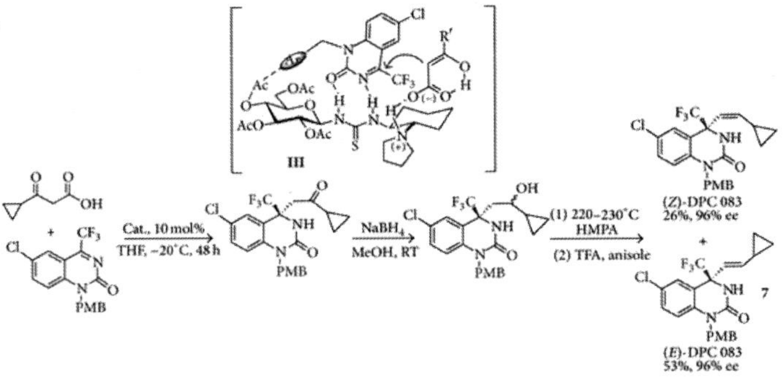

Scheme 5: Hydrogen-bonding assembly between organocatalyst, ketimine, and β-ketoacid in the preparation of the anti-HIV drug DPC 083.

Hydrogen bond-based organocatalysis also plays a primary role in the synthesis of low molecular weight drug candidates. The aza-Henry reaction (nitro-Mannich reaction) was used by Xu and coworkers [35] for the short asymmetric synthesis of the chiral piperidine derivative CP-99,994 8 (Scheme 6). The previous asymmetric syntheses of this potent neurokin-1 receptor antagonist were mainly based on the use of metal complexes as catalysts but suffered from several drawbacks, for example, low overall yield and enantioselectivity or a lengthy synthetic route. Notably the organocatalysed Takemoto's synthesis proceeded in five steps without the need to separate the diastereomeric intermediates that were cyclized as a mixture. The catalyst employed was a chiral thiourea IV, which served as an activator of both the nitroalkane and imine reactants. The transition state is relatively complex and is dominated by hydrogen-bonding interactions.

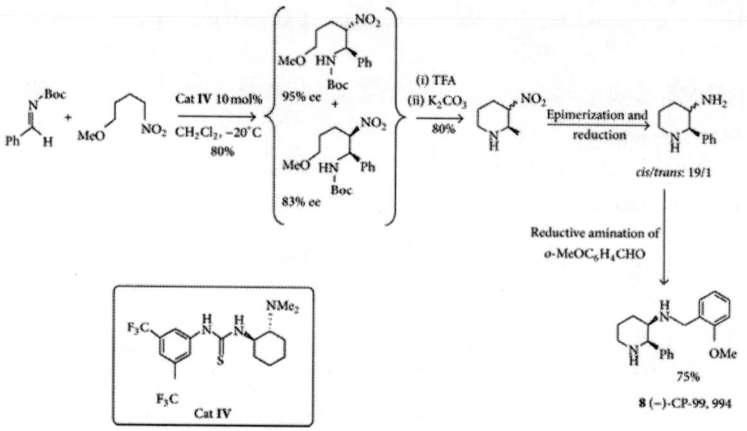

Scheme 6: Organ catalytic synthesis of CP-99,994 8, a neurokinin-1 receptor agonist.

The simultaneous donation of two hydrogen bonds has also proven to be a highly successful strategy for electrophilic activation in enzymes with an "oxyanion hole" having a postulated role in the stabilization of many high-energy tetrahedral intermediates [36]. It appears that living systems discovered and made use of these interactions in the ubiquitous useful ring-forming Diels-Alder reaction eons ago for the construction of complex natural products so that the prospect of discovering a Diels-Alderase mimic would be especially exciting. Following this concept and inspired by the antibody 13G5-catalyzed Diels-Alder cycloaddition of acrylamide with a carbamate (Scheme 7), taking place via a cooperative multiple hydrogen bond coordination to both diene and dienophile [37]; a catalytic asymmetric cycloaddition of 3-vinylindoles with activated dienophiles has been recently reported. The synthetic elaboration of vinyl indole derivatives via cycloaddition appears highly promising in that it leads (Figure 4) to fused poly-heterocyclic ring systems otherwise not easily accessible like carbazoles and pyridocarbazoles with antibiotic and antitumor activities.

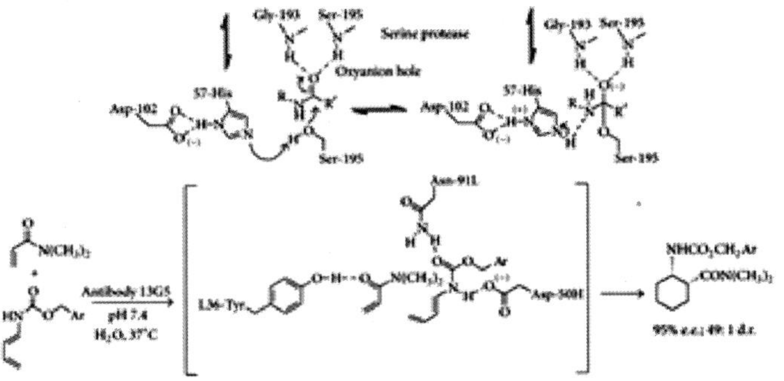

Scheme 7: Occurrence of the oxyanion hole in enzymatic processes.

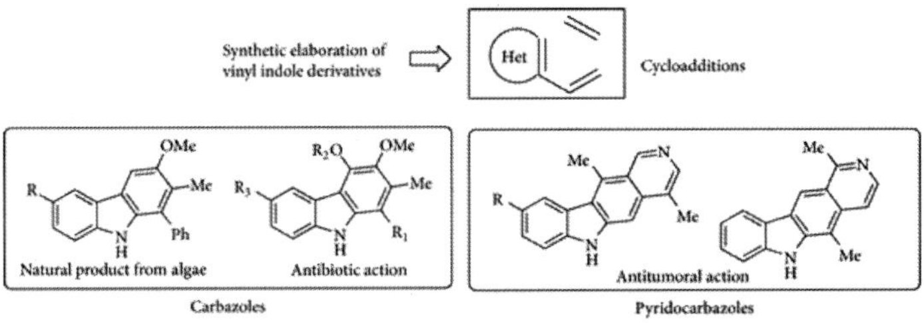

Figure 4: Biological activity of ring-fused indoles.

A scenario in which a suitable bifunctional acid-base organic catalyst (V)* coordinates through H-bond interactions both diene and dienophile leading (Scheme 8) to a highly organized transition state has been designed [38] delivering in very high yields and excellent enantioselectivities a wide range of indolines and tetracarbazoles common scaffolds in a variety of biologically active and pharmacologically important alkaloids [39–41]. The synthetic potential of the cycloadducts is exemplified by the access to indoline 9, to tetrahydrocarbazole 10 with potent activity against human papillomaviruses [42] and to a precursor [43] of tubifolidine 11 a Strychnos alkaloid previously prepared using a nine-step synthesis (Scheme 9).

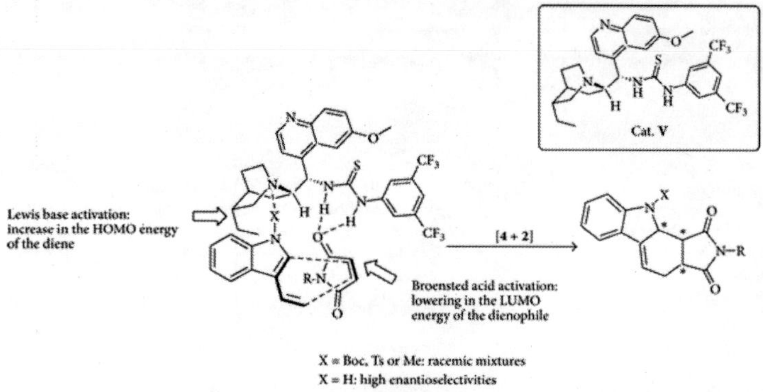

Lewis base activation: increase in the HOMO energy of the diene

Broensted acid activation: lowering in the LUMO energy of the dienophile

[4 + 2]

Cat. V

X = Boc, Ts or Me: racemic mixtures
X = H: high enantioselectivities

Scheme 8: Bifunctional activation in the Diels-Alder reaction of 3-vinylindoles.

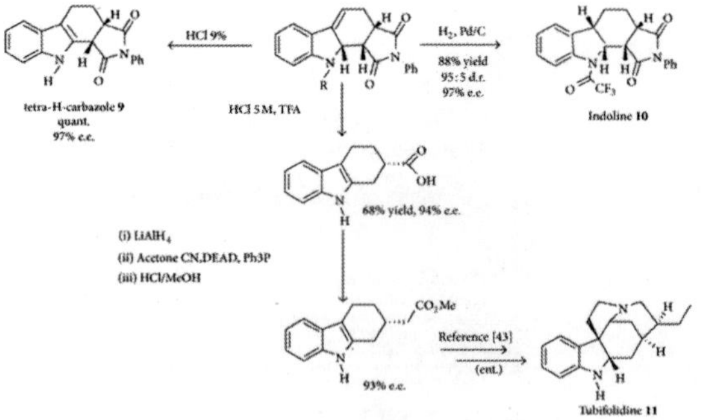

HCl 9%

H_2, Pd/C
88% yield
95 : 5 d.r.
97% e.e.

tetra-H-carbazole 9
quant.
97% e.e.

HCl 5 M, TFA

Indoline 10

68% yield, 94% e.e.

(i) $LiAlH_4$
(ii) Acetone CN, DEAD, Ph3P
(iii) HCl/MeOH

93% e.e.

Reference [43]
(ent.)

Tubifolidine 11

Scheme 9: Synthetic elaborations of the vinylindole cycloadducts.

The combination of hydrogen bond-based organocatalysis and cascade reactions or one-pot processes in the synthesis of therapeutics is powerful and can be illustrated by the synthesis of the alkaloid (−)-epibatidine, developed by the Takemoto's group and based on an enantioselective double Michael addition [44]. The bifunctional thiourea-based organocatalyst IV catalysed the first Michael addition of the γ,δ-unsaturated -ketoester 12 to the nitroalkene 13 and on addition of KOH the newly formed nitroalkane cyclized to form the polysubstituted cyclohexene 14 in a high yield and 75% ee (Scheme 10). The total synthesis of (−)-epibatidine 15 was achieved in further seven steps from 14. Though due to its high toxicity (200 times more potent than morphine) and lacking of selectivity on nicotinic receptors (−)-epibatidine cannot be considered a lead for pharmaceutical development; it has already opened the route to a wide series of more selective and promising derivatives.

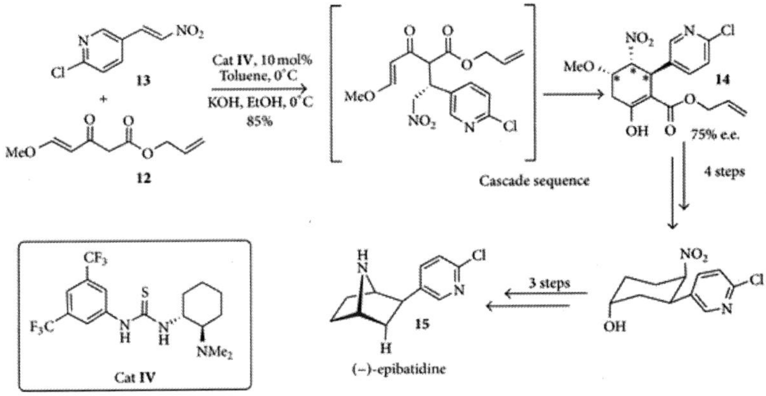

Scheme 10: Organocatalyzed cascade synthesis of (−)-epibatidine 15.

Other laboratory scale syntheses based on the use of thiourea-derived bifunctional organocatalysts have been reported leading to targets of interest in medicinal chemistry. Among them a further highly enantioselective (99% e.e.) synthesis of (R)-rolipram and of (3S-4R)-paroxetine (see Section 2.2.1.) has been accessed through the use of a combined thiourea-cinchona catalyst [45] using a highly enantioselective Michael addition of malonate nucleophiles as key steps. An indanol-thiourea organocatalyst resulted on the other hand very effectively in one of the first enantioselective Friedel-Crafts alkylations of indole with nitroalkenes leading after a synthetic elaboration of the alkylation products to the synthesis of 1,2,3,4-tetrahydro-β-carbolines [46] with anti-inflammatory and anti-arrhythmic activities.

Phase Transfer Catalysis

Phase transfer catalysis (PTC) has long been recognized as a versatile catalytic methodology for organic synthesis in both industry and academia. It features operational simplicity, typically mild reaction conditions, inexpensive and environmentally benign reagents and solvents, and relatively cheap catalysts that can be found in reasonable abundance [47]. Moreover, it has proven particularly viable for large- and industrial-scale applications. Chiral phase transfer catalysis has seen an explosive growth in the past couple of decades [48–50] and is still one of the hottest research areas in asymmetric noncovalent organocatalysis [51, 52]. The development through the years of various types of chiral phase transfer catalysts relying on the molecular design of both natural product-derived and purely synthetic quaternary ammonium salts delivered [53, 54] not only higher reactivity and stereoselectivity but also new synthetic opportunities [55]. So far a wide variety of highly enantioselective transformations catalyzed mainly by cinchona alkaloids or binaphthyl-derived quaternary ammonium salts have been introduced and applied to the asymmetric synthesis of biologically active compounds including a number of pharmaceuticals. Furthermore pharmaceutical companies have demonstrated the viability of asymmetric phase transfer reactions in the large-scale preparation of drugs.

Interestingly the first landmark example in the domain of chiral phase transfer organocatalysis was developed by Merck as early as in 1984 for the synthesis of a uricosuric drug (+)-indacrinone (MK-0197). In this work [56] the highly enantioselective alkylation of compound 16 was achieved using the cinchona alkaloid derivative V (obtained by N-alkylation of the quinuclidine core), NaOH as a base, and MeCl as the alkylating agent (Scheme 11). Using this approach intermediate 17, used for the synthesis of the indacrinone 18, could be accessed in high yield and enantiomeric purity on a pilot plant scale (~75 Kg); the cost of producing this enantiomer is significantly lower than the cost of producing the same molecule by a resolution process.

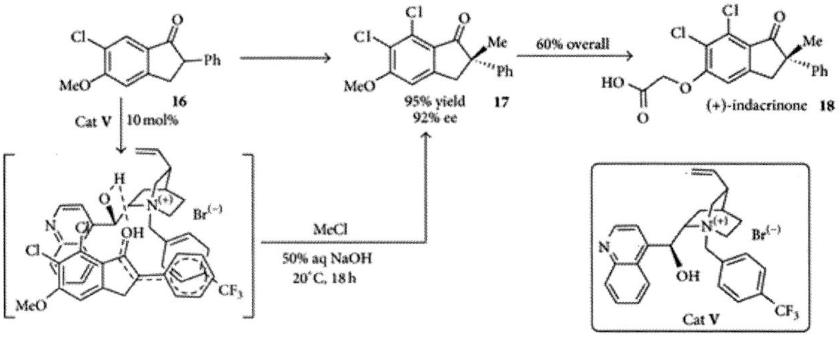

Scheme 11: Phase transfer catalysed synthesis of (+)-indacrinone 18.

Studies on the origin of the stereoselectivity substantiated the hypothesis of a tight ion pair transition state where the enolate anion and the cationic catalyst were held close to each other through π-interactions. Almost in the same period scientists from Merck demonstrated that cinchona derivatives such as VIcould catalyse the Michael addition of ketone 19 with methyl vinyl ketone (MVK) under mild conditions and crucially, at large scale [57] (Scheme 12) to give 20.

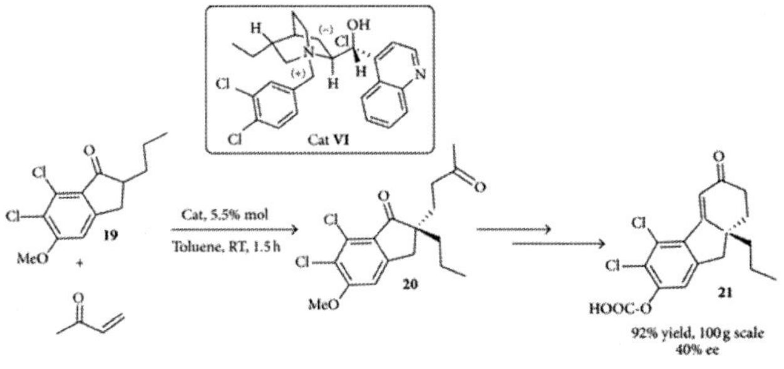

Scheme 12: Synthesis of a drug candidate for treatment of brain edema via PTC catalysis.

The ultimate goal of this study was the synthesis of drug candidate 21 (and analogues) for the treatment of brain edema and traumatic head injuries [58]. This reaction was carried out under various conditions and the operationally simple liquid/solid system gave excellent isolated yields at 100 g scale albeit with modest levels of enantioselectivity. These early examples showed the potential power of the asymmetric PTC reactions for industrial-oriented synthesis.

The learning generated in the previous examples was of great benefits for further developments of chiral phase transfer organocatalysis. An impressive use of the use of quaternary salts of cinchona alkaloids in phase transfer catalysis for the pilot scale production of drug candidates is shown in the development at Merck Sharp & Dohme of the asymmetric synthesis of an estrogen receptor -selective agonist [59] (Scheme 13). The base-catalysed Michael addition of the enolate of indanone 22 to MVK, in the presence of a (+)-cinchonine-derived quaternary ammonium phase transfer catalyst VII, gives diketone 23 in enantioenriched form. Robinson annulation then follows with construction of the cyclohexenone ring of tetrahydrofluorenone 24 that upon cyclization gives rise to the expected target 25. Overall the chemistry developed has been used to prepare >6 kg of the drug candidate in 18% overall yields and with >99% ee. The 2-naphtylmethylcinchoninium bromide catalyst VII selected on the basis of the 50% ee in the Michael addition step and on the bulk commercial availability of the required 2-naphtylmethyl bromide and the agitation rate were parameters critical to the success of this reaction.

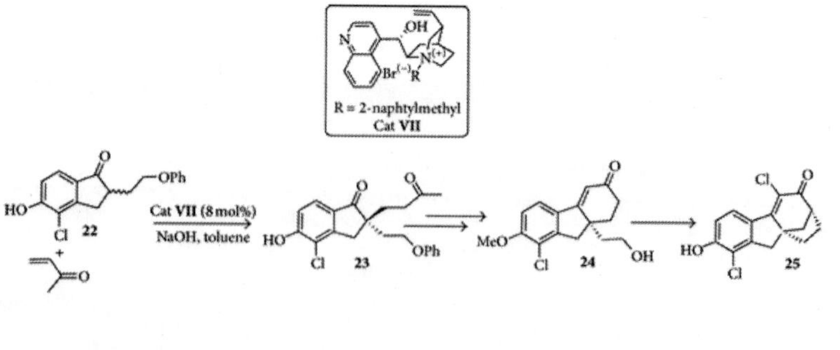

R = 2-naphtylmethyl
Cat VII

Cat VII (8 mol%)
NaOH, toluene

Scheme 13: Pilot-scale synthesis of an estrogen receptor-β.

In another more recent example the capability of chiral phase transfer catalysis based on quaternary ammonium salts VIII and ent-VIII-derived from cinchona alkaloids to induce highly enantioselective C–C bond forming reactions has been disclosed in the conjugate addition of nitroalkanes to 4-nitro-5-stirylisoxazoles, a valuable synthetic alternative to cinnamic esters [60] (Scheme 14). The transformation of the Michael adducts 26 into γ-nitro acids could be easily performed, and the subsequent Raney-Ni reduction gave the hydrochlorides of the GABA receptors (S)- and (R)-baclofen 4 thus outlining a short organocatalysed route alternative with respect to that outlined in Scheme 1.

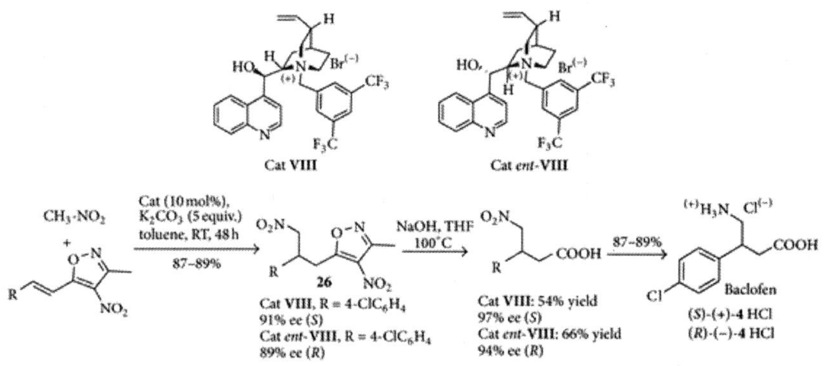

Scheme 14: Laboratory-scale synthesis of both the enantiomers of baclofen 4.

The accessibility of both the enantiomers in good yields and excellent enantioselectivities, the wide reaction scope, and the easy availability and the use of inexpensive organocatalysts outline major assets of this organocatalysed methodology.

Lewis and Brønsted Base Catalysis

Nucleophilic catalysts have had a wide role in the development of new synthetic methods [61]. In particular, the cinchona alkaloids catalyse many useful processes with high enantioselectivities [62]. They can be used as bases to deprotonate substrates with relatively acidic protons such as malonates, forming a contact pair between the resulting anion and the protonated amine. This interaction leads to a chiral environment around the anion and permits enantioselective reactions with electrophiles (Figure 5).

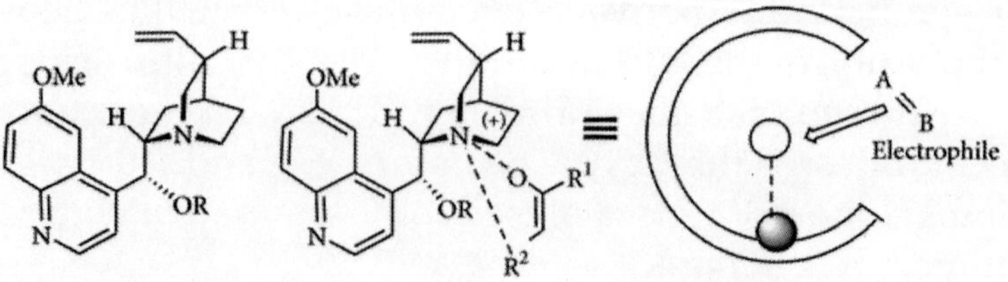

Figure 5: Cinchona alkaloids catalysis through chiral contact ion pair.

Since the seminal publication by Hiemstra and Wynberg [63] there have been different applications of this methodology with significantly improved catalysts [64]. Important in many of these processes is the ability to control the formation of quaternary centers with high enantiomeric excess [65]. The robustness and the easy availability of the commercially available cinchona derivatives attracted in the last decades increasing interest of both the academic and applied research. In medicinal chemistry relevant targets such as anticancer and antiparasitic agents were approached by using this methodology.

In the past 10 years, the number of chiral, nonracemic pharmaceuticals on the market was consistently increasing and many new single enantiomer drugs were produced to offer enhanced therapy and reduced toxicity. Organocatalysis emerged to be an effective way to reach this goal. A series of chiral 2-ethylthio-thiazolone derivatives 29 have been prepared (Scheme 15) by a straightforward enantioselective aza-Mannich addition of thiazolones 27 to N-tosylimines 28 catalyzed by a simple cinchona alkaloid (IX) as the chiral base with a 20 mol % of catalyst loading using diethyl ether as solvent [66]. The derivatives bearing a quaternary center were obtained in good yields and, in general, with high diastereo- and enantioselectivities. All the compounds, evaluated in five human cell cancer lines using MTT essay, caused a dose-dependent growth inhibitory effect on all the tested cancer lines. This study provides a foundation for further developments of new single enantiomer anticancer drugs.

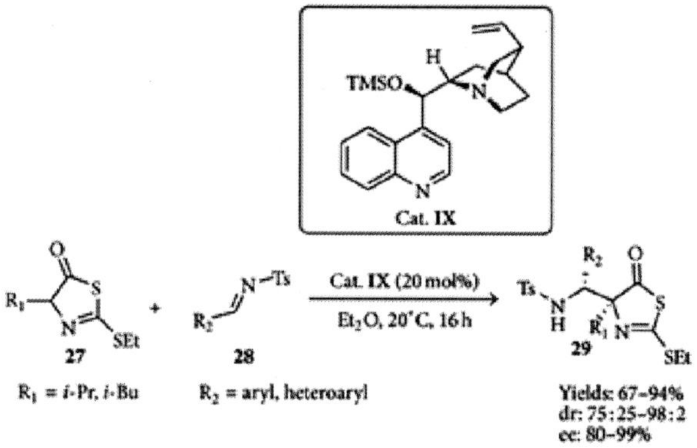

TMSO H N N Cat. IX

Cat. IX (20 mol%)
Et_2O, 20°C, 16 h

27 28 29

R_1 = *i*-Pr, *i*-Bu
R_2 = aryl, heteroaryl
Yields: 67–94%
dr: 75:25–98:2
ee: 80–99%

Scheme 15: Synthesis of anticancer thiazolone derivatives by organocatalytic aza-Mannich reaction.

Malaria is one of the most important diseases of the third world and the efficacy of the available drugs is limited by emerging resistance. In 2011, in an extensive effort to find unique chemotypes for the treatment of malaria, it has been found that dihydropyrimidinone-derived guanidine derivatives were the most promising [67]. These guanidine analogs 34 were synthesized in a multistep synthesis with commercially available and inexpensive (+)-cinchonine X and (−)-cinchonidine XI promoting the key organocatalytic step (Scheme 16).

In this step, the diketone derivative 30 was deprotonated by the nitrogen of the chiral base (cinchonine or cinchonidine) which attacks the imine formed in situ starting from 31 to give the corresponding intermediates 32 in high enantiomeric excesses. These were then cyclised into dihydropyrimidinones 33. Being the two organocatalysts pseudoenantiomers, both enantiomers of dihydropyrimidinones could be synthesized. Further treatment of 33 with Lawesson reagent, followed by sulphur alkylation and its substitution with different anilines led to a library of 96 guanidine derivatives 34.

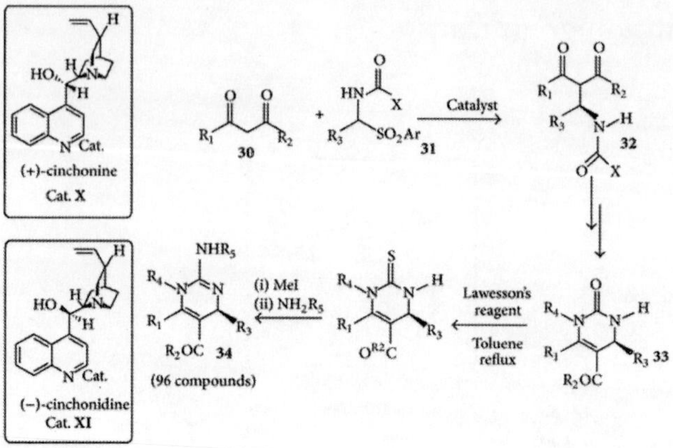

Scheme 16: Synthesis of a library of dihydropyrimidinones 34 anti-malarial derivatives by a cinchona alkaloid-driven key organocatalytic step.

Another quite impressive example of how simple and unmodified cinchona alkaloids can be used for the synthesis of medicinally important scaffolds is provided by the synthesis of (−)-uperzine A 37 currently being tested in clinical trials as a promising drug for the treatment of Alzheimer disease [68]. This reaction that can be considered as the first application of cascade reaction to the synthesis of targets in medicinal and natural product chemistry dates back to 1998 when the field of organocatalysis was just at its infancy. Huperzine-A, containing a challenging bridged tricyclic core, was obtained via a simple Michael/aldol cascade reaction sequence between a β-ketoester 35 and methacrolein (Scheme 17). The commercially available and inexpensive organocatalyst (−)-cinchonidine (XI) acts as a bifunctional organocatalyst. As a base it deprotonates 35 forming a chiral ion pair but the secondary alcohol function of the catalyst simultaneously activates a methacrolein molecule by forming a distinct hydrogen bond and incorporating it into the ionic complex. The Michael reaction, as the first step of the cascade reaction is thus initiated, followed by intramolecular aldol condensation. The tricyclic core 36 of (−)-huperzine A was formed with an overall yield of 60% and 64% enantiomeric excess (e.e.). The completion of the total synthesis starting from 36 required 5 further steps. It is worth noting that the synthesis of ent-37 could be achieved in the same way starting from cinchonine. Though to some extent disappointing for the modest enantioselectivity this procedure outlines a rapid one-pot entry to molecular complexity by using a simple metal-free, commercially available, and inexpensive air- and moisture-stable organocatalyst.

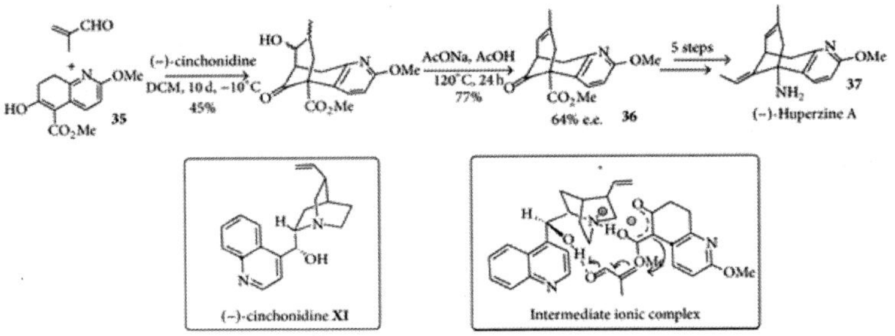

Scheme 17: Preparation of (−)-huperzine A by means ofan organocatalysed Michael/aldol cascade reaction sequence.

Brønsted Acid Catalysis

Recently chiral Brønsted acids have found widespread application in organocatalysis [69, 70]. For instance, in one of the most relevant processes the action of a Hantzsch ester, a biomimetic source of hydride, combines with that of chiral phosphoric acid as the catalyst. This can be considered as a metal-free simple H(+)-H(+) cascade reaction and has become a favourite application to the enantioselective reduction of nitrogen-containing heterocycles like pyridines or quinolines to the corresponding tetrahydroquinolines and tetrahydropyridines [71, 72]. This approach gives access to a variety of highly enantioenriched heterocycles that are privileged structures in natural products and drugs.

The preparation of fluoroquinolones reported by Rueping and coworkers [73] outlines the application of the transfer hydrogenation process to the synthesis of building blocks that have been utilized to complete the metal-free synthesis of drugs like (R)-flumequine (43) or (R)-levofloxacin (44) that display antibacterial activity towards a broad spectrum of bacteria [74, 75]. The readily available fluorinated quinoline 37 and benzoxazine 38 were reduced in the presence of Hantzsch esters 39 or 40 with only 1 mol% of the sterically demanding chiral phosphoric acid XII as catalyst to give the corresponding hydrogenated compounds 41 and 42 in very good yields and with excellent enantioselectivities (Scheme18).

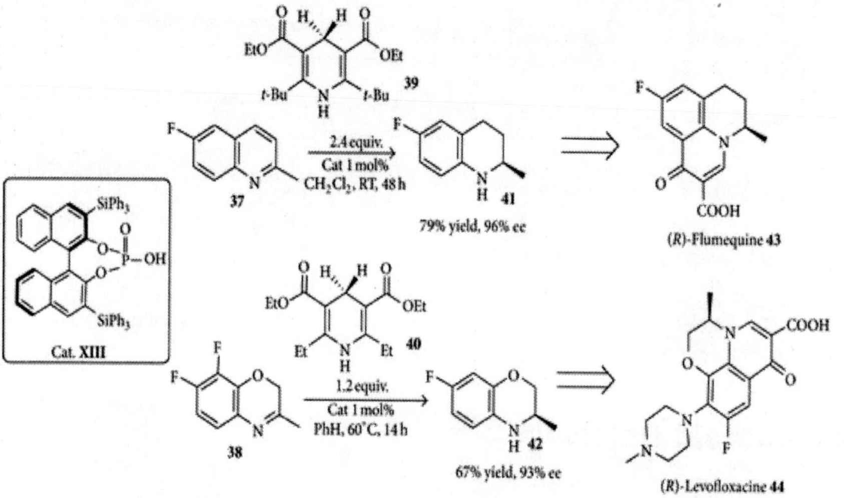

Scheme 18: Enantioselective transfer hydrogenation for the preparation of tri-cyclic fluoroquinolone antibacterial agents 43 and 44.

The synthesis of the two targets 43 and 44 was then accomplished in three more steps. Moreover through the use of only 1 mol% of the binaphthol phosphate catalysts XIV, a stepwise hydride transfer from the Hantzsch ester 45 to quinoline 46 afforded [76] the corresponding tetrahydroquinoline47 in excellent yields and enantioselectivities (Scheme 19). Mechanistically it has been assumed that this enantioselective cascade hydrogenation occurs in two cycles involving iminium ion an enamine species, respectively. A reductive N-methylation concludes a concise synthesis of (+)-galipinine 48 showing antimalarial activity on Plasmodium Falciparum for the chloroquine-resistant strains.

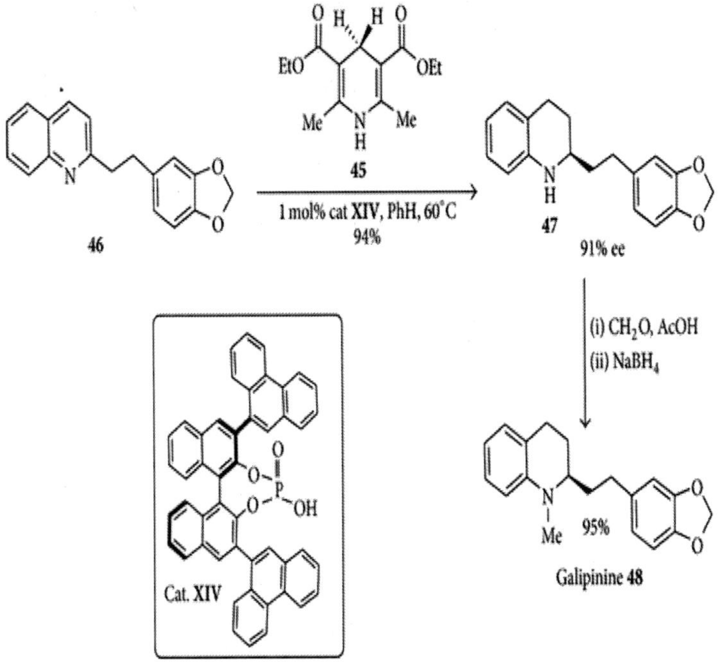

Scheme 19 Synthesis of (+)-galipinine via binolphosphoric acid-catalyzed enantioselective cascade reduction.

Another remarkable and to some extent different use of a chiral phosphoric acid in the synthesis of a drug candidate is represented by the one-pot acid-catalyzed three-component condensation of an aldehyde 49, a thiourea 50, and a β-ketoester 51 in an asymmetric Biginelli reaction to give the chiral 3,4-dihydropyrimidin-2-one derivatives 54 [77]. These scaffolds are privileged structures that depending on the substitution pattern exhibits a variety of important pharmacological properties like the inhibition of Hepatitis B virus replication. Here the chiral phosphoric acid XV catalyzes the Biginelli reaction by forming a chiral N-acyliminium phosphate ion pair 52 to which enantioselective addition of β-ketoesters 51 occurs to generate optically active 54 via the enantioenriched intermediate 53 (Scheme 20).

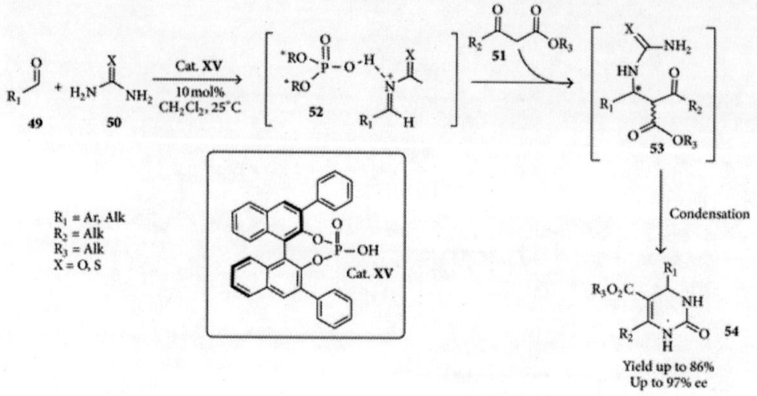

Scheme 20: Enantioselective chiral Brønsted acid-catalyzed three-component Biginelli reaction.

An asymmetric variant with an ytterbium-based catalyst for this Biginelli reaction was reported earlier [78], but the discovery of a metal-free synthesis by using Brønsted acid XV, which avoided contamination of the product with traces of metal, resulted in an important advancement. The phosphoric acid-based catalyst matched or even improved the level of conversion and stereoselectivity of the corresponding Lewis acid-catalyzed reaction, while maintaining the same substrate scope.

Covalent Organocatalysis

The area of amine-organocatalysed reactions is clearly dominated by secondary amines due to the versatility of possible combination of enamine (EN) and iminium (IM) activation. However, the primary amino function as a part of a chiral scaffold could be engaged as well in a number of synthetically appealing organocatalysed reactions. Several reviews on amino catalysis have recently appeared [79, 80].

Secondary Amine Organocatalysis via Enamines and Iminium Ions

The reaction that alerted the scientific community to the potential of organocatalysis was a proline-catalysed intramolecular aldol reaction reported almost simultaneously by two groups during the early 1970s [81, 82]. It was not until List et al. published a related intermolecular process [83] that secondary amine catalysis via enamine, inspired by Nature's aldolase enzymes, became en vogue in the domain of organocatalysed reactions. Since this report, there have been many subsequent publications of catalytic reactions via enamines. Proline-catalysed Mannich reactions [84], dihydroxylations [85], cross aldolizations [86], and aminations [87, 88] have held persistent interest in the area of asymmetric catalysis.

Mechanistically, this enamine catalysis might be better described as a bifunctional catalysis because the amine-containing catalyst (proline) typically interacts with a ketone substrate to form an enamine intermediate but simultaneously engages with an electrophilic reaction partner through either hydrogen bonding or electrostatic interaction (Scheme 21).

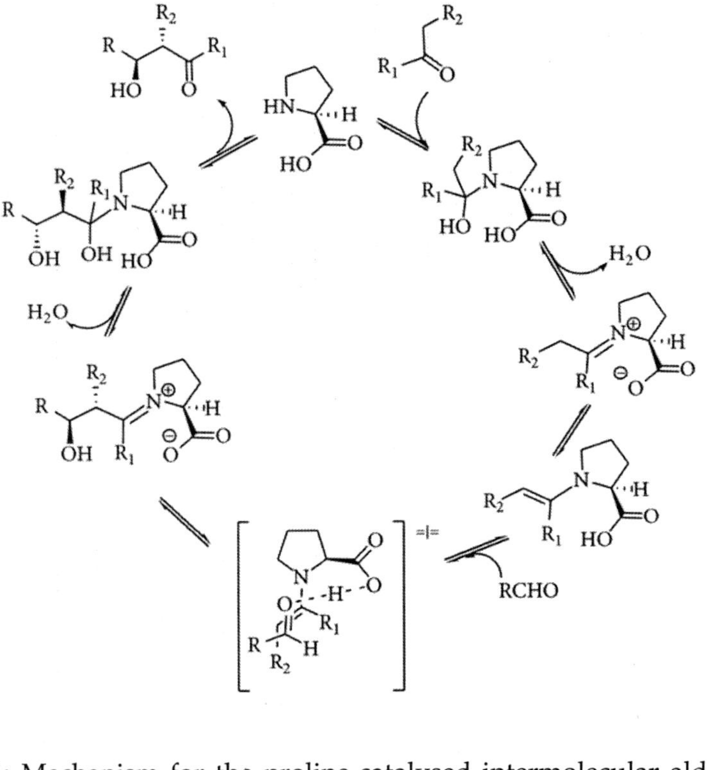

Scheme 21: Mechanism for the proline-catalysed intermolecular aldol reaction.

The capacity of chiral amines to function as enantioselective LUMO-lowering catalysts for a range of transformations that had traditionally employed Lewis acids has also been extensively used in organocatalysis. This strategy, termed iminium activation, was founded on the mechanistic postulate that the reversible formation of iminium ions from α,β-unsaturated aldehydes and chiral amines might emulate the equilibrium dynamics and π-orbital electronics that are inherent to Lewis acid catalysts, thereby providing a platform for designing organocatalytic processes (Scheme 22). The first generation catalyst to fulfil criteria such as efficient and easily reversible iminium ion formation, discrimination of the olefin π-face, and easy preparation was imidazolidinone XVI that in 2001 evolved in the more efficient imidazolidinone catalyst XVII (second generation). With its tailor-made family of imidazolidinone catalysts, iminium catalysis has been successfully applied to a broad range of chemical transformations including cycloadditions [89, 90], conjugate additions [91–93], hydrogenations [94], and cascade reactions [95]. The operational simplicity of these processes made them attractive alternatives to Lewis acid catalysis.

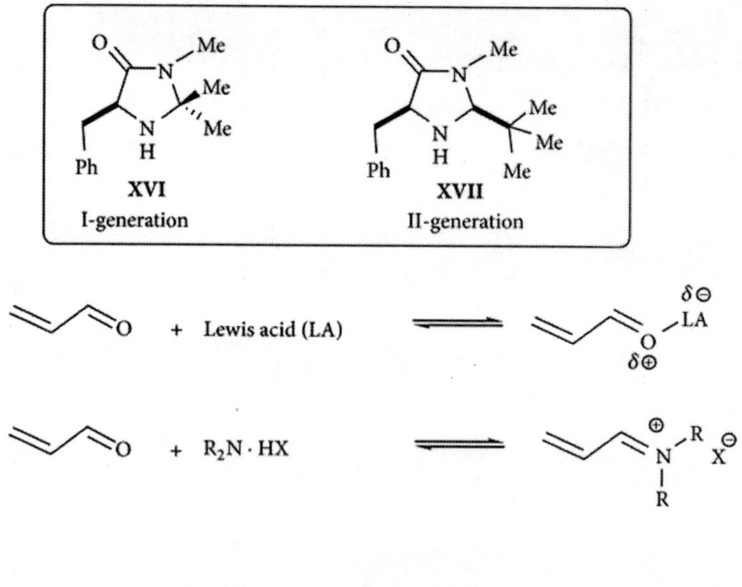

O Me N Me N H Me Ph
XVI
I-generation

O Me N Me N H Me Ph Me
XVII
II-generation

+ Lewis acid (LA) ⇌ δ⊖ LA O δ⊕

+ $R_2N \cdot HX$ ⇌ ⊕ N R R X⊖

Scheme 22: Iminium activation through LUMO lowering.

A number of drugs currently on the market have been approached with the enamine-iminium-based organocatalysis taking advantage by the simplicity of these inexpensive organocatalyst and by their high efficiency.

The case of warfarin is a very good example of the exceeding utility of organocatalytic methodologies in the assembly of relatively simple yet highly relevant molecules and many iminium-based organocatalysed processes have been designed for this aim. Warfarin is a vitamin K analogue, inhibiting vitamin K epoxide reductase. Its sodium salt, commercialised mainly under the trade names Coumadin and Marevan, is one of the most widely prescribed anticoagulants. Warfarin has been administered as a racemate for over fifty years; however, its two enantiomers display remarkably different pharmacological and pharmacokinetic profiles. Even if the S isomer shows higher activity, it is metabolised more rapidly than its less active R counterpart [96]. Thus, production of both (R)- and (S)-warfarin in enantiopure form might be of importance for a tailored patient treatment [97].

An obvious synthetic approach to warfarin is represented by the Michael addition of 4-hydroxycoumarin to benzylideneacetone, a reaction which is well posited for iminium ion catalysis through enone activation. Such an approach appears superior and more straightforward compared to the few reported catalytic asymmetric methods based on organometallic chemistry, which rely on more tortuous oxidation-reduction sequences with protecting groups usage [98, 99]. Accordingly, the feasibility of the organocatalytic strategy leading directly to warfarin has been well demonstrated, with the large number of reports well witnessing its success and appeal. A survey of the various organocatalysts engaged in this synthesis and of the related results is shown in Table 1.

Table 1: Catalytic asymmetric approaches to warfarin through iminium ion catalysis.

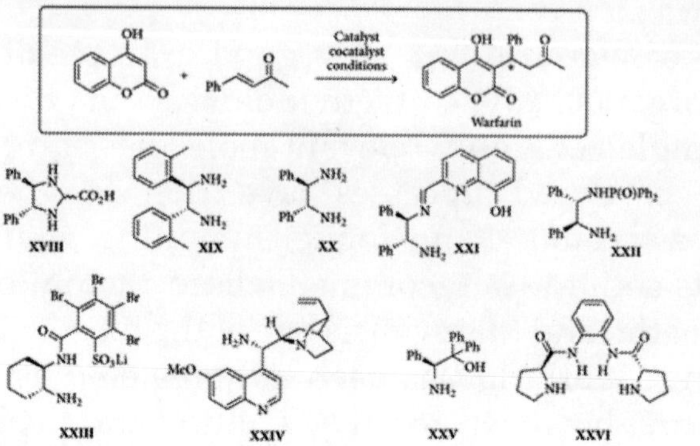

Entry	Catalyst	Cocatalyst	Conditions	Yield (%)	e.e. (%)	Ref.
1	**XVIII** (10 mol%)	—	CH_2Cl_2, RT, 150 h	96	82 (*R*)	[100, 101]
2	**XIX** (10 mol%)	AcOH (10 equiv.)	THF, RT, 24 h	99	92 (*R*)	[102]
3	**XX** (5 mol%)	$LiClO_4$ (5 mol%), AcOH (10 equiv.)	1,4-dioxane, RT, 24 h	61	92 (*R*)	[103]
4	*ent*-**XX** (2 mol%)	BzOH (4 mol%)	H_2O, RT, 12 h	78→ 50[a]	70 → >99[a] (*S*)	[104]
5	**XXI** (10 mol%)	Caproic acid (10 equiv.)	THF, RT, 36 h	96	91 (*S*)	[105]
6	**XXII** (10 mol%)	4-MeBzOH (20 mol%)	Toluene, RT, 72 h	99	94 (*R*)	[106]
7	**XXIII** (20 mol%)	—	THF/DMSO 4 : 1, RT, 48 h	99	42[b]	[107]
8	**XXIV** (20 mol%)	TFA (40 mol%)	CH_2Cl_2, 0°C, 96 h	88	96 (*S*)	[108]
9	**XXV** (10 mol%)	Succinic acid (10 mol%)	*n*-BuOH/H_2O, 40°C, 12 h	97	83 (*R*)	[109]
10	**XXVI** (20 mol%)	AcOH (1 equiv.)	THF, RT, 24 h	76	81[b]	[110]

The first approach to enantioenriched warfarin through iminium ion catalysis, disclosed in 2003, involved [100] the employment of the phenylenediamine-derived catalyst XVIII and provided after prolonged reaction time the target compound in good yield and moderate enantioselectivity, which could be improved to >99% by a simple crystallisation (Table 1, entry 1). It was also mentioned [101] that the reaction could be performed on a kg scale by employing a related catalyst structure, recoverable after reaction completion. The approach is the same as the commercial synthesis of Coumadin in a retrosynthetic sense, but crucially the presence of imidazolidine catalysts XVIII affords an enantioselective reaction whilst the commercial synthesis is racemic. A few years later it was discovered that the putative catalyst XVIII was undergoing a hydrolysis under the reaction conditions, delivering the corresponding free phenylenediamine, which was the actual catalytic species of the Michael addition. On this basis, the results obtained in this transformation could be improved [102] by employing a related diamine XIX in combination with a large amount of an acidic cocatalyst (acetic acid, entry 2). Subsequent efforts with simple primary diamine catalysts were directed [103] at improving the practical applicability of this reaction and included the discovery that the catalyst loading of phenylenediamineXX could be lowered by exploiting a combination of Lewis (lithium perchlorate) and Brønsted (acetic acid) acid cocatalysts in the reaction (entry 3) and that water as reaction medium under ultrasound activation [104] provided a fast protocol allowing the isolation of warfarin in essentially enantiopure form by simple filtration of the reaction mixture (entry 4). Variations in the phenylenediamine motif by functionalization of one of the amines have also been reported and include the monoamine XXI, generated in situ by acidic hydrolysis from a corresponding diimine [105], and the monophosphonamide XXII [106].

Both of these catalysts, employed in combination with carboxylic acids, provide the product warfarin in excellent enantioselectivity (entries5and6).Adifferentdiamine,namely,1,2-cyclohexanediamine, was employed for the preparation of catalyst XXIII bearing a lithium sulfate functionality on one of the amines, which however led [107] to rather poor results (entry 7). The lithium based catalyst XXIII was designed to exploit the mild Lewis acidity of lithium for the intramolecular activation of the imine formed between the catalyst and the Michael acceptor. Interestingly, computational studies showed that a related reaction (with cyclohexenone as Michael acceptor) follows an ene-type concerted pathway, with simultaneous C–C bond formation and proton transfer between the hydroxyl coumarin and the iminium ion activated olefin of the Michael acceptor (Figure 6).

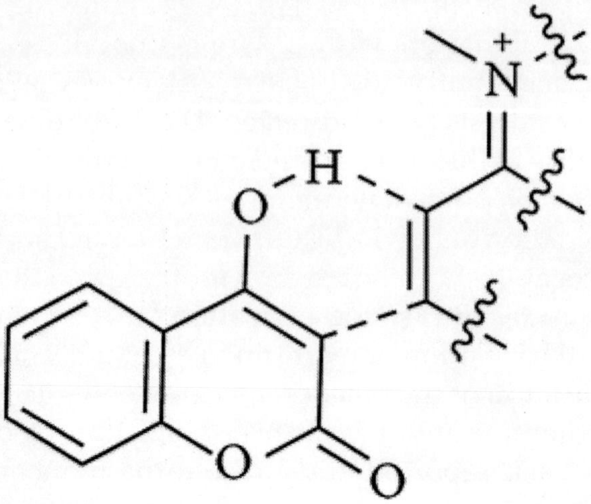

Figure 6: Reaction between coumarin and iminium ion in a concerted pathway.

A structurally distinct primary amine catalyst XXIV, derived from a Cinchona alkaloid, was also successfully applied in this reaction [108] with very good results at relatively high catalyst loadings (entry 8). Highlighting the high interest that this specific transformation has surged in the academic community, a recent publication reported the search for a more readily available catalytic system, considering that the preparation of the phenylenediamine cores of catalysts XVIII–XXII is rather troublesome and that the synthesis of the Cinchona derivative XXIV presents considerable safety issues (it is prepared through a Mitsunobu reaction which employs hazardous azides). To that purpose, the focus was set [109] on amino acid as starting materials, resulting in the disclosure of the diphenyl phenylglycinol XXV as a moderately efficient and easily available catalyst for the obtainment of warfarin in enantioenriched form (entry 9). The same report highlighted the better performances of primary amine compared to secondary amines in this reaction, reporting also the poor performances in this transformation of other popular secondary amine catalysts, such as proline or a MacMillan imidazolidinone. However, it was more recently demonstrated [110] that also a secondary amine, the dimeric proline-derived catalyst

XXV, could be applied in this Michael addition reaction with fairly good results (entry 10).

Simple iminium ion catalysed reactions have also been inserted into more complex multistep reaction sequences; some of them herein are discussed in detail leading to medicinally relevant compounds. In this context, the example of Telcagepant is significant, as an iminium ion catalysed reaction was selected as the key step for the establishment of the stereochemistry in an industrial-scale preparation of this molecule. Telcagepant [111] is an antagonist of the calcitonin gene-related neuropeptide and was considered highly promising for the treatment of migraine avoiding concomitant cardiovascular problems often generated by other antimigraine agents. Telcagepant can be retrosynthetically disconnected between a chiral 2-amino-5-aryl caprolactam and a 4-aminopiperidine units, assembled with concomitant urea formation (Scheme 23).

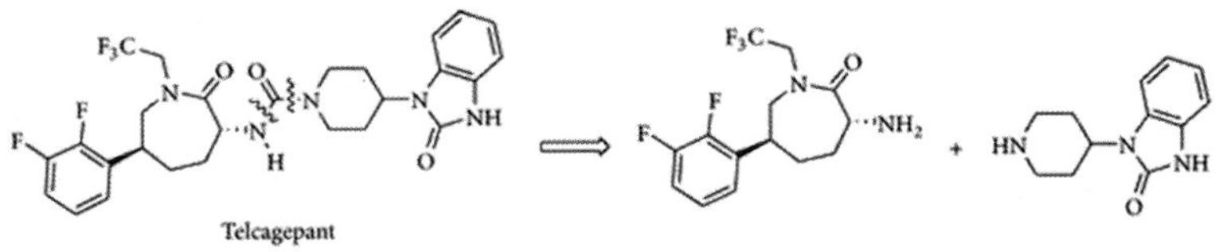

Telcagepant

Scheme 23: Retrosynthetic disconnection of Telcagepant.

A first-generation multi-kg (>500) process [112], relying on a dynamic kinetic resolution of the enantiomers through crystallisation, showcased several pitfalls in the preparation of the chiral caprolactam, thus calling for alternative synthetic strategies to this unit. To this end, an organocatalytic approach, namely, an iminium ion catalysed conjugate addition of nitromethane to a proper cinnamaldehyde [113], was judged as more promising than other asymmetric preparations based on transition metal catalysed reactions, such as ruthenium-catalysed hydrogenation [114] and rhodium based Hayashi-Miyaura addition [115]. Although previously reported with related substrates [116, 117] the organocatalytic step required a careful optimisation for its large scale implementation, as the formation of several by-products was observed under the reported conditions. In particular, to avoid the formation of acetals the usual alcoholic solvents were replaced with a THF/water solvent mixture. The employment of a carefully tailored additive mixture composed by boric and pivaloyl acid proved also to be necessary to achieve an acceptable reaction rate while minimising by-products formation. A final amelioration involved the employment of the crude mixture of the in situ silylated diphenylprolinolXXVII as catalyst [118, 119], thus avoiding its purification. Finally, the reaction could be performed on a pilot plant (>100 kg) affording the desired Michael adduct 56 in 73% assay yield and 95% enantiomeric excess (Scheme 24).

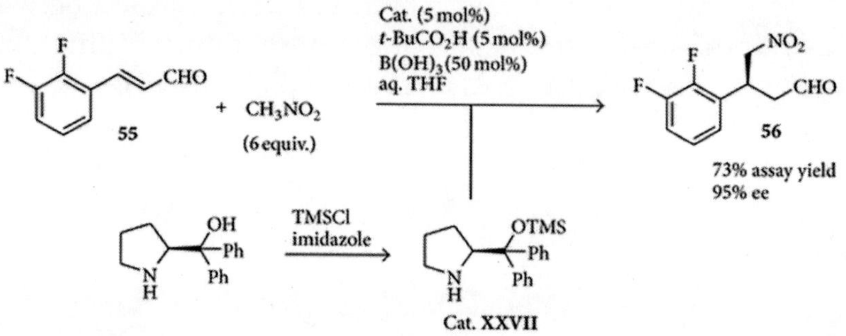

Scheme 24: Organocatalysed Michael addition.

Obviously, both the preparation of the starting aldehyde and the subsequent reaction steps leading to the caprolactam had also to be carefully optimised for the large scale synthesis. The first part of the overall sequence is depicted in Scheme 25 and involves the preparation of an allylic alcohol from difluorobenzene through lithiation, quenching with acrolein, acid promoted rearrangement, and TEMPO catalysed oxidation to give the substrate 55 to be submitted to the organocatalytic step. The aldehyde product 56 was not isolated, but treated directly in a Doebner-Knoevenagel type reaction with an in situ generated acetamido malonic acid (not isolated due to safety issues) in the presence of pyrrolidine. This decarboxylative Knoevenagel-type reaction, even if required extensive optimisation, was preferred over more established Wittig protocols based on atom economy reasoning. The resulting acid

57 was isolated and upgraded to >99% ee as a tributyl ammonium salt 58, obtained in an impressive 48% overall yield starting from difluorobenzene.

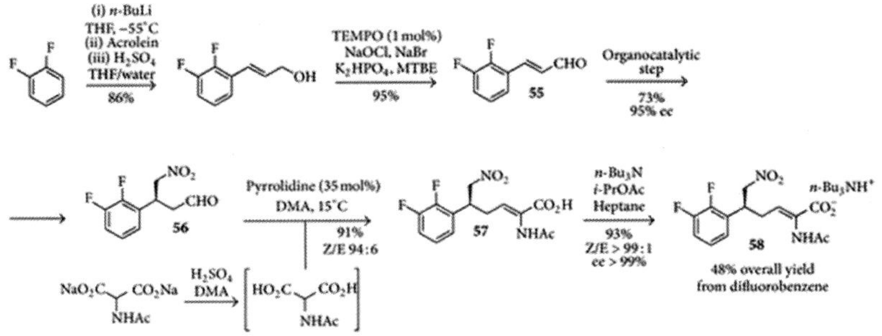

Scheme 25: Large scale preparation of 58.

Hydrogenation of this enamide intermediate (57) with concomitant nitro group reduction proved to be very challenging, due to the unwanted formation of desfluoro products, which could not be present in more than tiny (0.2%) amounts for the successful employment of this synthesis for telcagepant production. After several attempts (Scheme 26), the hydrogenation reduction showed the desired features when performed on the acid in the presence of lithium salts, in isopropanol under acidic conditions, conditions which also promoted esterification of the carboxylate. Trifluoroethyl alkylation with a triflate gave the N-alkylated compound 59, whose ester was hydrolysed to the carboxylic acid.

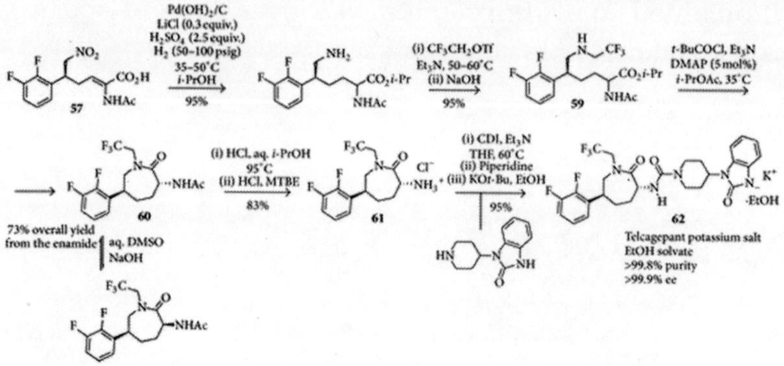

Scheme 26: Second part of the synthesis of Telcagepant.

Cyclisation involved the employment of a mixed anhydride, possible for the decreased nucleophilicity of the trifluoroethyl nitrogen, and was followed by a base promoted dynamic crystallisation process in aqueous NaOH/DMSO mixture, which allowed the isolation of the 2-acetamido caprolactam 60 in 73% overall yield from the enamide. Acidic hydrolysis of the acetamide gave the amine 61, which was engaged in urea formation by treatment with carbonyl diimidazole and the corresponding piperidine. Deprotonation of the benzamide hydrogen in ethanol furnished the potassium salt of telcagepant 62 as an ethanol solvate, with >99.8% purity and >99.9% ee. This environmentally responsible synthesis contains all of the elements required for a manufacturing process and prepares telcagepant 62 with the high quality required for pharmaceutical use.

The key role played by proline-derived organocatalysts in medicinal chemistry is also witnessed by the recent synthesis of maraviroc. Human immunodeficiency virus (HIV) is a pandemic that was first recognized in 1981. In 2008 HIV caused the death of 2 million people due to the acquired immune deficiency syndrome (AIDS). Due to high medical need the exciting prospect of a new class of drugs for HIV led to the rapid development of the academic and industrial research in this field. Maraviroc (UK-427,857) (69) is a chemokine receptor 5 (CCR-5) receptor antagonist that is currently developed at Pfizer for the treatment of HIV. A robust, plant-suitable synthesis of 69 was urgently required to manufacture larger quantities and the process research [120, 121] of a commercialisable route has been recently developed.

The chiral β-amino aldehyde 65 a key intermediate in synthesis of 68 was produced in the industrial process in an 8.3% overall yield in a multistep sequence in which the enantioenriched β-amino ester 64was obtained by tartaric acid resolution of the racemic counterpart. Moreover the resolution of the β-amino ester was only moderately efficient with two recrystallisations required to achieve the desired enantiomeric excess of >95% (Scheme 27, route a).

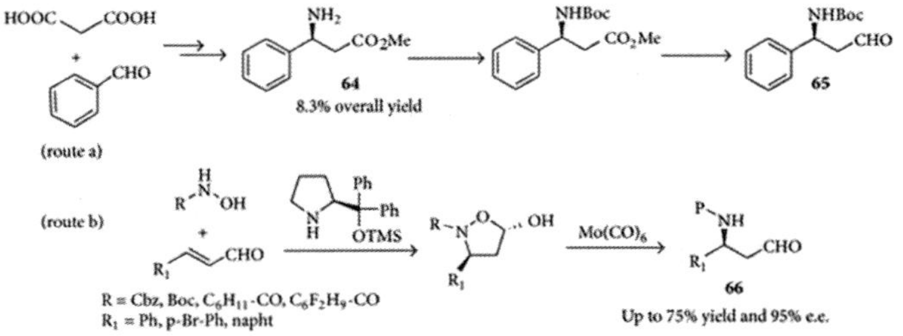

Scheme 27: Resolution-based (route a) and organocatalysed (route b) asymmetric synthesis of β-amino aldehydes.

An interesting perspective for introducing improvements in this process is offered by organocatalysis. In a recent paper [122] the Córdova group has proposed the asymmetric synthesis of maraviroc and its analogues via an organocatalysed highly enantioselective synthesis of the key chiral β-amino aldehyde 66fragment. The assembly of 66 occurs via a two-step protocol involving catalytic enantioselective tandem aza-Michael hemiacetal formation between hydroxylamine and enal followed by N–O cleavage (Scheme27, route b).

The use of (S)-diphenylprolinol silyl ether XXVII as the catalyst appears to be very helpful in providing the β-amino aldehydes 66 in yields up to 75% and remarkably good (92–95%) enantioselectivities. Moreover the tolerance of the organocatalysed reaction to a wide range of protecting groups at nitrogen opens the route to structural diversity leading to different analogues of maraviroc of potential interest for scouting new medicinal targets.

Following the organocatalysed methodology maraviroc 69 (Figure 7) could be obtained (3 steps) by reaction of -amino aldehyde 67 and tropane 68 in 53% overall yield with 80% e.e. and (3 steps) 45% overall yield with 92% e.e. by varying the reaction conditions (Scheme 28). It is also worth noting that the reaction presented herein can be scaled up. Thus the multigram-scale reaction between cinnamic aldehyde (31.4 g) and Cbz-protected hydroxylamine in the presence of 7 mol% of (S)-proline-derived catalyst gave the corresponding 5-hydroxyoxazolidinone precursor of the of β-amino aldehyde 67 in 91% yield (65 g) and 99% e.e.

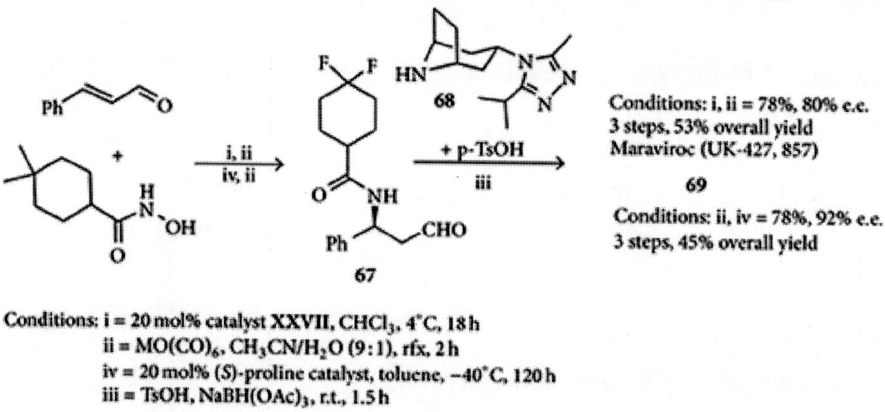
Ph
O
H
N
OH
O
i, ii
iv, ii
F
F
O
NH
Ph
CHO
67
HN
68
N
N
N
+ p-TsOH
iii
Conditions: i, ii = 78%, 80% e.e.
3 steps, 53% overall yield
Maraviroc (UK-427, 857)
69
Conditions: ii, iv = 78%, 92% e.e.
3 steps, 45% overall yield
Conditions: i = 20 mol% catalyst XXVII, CHCl3, 4°C, 18 h
ii = MO(CO)6, CH3CN/H2O (9:1), rfx, 2 h
iv = 20 mol% (S)-proline catalyst, toluene, −40°C, 120 h
iii = TsOH, NaBH(OAc)3, r.t., 1.5 h

Scheme 28: Silylated prolinol-catalysed enantioselective synthesis of Maraviroc.

Iminium ion organocatalysis appears to be key step in the synthesis of the antidepressants (−)-paroxetine (72). This molecule originally marketed by GlaxoSmithKline as Seroxat, a blockbuster selective serotonin reuptake inhibitor for the treatment of depression and anxiety related disorders, has previously been focused on enzymatic asymmetric desymmetrization [123, 124] chiral auxiliary-assisted [125], or asymmetric, deprotonation reactions [126]. In the previous methodologies the total synthesis of this chiral compound consisted of approximately 12–14 steps. A novel approach, based on organocatalysis employing a chiral proline-derived catalyst led to a number of brief (formal) syntheses of paroxetine.

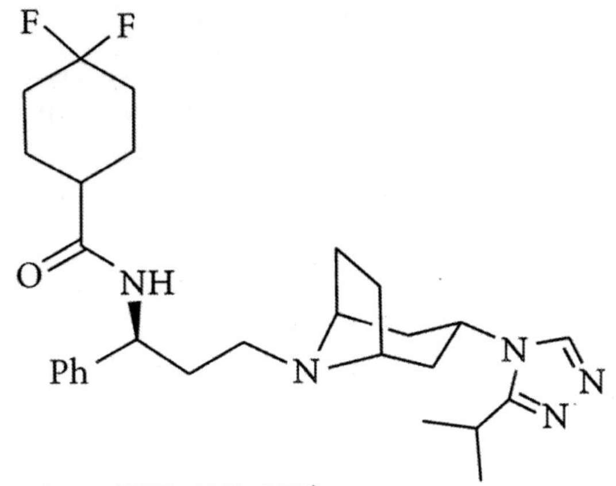

Maraviroc (UK-427, 857)
69

Figure 7: The structure of maraviroc (69).

Using simple building blocks like fluorocinnamaldehyde 70 and malonate derivative 71 with the silylated prolinol XXVII as the organocatalyst, Michael reaction was shown (Scheme 29, route a) to proceed in good yields and stereoselectivity [127] to afford the target product 72 in a three-step sequence. A previous report [128] leading to (−)-paroxetine in six steps overall and based on organocatalyst XVIII had also shown the possibility of performing a potentially scalable Michael addition (Scheme 29, route b).

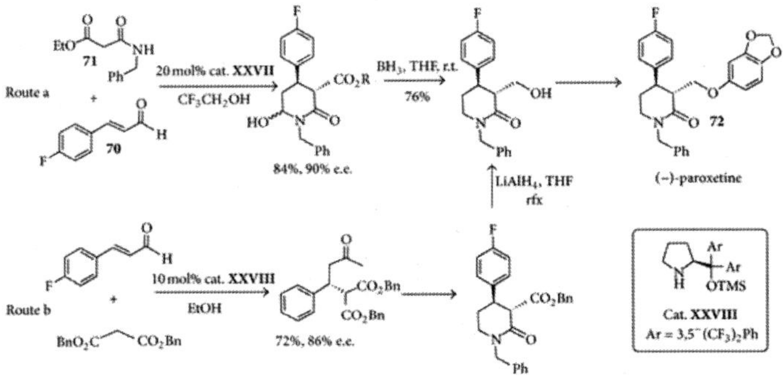

Scheme 29: Formal syntheses of (−)-paroxetine via organocatalysis.

In both cases, the nature of the solvent was highlighted as a key parameter, and polar-protic solvents were required. Whereas the latter example utilised a much more industrially friendly solvent (ethanol versus trifluoroethanol), the reaction times were greater, running for multiple days. Regardless, the potential step saving in these routes is significant taking into account that current industrial syntheses are typically 10–15 steps. Moreover the simplicity of the chemistry involved makes these organocatalysed processes suitable candidates or further scale-up activities.

The spirocyclic motif is featured in a number of natural products as well as medicinally relevant compounds [129–131]. The stereocontrolled construction of the spirocyclic indole core, of high interest for the synthesis of biologically active compounds, poses a challenging synthetic problem mainly in connection with the installation of the spiroquaternary stereocenter. Only a few asymmetric transformations based on cycloaddition processes [132–134] or on the intramolecular Heck reaction [135, 136] have proven successful for achieving this goal. The asymmetric cascade organocatalysis exploiting the ability of chiral amines to combine two modes of activation of carbonyl compounds viaiminium and enamine catalysis into one mechanistic scheme allows the direct, one-step synthesis of complex spirocyclic oxindoles having multiple stereocenters.

Among them the synthesis of spirooxindole 77 recently patented by Hoffmann-La Roche [137], a specific and potent inhibitor of MDM2-p53 interaction and innovative target for the discovery of anticancer agents [138], has been recently reported [139] which highlights the potential of organocascades in building complex structures of interest in medicinal chemistry (Scheme 30). In the one-pot double organocascade compound 73 would first act as a Michael acceptor, intercepting the nucleophilic diamine intermediate 75 generated by condensation of the catalyst XXIX with the α,β-unsaturated ketone 74. The resulting Michael addition product 76 would then selectively engage itself in an intramolecular iminium catalysed conjugate addition to afford the spirooxindole 77. The target product could be obtained with 99% d.e. and e.e. after a single crystallization.

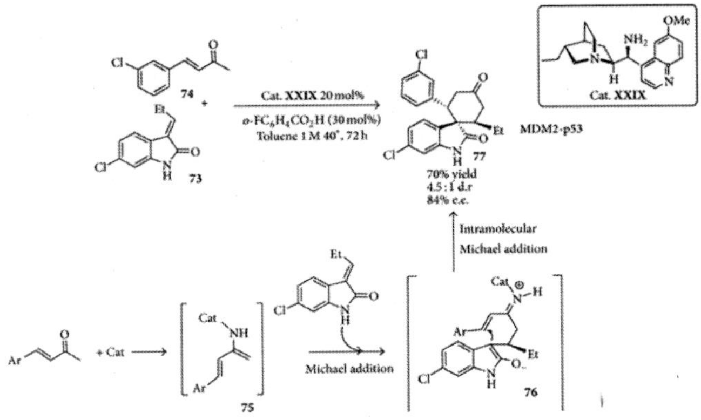

Scheme 30: Tandem double Michael addition via enamine-iminium activation sequence toward spirocyclic oxindole.

Influenza viruses pose a serious threat to world public health. Two of the drugs currently used to treat influenza patients are (−)-oseltamivir phosphate (Tamiflu) [140] and zanamivir (Relenza) [141]. Enamine catalysed transformations were exploited in the realisation of synthetic sequences leading to (−)-oseltamivir. The phosphate salt of this molecule, commercialised as Tamiflu, is a neuraminidase inhibitor useful as antiviral agent for the treatment and the prevention of influenza infections. Due to its specific biological action, (−)-oseltamivir is considered to be very general against all types of influenza viruses, including the avian H5N1, which has a mortality rate close to 50%. In the mid-2010s, fear for a potentially disastrous pandemic spread of a mutation of this virus amongst humans has prompted several nations to plan the accumulation of massive stocks of this drug. The production of this molecule is being based on a multistep synthesis employing shikimic acid as starting material. Both the variable availability of natural shikimic acid and the relative length of the synthesis put great pressure on industrialists and academics to develop very quickly alternative synthetic routes, in order to satisfy the exponentially increasing requests [142]. Asymmetric catalysis was considered a very attractive method, as it would not rely heavily on natural chiral sources. Accordingly, several syntheses based on enantioselective Lewis acid catalysis have been reported [143], starting already from 2006 [144,

145]. Enamine catalysis was later appreciated as a useful alternative and was consequently applied as the key stereodetermining step in a few syntheses of (−)-oseltamivir. Interestingly, many of the reported protocols went beyond the study of the simple reaction steps but increased the overall efficiency by combining many of the steps in a one pot fashion.

Essentially, two complementary organocatalytic strategies to (−)-oseltamivir have been developed, both based on the addition of an α-alkoxy aldehyde (pentan-3-yloxyacetaldehyde) to a nitroalkene. The first approach, disclosed in 2009 [146], involved the addition of this aldehyde to a 2-ester substituted nitroalkene, catalysed by a silyl-protected D-prolinol derivative. This constituted the first reaction of a nine-step synthesis leading to the target compound (−)-oseltamivir, in which all steps were carefully optimised and designed for their combination in one-pot sequences.

The whole sequence could be in fact carried out in only three one-pot operations, which soon after [147] were reduced to two in the second generation synthesis depicted in Scheme 31. The synthesis starts with the mentioned organocatalytic reaction. The nitroaldehyde product 78, generated with high enantioselectivity, reacts then with a vinyl phosphonate in the presence of a base, giving a nitro-Michael Horner-Wadsworth-Emmons sequential reaction which however leads to a mixture of different products 79 a–c.

Solvent evaporation followed by stirring the basic mixture in ethanol triggers retroaldol elimination processes followed by olefination rendering very nicely a single cyclohexene product 80, which upon treatment with a thiophenol gives a cyclic five-substituted cyclohexane 81 featuring the desired stereochemistry, isolated in overall 56% yield over the three steps. The introduction of the thiophenol is necessary to control the relative configuration at the α-nitro chiral centre through equilibration. This intermediate (81) is converted into the target (−)-oseltamivir 84 through a six-step one-pot sequence, starting withtert-butyl ester cleavage with trifluoroacetic acid, leading to an acid which is converted first in a chloride and then in an acylazide 82.

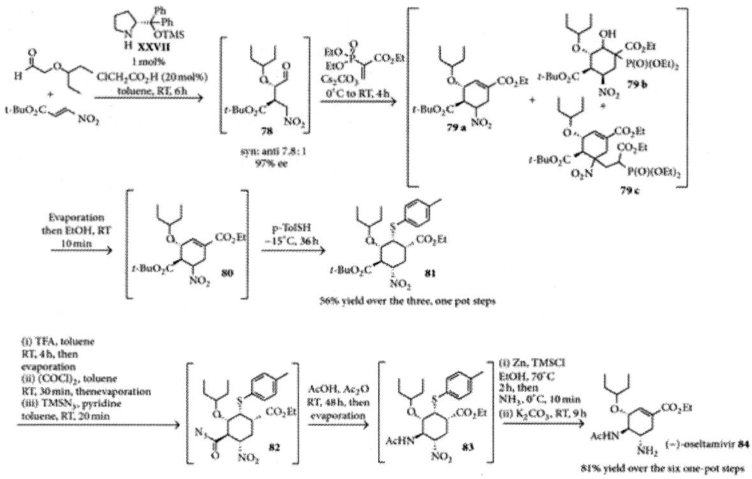

Scheme 31: Two-pot organocatalysed synthesis of (−)-oseltamivir 84.

This azide undergoes a Curtius rearrangement with concomitant acetylation, which serves to install the desired acetamide functionality (83). Nitro group reduction with zinc, ammonia bubbling, and elimination of the stereodirecting thiophenol through a retro-Michael addition finally leads to the target oseltamivir product 84. It is worth noting that the metal-based reagents employed in this synthesis contain either alkali-metal ions or nontoxic Zn. Thus this procedure is suitable for large-scale preparation.

A second approach to this target molecule through organocatalysis aimed at avoiding the hazardous azide addition and Curtius rearrangement steps, by introducing the amide functionality already at the beginning of the synthesis. To this end, the readily available 2-Z-acetamido nitroethene 85 was engaged in an enamine catalysed reaction with the same aldehyde previously employed. By this approach, disclosed in 2010 [148], the target oseltamivir 84 could be obtained after a few more steps based on the previous synthesis. Due to its importance, this organocatalytic reaction and the subsequent steps were then the subject of thorough studies directed at their generalisation [149, 150] and careful optimisation [151], which culminated very recently in the disclosure of a synthesis of oseltamivir proceeding in one-pot from the aldehyde and this nitroalkene [152]. Compared to the previous sequence reported in Scheme 31, this synthesis (Scheme 32) not only obviates the need of introducing the acetamide functionality with azide chemistry, but also avoids evaporation steps and the usage of toxic chlorinated solvents, allowing the obtainment of oseltamivir even on gram scale with a remarkable 28% yield over the five steps without any intermediate work-up or isolation.

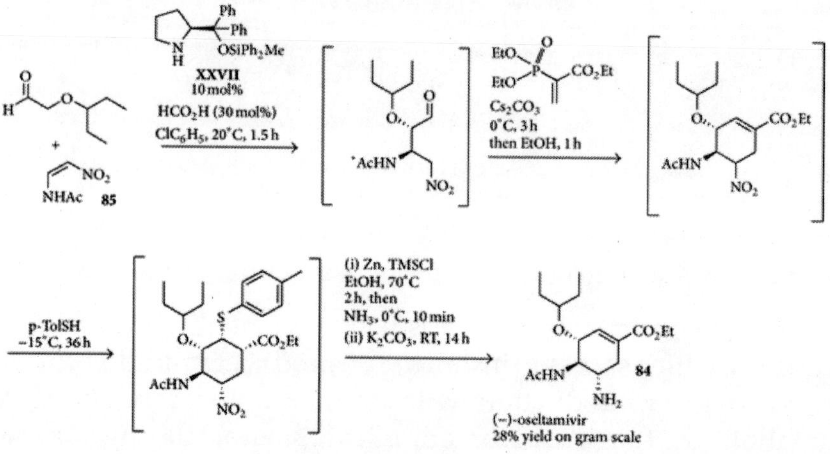

Scheme 32: Scalable organocatalysed short synthesis of (−)-oseltamivir 84.

In the same year a nine-step synthesis of (−)-oseltamivir has been proposed [153] in which a novel barium-catalysed asymmetric Diels-Alder-type reaction gives access to the cyclohexene framework of Tamiflu in the key enantioselective reaction path with a Pd-catalysed allylic occurring in the following steps of the synthetic sequence. Though of relevant interest for employing a nontoxic metal and for being effective at 60 g scale, the advantage of this synthesis with respect to the organocatalysed methodology is to some extent diminished by the need of isolating all the intermediates and by the use of a potentially explosive reagent such as diphenyl phosphoryl azide.

Concluding Remarks

Featuring aspects of medicinal chemistry are the strong reliability of the processes, the use of cheap and commercial catalysts, and short synthetic sequences. Moreover, both enantiomers of the catalysts should be easily available in order to obtain both enantiomers of the products and allow analysing them from a biological viewpoint. As herein highlighted, enantioselective organocatalysis plays nowadays a major role, alongside metal-catalysed processes, in chemical synthesis because together these complementary disciplines have revolutionized the way the synthesis is carried out. A point that should not be underestimated is the attractiveness of the operational simplicity of organocatalytic reactions: rigorous exclusion of oxygen and moisture is usually not required, and potential toxic metal contamination is avoided. The success and the usefulness of the organocatalytic approach are well demonstrated by the several examples in which different organocatalytic strategies, involving various activation modes and/or catalyst structures, have been successfully applied to the same target compound, as herein exemplified in the case of baclofen or rolipram.

The application of organocatalysis to the synthesis of molecules of interest in various fields including medicinal chemistry is however still facing problems such as the relatively high catalyst loading, the long reaction time, and also the difficult recyclability of the organocatalyst. To give a partial solution, significant advances have been reported in recent times describing polymer supported organic catalysts, easily recovered after the reaction has taken place by filtration [154–156]. In contrast with immobilised metal complexes (via solid-supported bound ligand), leaching problems are much less critical when using organocatalysts immobilised by covalent bonding to the solid support. Even if costs considerations, as well as decrease of catalyst activity and selectivity, are issues not fully solved yet, several remarkable examples have been reported, which show great promise for future development.

Conflict of Interests

The authors declare that there is no conflict of interests regarding the publication of this paper.

ACKNOWLEDGMENTS

The author expresses the warmest thanks to Dr. Luca Bernardi for his precious suggestions and reading of the paper.

REFERENCES

1. P. I. Dalko and L. Moisan, "In the golden age of organocatalysis," Angewandte Chemie International Edition, vol. 43, no. 39, pp. 5138–5175, 2004. View at Publisher · View at Google Scholar
2. B. List, "The ying and yang of asymmetric aminocatalysis," Chemical Communications, pp. 819–824, 2006.
3. B. List, "Organocatalysis: a complementary catalysis strategy advances organic synthesis," Advanced Synthesis & Catalysis, vol. 346, no. 9-10, p. 1021, 2004. View at Publisher · View at Google Scholar
4. M. Nielsen, D. Worgull, T. Zwifel, B. Gschwend, S. Bertelsen, and K. A. Jorgensen, "Mechanisms in aminocatalysis," Chemical Communications, vol. 47, no. 2, pp. 632–649, 2011. View at Publisher · View at Google Scholar.
5. A. Grossmann and D. Enders, "N-heterocyclic carbene catalyzed domino reactions," Angewandte Chemie International Edition, vol. 51, no. 2, pp. 314–325, 2012. View at Publisher · View at Google Scholar
6. P. M. Pihko, Hydrogen Bonding in Organic Synthesis, Wiley-VCH, Weinheim, Germany, 2009.
7. K. Maruoka, Asymmetric Phase Transfer Catalysis, Wiley-VCH, Weinheim, Germany, 2008.
8. E. N. Jacobsen and D. W. C. McMillan, "Organocatalysis," Proceedings of the National Academy of Sciences of the United States of America, vol. 107, no. 48, pp. 20618–20619, 2010. View at Publisher · View at Google Scholar
9. D. W. C. McMillan, "Commentary the advent and development of organocatalysis," Nature, vol. 455, pp. 304–308, 2008. View at Publisher · View at Google Scholar
10. B. List, "Biocatalysis and organocatalysis: asymmetric synthesis inspired by nature," inAsymmetric Synthesis: The Essentials, M. Christmann and S. Brase, Eds., pp. 161–165, Wiley-VCH, Weinheim, Germany, 2007.
11. P. I. Dalko and L. Moisan, "In the golden age of organocatalysis," Angewandte Chemie International Edition, vol. 43, no. 39, pp. 5138–

5175, 2004. View at Publisher · View at Google Scholar · View at Scopus.

12. T. Newhouse, P. S. Baran, and R. W. Hoffmann, "The economies of synthesis," Chemical Society Reviews, vol. 38, no. 11, pp. 3010–3021, 2009. View at Publisher · View at Google Scholar
13. D. Enders, M. R. M. Hüttl, C. Grondal, and G. Raabe, "Control of four stereocentres in a triple cascade organocatalytic reaction," Nature, vol. 441, no. 7095, pp. 861–863, 2006. View at Publisher · View at Google Scholar · View at Scopus.
14. A. Carlone, S. Cabrera, M. Marigo, and K. A. Jørgensen, "A new approach for an organocatalytic multicomponent domino asymmetric reaction," Angewandte Chemie International Edition, vol. 46, no. 7, pp. 1101–1104, 2007. View at Publisher · View at Google Scholar
15. M. Eichelbaum, B. Testa, and A. Somogyi, Eds., Stereochemical Aspects of Drug Action and Disposition, Spriger, Heidelberg, Germany, 2003.
16. K. C. Nicolau, D. J. Edmonds, and P. G. Bulger, "Cascade reactions in total synthesis," Angewandte Chemie International Edition, vol. 45, no. 43, pp. 7134–7186, 2006. View at Publisher · View at Google Scholar
17. C. Grondal, M. Jeanty, and D. Enders, "Organocatalytic cascade reactions as a new tool in total synthesis," Nature Chemistry, vol. 2, no. 3, pp. 167–178, 2010. View at Publisher · View at Google Scholar · View at Scopus
18. R. M. de Figueiredo and M. Christman, "Organocatalytic synthesis of drugs and bioactive natural products," European Journal of Organic Chemistry, no. 16, pp. 2575–2600, 2007.
19. J. Alemán and S. Cabrera, "Applications of asymmetric organocatalysis in medicinal chemistry," Chemical Society Reviews, vol. 42, no. 2, pp. 774–793, 2013. View at Publisher · View at Google Scholar
20. H. U. Blaser and E. Schmidt, Asymmetric Catalysis on Industrial Scale: Challenges, Approaches and Solutions, Wiley-VCH, Weinheim, Germany, 2004.
21. H. Gröger, "Asymmetric organocatalysis on a technical scale: current status and future challenges," Organocatalysis, vol. 2007/2, pp. 227–258, 2008. View at Publisher · View at Google Scholar
22. C. A. Busacca, D. R. Fandrick, J. J. Song, and C. H. Senanayake, "The growing impact of catalysis in the pharmaceutical industry," Advanced Synthesis and Catalysis, vol. 353, no. 11-12, pp. 1825–1864, 2011. View at Publisher · View at Google Scholar · View at Scopus
23. G. P. Howell, "Asymmetric and diastereoselective conjugate addition reactions: C–C bond formation at large scale," Organic Process Research & Development, vol. 16, no. 7, pp. 1258–1272, 2012. View at

Publisher · View at Google Scholar

24. M. S. Sigman and E. N. Jacobsen, "Schiff base catalysts for the asymmetric strecker reaction identified and optimized from parallel synthetic libraries," Journal of the American Chemical Society, vol. 120, no. 19, pp. 4901–4902, 1998. View at Publisher · View at Google Scholar · View at Scopus
25. E. J. Corey and M. J. Grogan, "Enantioselective synthesis of α-amino nitriles from N-benzhydryl imines and HCN with a chiral bicyclic guanidine as catalyst," Organic Letters, vol. 1, no. 1, pp. 157–160, 1999. View at Publisher · View at Google Scholar
26. E. A. C. Davie, S. M. Mennen, Y. Xu, and S. J. Miller, "Asymmetric catalysis mediated by synthetic peptides," Chemical Reviews, vol. 107, no. 12, pp. 5759–5812, 2007. View at Publisher ·View at Google Scholar · View at Scopus
27. L. Bernardi, M. Fochi, M. Comes Franchini, and A. Ricci, "Bioinspired organocatalytic asymmetric reactions," Organic and Biomolecular Chemistry, vol. 10, no. 15, pp. 2911–2922, 2012.View at Publisher · View at Google Scholar · View at Scopus
28. M. M. Benning, T. Haller, J. A. Gerlt, and H. M. Holden, "New reactions in the crotonase superfamily: structure of methylmalonyl CoA decarboxylase from Escherichia coli," Biochemistry, vol. 39, no. 16, pp. 4630–4639, 2000. View at Publisher · View at Google Scholar · View at Scopus
29. S. Nakamura, "Catalytic enantioselective decarboxylative reactions using organocatalysts," Organic & Biomolecular Chemistry, vol. 12, pp. 394–405, 2014. View at Publisher · View at Google Scholar
30. J. Lubkoll and H. Wennemers, "Mimicry of polyketide synthases-enantioselective 1,4-addition reactions of malonic acid half-thioesters to nitroolefins," Angewandte Chemie International Edition, vol. 46, no. 36, pp. 6841–6844, 2007. View at Publisher · View at Google Scholar · View at Scopus
31. H. Y. Bae, S. Some, J. Y. - Kim et al., "Organocatalytic enantioselective Michael-addition of malonic acid half-thioesters to β-nitroolefins: from mimicry of polyketide synthases to scalable synthesis of γ-amino acids," Advanced Synthesis & Catalysis, vol. 353, no. 17, pp. 3196–3202, 2011. View at Publisher · View at Google Scholar
32. J. Becht, O. Meyer, and G. Helmchen, "Enantioselective syntheses of (-)-(R)-rolipram, (-)-(R)-baclofen and other GABA analogues via rhodium-catalyzed conjugate addition of arylboronic acids," Synthesis, no. 18, pp. 2805–2810, 2003. View at Scopus
33. D. M. Barnes, S. J. Wittenberger, J. Zhang et al., "Development of

a catalytic enantioselective conjugate addition of 1,3-dicarbonyl compounds to nitroalkenes for the synthesis of endothelin-A antagonist ABT-546. Scope, mechanism, and further application to the synthesis of the antidepressant rolipram," Journal of the American Chemical Society, vol. 124, no. 44, pp. 13097–13105, 2002. View at Publisher · View at Google Scholar · View at Scopus

34. H. N. Yuan, S. Wang, J. Nie, W. Meng, Q. Yao, and J. A. Ma, "Hydrogen-bond-directed enantioselective decarboxylative mannich reaction of β-ketoacids with ketimines: application to the synthesis of anti-HIV drug DPC 083," Angewandte Chemie International Edition, vol. 52, no. 14, pp. 3869–3873, 2013. View at Publisher · View at Google Scholar

35. X. Xu, T. Furukawa, T. Okino, H. Miyabe, and Y. Takemoto, "Bifunctional-thiourea-catalyzed diastereo- And enantioselective Aza-Henry reaction," Chemistry, vol. 12, no. 2, pp. 466–476, 2005. View at Publisher · View at Google Scholar · View at Scopus

36. Y. Zhang, J. Kua, and J. A. McCammon, "Role of the catalytic triad and oxyanion hole in acetylcholinesterase catalysis: an ab initio QM/MM study," Journal of the American Chemical Society, vol. 124, no. 35, pp. 10572–10577, 2002. View at Publisher · View at Google Scholar ·View at Scopus

37. J. T. Yli-Kauhaluoma, J. A. Ashley, L. C. Lo Chih-Hung, L. Tucker, M. M. Wolfe, and K. D. Janda, "Anti-metallocene antibodies: a new approach to enantioselective catalysis of the diels-alder reaction," Journal of the American Chemical Society, vol. 117, no. 27, pp. 7041–7047, 1995. View at Scopus

38. C. Gioia, A. Hauville, L. Bernardi, F. Fini, and A. Ricci, "Organocatalytic asymmetric diels-Alder reactions of 3-vinylindoles," Angewandte Chemie International Edition, vol. 47, no. 48, pp. 9236–9239, 2008. View at Publisher · View at Google Scholar · View at Scopus

39. M. Hesse, Alkaloids, Nature Curse or Blessing?Wiley-VCH, New York, NY, USA, 2002.

40. J. E. Saxton, "Recent progress in the chemistry of the monoterpenoid indole alkaloids," Natural Product Reports, vol. 14, no. 6, pp. 559–590, 1997. View at Publisher · View at Google Scholar

41. G. Abbiati, V. Canevari, D. Facoetti, and E. Rossi, "Diels-alder reactions of 2-vinylindoles with open-chain C=C dienophiles," European Journal of Organic Chemistry, vol. 2007, no. 3, pp. 517–525, 2007.

42. K. S. Gunmudsson, P. R. Sebahar, L. D'Aurora Richardson et al., "Substituted tetrahydrocarbazoles with potent activity against human papillomaviruses," Bioorganic & Medicinal Chemistry Letters, vol. 19, no. 13, pp. 3489–3492, 2009. View at Publisher · View at Google Scholar

43. S. Shimizu, K. Ohori, T. Arai, H. Sasai, and M. Shibasaki, "A catalytic asymmetric synthesis of tubifolidine," Journal of Organic Chemistry, vol. 63, no. 21, pp. 7547–7551, 1998. View at Scopus

44. Y. Hoashi, T. Yabuta, and Y. Takemoto, "Bifunctional thiourea-catalyzed enantioselective double Michael reaction of γ,δ-unsaturated β-ketoester to nitroalkene: asymmetric synthesis of (-)-epibatidine," Tetrahedron Letters, vol. 45, no. 50, pp. 9185–9188, 2004. View at Publisher · View at Google Scholar · View at Scopus

45. P. S. Haynes, P. A. Stupple, and D. J. Dixon, "Organocatalytic asymmetric total synthesis of (R)-rolipram and formal synthesis of (3S,4R)-paroxetine," Organic Letters, vol. 10, no. 7, pp. 1389–1391, 2008. View at Publisher · View at Google Scholar

46. R. P. Herrera, V. Sgarzani, L. Bernardi, and A. Ricci, "Catalytic enantioselective Friedel-Crafts alkylation of indoles with nitroalkenes by using a simple thiourea organocatalyst," Angewandte Chemie International Edition, vol. 44, no. 40, pp. 6576–6579, 2005. View at Publisher · View at Google Scholar · View at Scopus

47. K. Maruoka, "Practical aspects of recent asymmetric phase-transfer catalysis," Organic Process Research & Development, vol. 12, no. 4, pp. 679–697, 2008. View at Publisher · View at Google Scholar

48. T. Shioiri, "Chiral phase transfer catalysis," in Handbook of Phase Transfer Catalysis, Y. Sasson and R. Nuemann, Eds., chapter 14, pp. 462–479, Blackie Academic & Professional, London, UK, 1997.

49. A.Nelson, "Asymmetric phase-transfer catalysis," Angewandte Chemie International Edition, vol. 38, no. 11, pp. 1583–1585, 1999.

50. K. Maruoka and T. Ooi, "Enantioselective amino acid synthesis by chiral phase-transfer catalysis,"Chemical Reviews, vol. 103, no. 8, pp. 3013–3028, 2003. View at Publisher · View at Google Scholar · View at Scopus

51. M. J. O'Donnell, "The enantioselective synthesis of α-amino acids by phase-transfer catalysis with achiral schiff base esters," Accounts of Chemical Research, vol. 37, no. 8, pp. 506–517, 2004. View at Publisher · View at Google Scholar

52. B. Lygo and B. I. Andrews, "Asymmetric phase-transfer catalysis utilizing chiral quaternary ammonium salts: asymmetric alkylation of glycine imines," Accounts of Chemical Research, vol. 37, no. 8, pp. 518–525, 2004. View at Publisher · View at Google Scholar

53. S. Shirakawa and K. Maruoka, in Catalytic Asymmetric Synthesis, I. Ojima, Ed., chapter 2C, p. 95, Wiley, Hoboken, NJ, USA, 3rd edition, 2010.

54. K. Maruoka, "Highly practical amino acid and alkaloid synthesis using designer chiral phase transfer catalysts as high-performance organocatalysts," The Chemical Record, vol. 10, no. 5, pp. 254–259, 2010. View at Publisher · View at Google Scholar

55. S. Shirakawa and K. Maruoka, "Recent developments in asymmetric phase-transfer reactions,"Angewandte Chemie International Edition, vol. 52, no. 16, pp. 4312–4348, 2013. View at Publisher · View at Google Scholar

56. U. H. Dolling, P. Davis, and E. J. J. Grabowski, "Efficient catalytic asymmetric alkylations. 1. Enantioselective synthesis of (+)-indacrinone via chiral phase-transfer catalysis," Journal of the American Chemical Society, vol. 106, no. 2, pp. 446–447, 1984. View at Publisher · View at Google Scholar

57. R. S. E. Conn, A. V. Lovell, S. Karaday, and L. M. Weinstock, "Chiral Michael addition: methyl vinyl ketone addition catalyzed by Cinchona alkaloid derivatives," Journal of Organic Chemistry, vol. 51, no. 24, pp. 4710–4711, 1986. View at Publisher · View at Google Scholar

58. E. J. Cragoe Jr., O. W. Woltersdorf Jr., N. P. Gould et al., "Agents for the treatment of brain edema. 2. [(2,3,9,9a-tetrahydro-3-oxo-9a-substituted-1H-fluoren-7-yl)oxy]alkanoic acids and some of their analogues," Journal of Medicinal Chemistry, vol. 29, no. 5, pp. 825–841, 1986. View at Scopus

59. J. P. Scott, M. S. Ashwood, K. M. J. Brands et al., "Development of a phase transfer catalyzed asymmetric synthesis for an estrogen receptor beta selective agonist," Organic Process Research and Development, vol. 12, no. 4, pp. 723–730, 2008. View at Publisher · View at Google Scholar ·View at Scopus.

60. A.Baschieri, L. Bernardi, A. Ricci, S. Suresh, and M. F. A. Adamo, "Catalytic asymmetric conjugate addition of nitroalkanes to 4-nitro5-styrylisoxazoles," Angewandte Chemie International Edition, vol. 48, no. 49, pp. 9342–9345, 2009. View at Publisher · View at Google Scholar · View at Scopus

61. S. France, D. J. Guerin, S. J. Miller, and T. Letchka, "Nucleophilic chiral amines as catalysts in asymmetric synthesis," Chemical Reviews, vol. 103, no. 8, pp. 2985–3012, 2003. View at Publisher· View at Google Scholar

62. T. Marcelli and H. Hiemstra, "Cinchona alkaloids in asymmetric organocatalysis," Synthesis, no. 8, Article ID E26109SS, pp. 1229–1279, 2010. View at Publisher · View at Google Scholar · View at Scopus

63. H. Hiemstra and H. Wynberg, "Addition of aromatic thiols to conjugated cycloalkenones, catalyzed by chiral .beta.-hydroxy amines. A mechanistic study of homogeneous catalytic asymmetric

synthesis," Journal of the American Chemical Society, vol. 103, no. 2, pp. 417–430, 1981. View at Publisher · View at Google Scholar

64. M. Bella and K. A. Jørgensen, "Organocatalytic enantioselective conjugate addition to alkynones,"Journal of the American Chemical Society, vol. 126, no. 18, pp. 5672–5673, 2004. View at Publisher · View at Google Scholar
65. H. Li, Y. Wang, L. Tang et al., "Stereocontrolled creation of adjacent quaternary and tertiary stereocenters by a catalytic conjugate addition," Angewandte Chemie International Edition, vol. 44, no. 1, pp. 105–108, 2004.
66. X. D. Liu, L. J. Deng, H. J. Song, H. Z. Jia, and R. Wang, "Asymmetric Aza-Mannich addition: synthesis of modified chiral 2-(Ethylthio)-thiazolone derivatives with anticancer potency,"Organic Letters, vol. 13, no. 6, pp. 1494–1497, 2011. View at Publisher · View at Google Scholar
67. L. E. Brown, K. C. Cheng, W. Wei et al., "Discovery of new antimalarial chemotypes through chemical methodology and library development," Proceedings of the National Academy of Sciences of the United States of America, vol. 108, no. 17, pp. 6775–6780, 2011. View at Publisher ·View at Google Scholar · View at Scopus
68. S. Kaneko, T. Yoshino, T. Katoh, and S. Terashima, "Synthetic studies of huperzine A and its fluorinated analogues. 1. Novel asymmetric syntheses of an enantiomeric pair of huperzine A,"Tetrahedron, vol. 54, no. 21, pp. 5471–5484, 1998. View at Publisher · View at Google Scholar ·View at Scopus
69. T. Akiyama, J. Itoh, K. Yokota, and K. Fuchibe, "Enantioselective mannich-type reaction catalyzed by a chiral Brønsted acid," Angewandte Chemie International Edition, vol. 43, no. 12, pp. 1566–1568, 2004. View at Publisher · View at Google Scholar · View at Scopus
70. D. Uraguchi and M. Terada, "Chiral Brønsted acid-catalyzed direct mannich reactions via electrophilic activation," Journal of the American Chemical Society, vol. 126, no. 17, pp. 5356–5357, 2004. View at Publisher · View at Google Scholar · View at Scopus
71. S. Hoffmann, A. M. Seyad, and B. List, "A powerful Brønsted acid catalyst for the organocatalytic asymmetric transfer hydrogenation of imines," Angewandte Chemie International Edition, vol. 44, no. 45, pp. 7424–7427, 2005. View at Publisher · View at Google Scholar
72. M. Rueping, J. Dufour, and F. R. Schoepke, "Advances in catalytic metal-free reductions: from bio-inspired concepts to applications in the organocatalytic synthesis of pharmaceuticals and natural products," Green Chemistry, vol. 13, no. 5, pp. 1084–1105, 2011. View at Publisher · View at Google Scholar · View at Scopus

73. M. Rueping, M. Stoeckel, E. Sugiono, and T. Theissmann, "Asymmetric metal-free synthesis of fluoroquinolones by organocatalytic hydrogenation," Tetrahedron, vol. 66, no. 33, pp. 6565–6568, 2010. View at Publisher · View at Google Scholar · View at Scopus

74. Hayakawa, S. Atarashi, S. Yokohama, M. Imamura, K. Sakano, and M. Furukawa, "Synthesis and antibacterial activities of optically active ofloxacin," Antimicrobial Agents and Chemotherapy, vol. 29, no. 1, pp. 163–164, 1986. View at Publisher · View at Google Scholar

75. D. Seiyaku, "The s-(-) isomer of 7,8-difluoro-2,3-dihydro-3-methyl-4H-1,4-benzoxazine can be utilized in the synthesis of the optically active form of Ofloxacin known as Levofloxacin. Levofloxacin is 8 to 128 times more active than Ofloxacin depending upon the bacteria tested,"Drugs Future, vol. 17, no. 7, pp. 559–563, 1992.

76. M. Rueping, A. P. Antonchick, and T. Theissmann, "A highly enantioselective Brønsted acid catalyzed cascade reaction: organocatalytic transfer hydrogenation of quinolines and their application in the synthesis of alkaloids," Angewandte Chemie International Edition, vol. 45, no. 22, pp. 3683–3686, 2006. View at Publisher · View at Google Scholar · View at Scopus

77. X. Chen, X. Xu, H. Liu, L. Cun, and L. Gong, "Highly enantioselective organocatalytic Biginelli reaction," Journal of the American Chemical Society, vol. 128, no. 46, pp. 14802–14803, 2006.View at Publisher · View at Google Scholar · View at Scopus

78. Y. Huang, F. Yang, and C. Zhu, "Highly enantioseletive biginelli reaction using a new chiral ytterbium catalyst: asymmetric synthesis of dihydropyrimidines," Journal of the American Chemical Society, vol. 127, no. 47, pp. 16386–16387, 2005. View at Publisher · View at Google Scholar

79. P. Melchiorre, M. Marigo, A. Carlone, and G. Bartoli, "Asymmetric aminocatalysis-gold rush in organic chemistry," Angewandte Chemie International Edition, vol. 47, no. 33, pp. 6138–6171, 2008. View at Publisher · View at Google Scholar · View at Scopus

80. C. F. Barbas III, "Organocatalysis lost: modern chemistry, ancient chemistry, and an unseen biosynthetic apparatus," Angewandte Chemie International Edition, vol. 47, no. 1, pp. 42–47, 2008. View at Publisher · View at Google Scholar

81. Z. G. Hajos and D. R. Parrish, "Asymmetric synthesis of bicyclic intermediates of natural product chemistry," Journal of Organic Chemistry, vol. 39, no. 12, pp. 1615–1621, 1974. View at Publisher· View at Google Scholar

82. U. Eder, G. Sauer, and R. Wiechert, "New type of asymmetric

cyclization to optically active steroid CD partial structures," Angewandte Chemie International Edition in English, vol. 10, no. 7, pp. 496–497, 1971. View at Publisher · View at Google Scholar

83. B. List, R. A. Lerner, and C. F. Barbas III, "Proline-catalyzed direct asymmetric aldol reactions,"Journal of the American Chemical Society, vol. 122, no. 10, pp. 2395–2396, 2000. View at Publisher · View at Google Scholar · View at Scopus

84. W. Notz, F. Tanaka, and C. F. Barbas III, "Enamine-based organocatalysis with proline and diamines: the development of direct catalytic asymmetric aldol, mannich, Michael, and diels-alder reactions," Accounts of Chemical Research, vol. 37, no. 8, pp. 580–591, 2004. View at Publisher ·View at Google Scholar

85. W. Notz and B. List, "Catalytic asymmetric synthesis of anti-1,2-diols," Journal of the American Chemical Society, vol. 122, no. 30, pp. 9336–7387, 2000. View at Publisher · View at Google Scholar

86. A.B. Northrup and D. W. C. MacMillan, "The first direct and enantioselective cross-aldol reaction of aldehydes," Journal of the American Chemical Society, vol. 124, no. 24, pp. 6798–6799, 2002. View at Publisher · View at Google Scholar · View at Scopus

87. B. List, "Direct catalytic asymmetric α-amination of aldehydes," Journal of the American Chemical Society, vol. 124, no. 20, pp. 5656–5657, 2002. View at Publisher · View at Google Scholar

88. N. Kumaragurubaran, K. Juhl, W. Zhuang, A. Bøgevig, and K. A. Jørgensen, "Direct l-proline-catalyzed asymmetric α-amination of ketones," Journal of the American Chemical Society, vol. 124, no. 22, pp. 6254–6255, 2002. View at Publisher · View at Google Scholar.

89. A. B. Northrup and D. W. C. MacMillan, "The first general enantioselective catalytic Diels-Alder reaction with simple (α,β-unsaturated ketones," Journal of the American Chemical Society, vol. 124, no. 11, pp. 2458–2460, 2002. View at Publisher · View at Google Scholar · View at Scopus

90. R. M. Wilson, W. S. Jen, and D. W. C. MacMillan, "Enantioselective organocatalytic intramolecular Diels-Alder reactions. The asymmetric synthesis of solanapyrone D," Journal of the American Chemical Society, vol. 127, no. 33, pp. 11616–11617, 2005. View at Publisher · View at Google Scholar · View at Scopus

91. F. Austin and D. W. C. MacMillan, "Enantioselective organocatalytic indole alkylations. Design of a new and highly effective chiral amine for iminium catalysis," Journal of the American Chemical Society, vol. 124, no. 7, pp. 1172–1173, 2002. View at Publisher · View at Google Scholar

92. N. A. Paras and D. W. C. MacMillan, "The enantioselective

organocatalytic 1,4-addition of electron-rich benzenes to α,β-unsaturated aldehydes," Journal of the American Chemical Society, vol. 124, no. 27, pp. 7894–7895, 2002. View at Publisher · View at Google Scholar · View at Scopus

93. S. P. Brown, N. C. Goodwin, and D. W. C. MacMillan, "The first enantioselective organocatalytic Mukaiyama-Michael reaction: a direct method for the synthesis of enantioenriched γ-butenolide architecture," Journal of the American Chemical Society, vol. 125, no. 5, pp. 1192–1194, 2003.View at Publisher · View at Google Scholar · View at Scopus
94. S. G. Ouellet, J. B. Tuttle, and D. W. C. MacMillan, "Enantioselective organocatalytic hydride reduction," Journal of the American Chemical Society, vol. 127, no. 1, pp. 32–33, 2005. View at Publisher · View at Google Scholar · View at Scopus
95. F. Austin, S. G. Kim, C. J. Sinz, W. J. Xiao, and D. W. C. MacMillan, "Enantioselective organocatalytic construction of pyrroloindolines by a cascade addition-cyclization strategy: synthesis of (-)-flustramine B," Proceedings of the National Academy of Sciences of the United States of America, vol. 101, no. 15, pp. 5482–5487, 2004. View at Publisher · View at Google Scholar
96. H. P. Rang, M. M. Dale, J. M. Ritter, and P. Gardner, Pharmacology, Churchill Livingstone, Philadelphia, Pa, USA, 4th edition, 2001.
97. H. J. Bardsley and A. K. Daly, "The therapeutic use of r-warfarin as anticoagulant," Patent Cooperation Treaty International Application WO, 0043003, 2000.
98. A. Robinson and H. Y. Li, "The first practical asymmetric synthesis of R and S-Warfarin,"Tetrahedron Letters, vol. 37, no. 46, pp. 8321–8324, 1996. View at Publisher · View at Google Scholar
99. Y. Tsuchiya, Y. Hamashima, and M. Sodeoka, "A new entry to Pd–H chemistry: catalytic asymmetric conjugate reduction of enones with EtOH and a highly enantioselective synthesis of warfarin," Organic Letters, vol. 8, no. 21, pp. 4851–4854, 2006. View at Publisher View at Google Scholar View at Scopus
100. N. Halland, T. Hansen, and K. A. Jørgensen, "Organocatalytic asymmetric Michael reaction of cyclic 1,3-dicarbonyl compounds and α ,β-unsaturated ketones-a highly atom-economic catalytic one-step formation of optically active warfarin anticoagulant," Angewandte Chemie International Edition, vol. 42, no. 40, pp. 4955–4957, 2003. View at Publisher · View at Google Scholar
101. N. Halland, K. A. Jørgensen, and T. Hansen, Patent Cooperation Treaty International Application, WO 03/050105, 9, 413, 2003.

102. H. Kim, C. Yen, P. Preston, and J. Chin, "Substrate-directed stereoselectivity in vicinal diamine-catalyzed synthesis of warfarin," Organic Letters, vol. 8, no. 23, pp. 5239–5242, 2006. View at Publisher · View at Google Scholar · View at Scopus

103. H. Yang, L. Li, K. Jiang, J. Jiang, G. Lai, and L. Xu, "Highly enantioselective synthesis of warfarin and its analogs by means of cooperative LiClO4/DPEN-catalyzed Michael reaction: enantioselectivity enhancement and mechanism," Tetrahedron, vol. 66, no. 51, pp. 9708–9713, 2010. View at Publisher · View at Google Scholar · View at Scopus

104. Rogozińska, A. Adamkiewicz, and J. Mlynarsky, "Efficient "on water" organocatalytic protocol for the synthesis of optically pure warfarin anticoagulant," Green Chemistry, vol. 13, no. 5, pp. 1155–1157, 2011. View at Publisher · View at Google Scholar

105. X. Zhu, A. Lin, Y. Shi, J. Guo, C. Zhu, and Y. Cheng, "Enantioselective synthesis of polycyclic coumarin derivatives catalyzed by an in situ formed primary amine-imine catalyst," Organic Letters, vol. 13, no. 16, pp. 4382–4385, 2011. View at Publisher · View at Google Scholar · View at Scopus

106. J. Dong and D. M. Du, "Highly enantioselective synthesis of warfarin and its analogs catalysed by primary amine-phosphinamide bifunctional catalysts," Organic & Biomolecular Chemistry, vol. 10, no. 40, pp. 8125–8131, 2012. View at Publisher · View at Google Scholar

107. Leven, J. M. Neudörfl, and B. Goldfuss, "Metal-mediated aminocatalysis provides mild conditions: enantioselective Michael addition mediated by primary amino catalysts and alkali-metal ions," Beilstein Journal of Organic Chemistry, vol. 9, pp. 155–165, 2013. View at Publisher ·View at Google Scholar

108. J. W. Xie, L. Yue, W. Chen et al., "Highly enantioselective Michael addition of cyclic 1,3-dicarbonyl compounds to α,β-unsaturated ketones," Organic Letters, vol. 9, no. 3, pp. 413–415, 2007. View at Publisher · View at Google Scholar

109. T. E. Kristensen, K. Vestli, F. K. Hansen, and T. Hansen, "New phenylglycine-derived primary amine organocatalysts for the preparation of optically active warfarin," European Journal of Organic Chemistry, vol. 2009, no. 30, pp. 5185–5191, 2009. View at Publisher · View at Google Scholar

110. Z. H. Dong, L. J. Wang, X. H. Chen, X. H. Liu, L. L. Lin, and X. M. Feng, "Organocatalytic enantioselective Michael addition of 4-hydroxycoumarin to α ,β-unsaturated ketones: a simple synthesis of warfarin," European Journal of Organic Chemistry, no. 30, pp. 5192–5197, 2009. View at Publisher · View at Google Scholar

111. D. V. Paone, A. W. Shaw, D. N. Nguyen et al., "Potent, orally bioavailable calcitonin gene-related peptide receptor antagonists for the treatment of migraine: discovery of N-(3R,6S)-6-(2,3-difluorophenyl)-2-oxo-1-(2,2,2-trifluoroethyl)azepan-3-yl-4-(2-oxo-2, 3-dihydro-1H-imidazo[4,5-b]pyridin-1-yl)piperidine-1-carboxamide (MK-0974)," Journal of Medicinal Chemistry, vol. 50, no. 23, pp. 5564–5567, 2007. View at Publisher · View at Google Scholar ·View at Scopus

112. Palucki, I. Davies, D. Steinhuebel, and J. Rosen, Patent Cooperation Treaty International Application WO2007120589, 2007.

113. F. Xu, M. Zacuto, N. Yoshikawa et al., "Asymmetric synthesis of telcagepant, a CGRP receptor antagonist for the treatment of migraine," Journal of Organic Chemistry, vol. 75, no. 22, pp. 7829–7841, 2010. View at Publisher · View at Google Scholar · View at Scopus

114. D. P. Steinhuebel, S. W. Krska, A. Alorati et al., "Asymmetric hydrogenation of protected allylic amines," Organic Letters, vol. 12, no. 18, pp. 4201–4203, 2010. View at Publisher · View at Google Scholar · View at Scopus

115. C. S. Burgey, D. V. Paone, A. W. Shaw et al., "Synthesis of the (3R,6S)-3-amino-6-(2,3-difluorophenyl)azepan-2-one of telcagepant (MK-0974), a calcitonin gene-related peptide receptor antagonist for the treatment of migraine headache," Organic Letters, vol. 10, no. 15, pp. 3235–3238, 2008. View at Publisher · View at Google Scholar · View at Scopus

116. H. Gotoh, H. Ishikawa, and Y. Hayashi, "Diphenylprolinol silyl ether as catalyst of an asymmetric, catalytic, and direct Michael reaction of nitroalkanes with α,β-unsaturated aldehydes," Organic Letters, vol. 9, no. 25, pp. 5307–5309, 2007. View at Publisher · View at Google Scholar

117. Y. Wang, P. Li, X. Liang, T. Y. Zhang, and J. Ye, "An efficient enantioselective method for asymmetric Michael addition of nitroalkanes to α,β-unsaturated aldehydes," Chemical Communications, no. 10, pp. 1232–1234, 2008. View at Publisher · View at Google Scholar

118. Y. Hayashi, H. Gotoh, T. Hayashi, and M. Shoji, "Diphenylprolinol silyl ethers as efficient organocatalysts for the asymmetric Michael reaction of aldehydes and nitroalkenes," Angewandte Chemie International Edition, vol. 44, no. 27, pp. 4212–4215, 2005. View at Publisher · View at Google Scholar · View at Scopus

119. M. Marigo, T. C. Wabnitz, D. Fielenbach, and K. A. Jørgensen, "Enantioselective organocatalyzed sulfenylation of aldehydes," Angewandte Chemie International Edition, vol. 44, no. 5, pp. 794–797, 2005. View at Publisher · View at Google Scholar

120. J. Åhman, M. Birch, S. J. Haycock-Lewandowski, J. Long, and A. Wilder, "Process research and scale-up of a commercialisable route to maraviroc (UK-427,857), a CCR-5 receptor antagonist,"Organic Process Research & Development, vol. 12, no. 6, pp. 1104–1113, 2008. View at Publisher ·View at Google Scholar

121. S. J. Haycock-Lewandowski, A. Wilder, and J. Åhman, "Development of a bulk enabling route to maraviroc (UK-427,857), a CCR-5 receptor antagonist," Organic Process Research & Development, vol. 12, no. 6, pp. 1094–1103, 2008. View at Publisher · View at Google Scholar

122. G.-L. Zhao, S. Lin, A. Korotvička, L. Deiana, M. Kullberg, and A. Córdova, "Asymmetric synthesis of maraviroc (UK-427,857)," Advanced Synthesis & Catalysis, vol. 352, no. 13, pp. 2291–2298, 2010. View at Publisher · View at Google Scholar

123. M. S. Yu, I. Lantos, Z. Peng, J. Yu, and T. Cacchio, "Asymmetric synthesis of (-)-paroxetine using PLE hydrolysis," Tetrahedron Letters, vol. 41, no. 30, pp. 5647–5651, 2000. View at Scopus

124. J. M. Palomo, G. Fernández-Lorente, C. Mateo, R. Fernández-Lafuente, and J. M. Grison, "Enzymatic resolution of (±)-trans-4-(4′-fluorophenyl)-6-oxo-piperidin-3-ethyl carboxylate, an intermediate in the synthesis of (−)-Paroxetine," Tetrahedron, vol. 13, no. 21, pp. 2375–2361, 2002. View at Publisher · View at Google Scholar

125. M. Amat, J. Bosch, J. Hidalgo et al., "Synthesis of enantiopure trans-3,4-disubstituted piperidines. An enantiodivergent synthesis of (+)- and (-)-paroxetine," Journal of Organic Chemistry, vol. 65, no. 10, pp. 3074–3084, 2000. View at Publisher · View at Google Scholar · View at Scopus

126. T. A. Johnson, D. O. Jang, W. Slafer, M. D. Curtius, and P. Beak, "Asymmetric carbon-carbon bond formations in conjugate additions of lithiated N-Boc allylic and benzylic amines to nitroalkenes: enantioselective synthesis of substituted piperidines, pyrrolidines, and pyrimidinones," Journal of the American Chemical Society, vol. 124, no. 39, pp. 11689–11698, 2002. View at Publisher · View at Google Scholar

127. G. Valero, J. Schimer, I. Cisarova, J. Vesely, A. Moyano, and R. Rios, "Highly enantioselective organocatalytic synthesis of piperidines. Formal synthesis of (-)-Paroxetine," Tetrahedron Letters, vol. 50, no. 17, pp. 1943–1946, 2009. View at Publisher · View at Google Scholar · View at Scopus

128. S. Brandau, A. Landa, J. Franzen, M. Marigo, and K. A. Jorgensen, "Organocatalytic conjugate addition of malonates to α ,β-unsaturated aldehydes: asymmetric formal synthesis of (-)-paroxetine, chiral

lactams, and lactones," Angewandte Chemie International Edition, vol. 45, no. 26, pp. 4305–4309, 2006. View at Publisher · View at Google Scholar

129. C. Lagisetti, A. Pourpak, Q. Jiang et al., "Antitumor compounds based on a natural product consensus pharmacophore," Journal of Medicinal Chemistry, vol. 51, no. 19, pp. 6220–6224, 2008.View at Publisher · View at Google Scholar · View at Scopus

130. X. Jiang, Y. Cao, Y. Wang, L. Liu, F. Shen, and R. Wang, "A unique approach to the concise synthesis of highly optically active spirooxazolines and the discovery of a more potent oxindole-type phytoalexin analogue," Journal of the American Chemical Society, vol. 132, no. 43, pp. 15328–15333, 2010. View at Publisher · View at Google Scholar · View at Scopus

131. S. Kotha, A. C. Deb, K. Lahiri, and E. Manivannan, "Selected synthetic strategies to spirocyclics,"Synthesis, no. 2, pp. 165–193, 2009. View at Publisher · View at Google Scholar · View at Scopus

132. E. J. Corey, "Catalytic enantioselective diels-alder reactions: methods, mechanistic fundamentals, pathways, and applications," Angewandte Chemie International Edition, vol. 114, pp. 1724–1741, 2002.

133. R. Sebahar and R. M. Williams, "The asymmetric total synthesis of (+)- and (-)-spirotryprostatin B," Journal of the American Chemical Society, vol. 122, no. 23, pp. 5666–5667, 2000. View at Publisher · View at Google Scholar · View at Scopus

134. B. M. Trost, N. Cramer, and S. M. Silverman, "Enantioselective construction of spirocyclic oxindolic cyclopentanes by palladium-catalyzed trimethylenemethane-[3+2]-cycloaddition,"Journal of the American Chemical Society, vol. 129, no. 41, pp. 12396–12397, 2007. View at Publisher · View at Google Scholar · View at Scopus

135. A.B. Dounay and L. E. Overman, "The asymmetric intramolecular heck reaction in natural product total synthesis," Chemical Reviews, vol. 103, no. 8, pp. 2945–2963, 2003. View at Publisher · View at Google Scholar · View at Scopus

136. Madin, C. J. O'Donnell, T. Oh, D. W. Old, L. E. Overman, and M. J. Sharp, "Use of the intramolecular heck reaction for forming congested quaternary carbon stereocenters. Stereocontrolled total synthesis of (±)-gelsemine," Journal of the American Chemical Society, vol. 127, no. 51, pp. 18054–18065, 2005. View at Publisher · View at Google Scholar · View at Scopus

137. J. J. Liu and Z. Zhang, Patent Cooperation Treaty International Application WO, 2008/055812, 2008.

138. Chène, "Inhibiting the p53-MDM2 interaction: an important target for

cancer therapy," Nature Reviews Cancer, vol. 3, pp. 102–109, 2003. View at Publisher · View at Google Scholar

139. G. Bencivenni, L. Wu, A. Mazzanti et al., "Targeting structural and stereochemical complexity by organocascade catalysis: construction of spirocyclic oxindoles having multiple stereocenters,"Angewandte Chemie International Edition, vol. 48, no. 39, pp. 7200–7203, 2009. View at Publisher · View at Google Scholar · View at Scopus

140. U. Kim, W. Lew, M. A. Williams et al., "Influenza neuraminidase inhibitors possessing a novel hydrophobic interaction in the enzyme active site: design, synthesis, and structural analysis of carbocyclic sialic acid analogues with potent anti-influenza activity," Journal of the American Chemical Society, vol. 119, no. 4, pp. 681–690, 1997. View at Publisher · View at Google Scholar ·View at Scopus

141. M. Von Itzstein, W.-Y. Wu, G. B. Kok et al., "Rational design of potent sialidase-based inhibitors of influenza virus replication," Nature, vol. 363, no. 6428, pp. 418–423, 1993. View at Publisher ·View at Google Scholar · View at Scopus

142. V. Farina and J. D. Brown, "Tamiflu: the supply problem," Angewandte Chemie International Edition, vol. 45, no. 44, pp. 7330–7334, 2006. View at Publisher · View at Google Scholar

143. M. Shibasaki and M. Kanai, "Synthetic strategies for oseltamivir phosphate," European Journal of Organic Chemistry, vol. 2008, no. 11, pp. 1839–1850, 2008. View at Publisher · View at Google Scholar

144. Y. Y. Yeung, S. Hong, and E. J. Corey, "A short enantioselective pathway for the synthesis of the anti-influenza neuramidase inhibitor oseltamivir from 1,3-butadiene and acrylic acid," Journal of the American Chemical Society, vol. 128, no. 19, pp. 6310–6311, 2006. View at Publisher · View at Google Scholar

145. Y. Fukuta, T. Mita, N. Fukuda, M. Kanai, and M. Shibasaki, "De novo synthesis of tamiflu via a catalytic asymmetric ring-opening of meso-aziridines with TMSN3," Journal of the American Chemical Society, vol. 128, no. 19, pp. 6312–6313, 2006. View at Publisher · View at Google Scholar · View at Scopus

146. H. Ishikawa, T. Suzuki, and Y. Hayashi, "High-yielding synthesis of the anti-influenza neuramidase inhibitor (-)-oseltamivir by three "one-pot" operations," Angewandte Chemie International Edition, vol. 48, no. 7, pp. 1304–1307, 2009. View at Publisher · View at Google Scholar

147. H. Ishikawa, T. Suzuki, H. Orita, T. Uchimaru, and Y. Hayashi, "High-yielding synthesis of the anti-influenza neuraminidase inhibitor (-)-oseltamivir by two "one-pot" sequences," Chemistry, vol. 16, no.

42, pp. 12616–12626, 2010. View at Publisher · View at Google Scholar · View at Scopus

148. S. Zhu, S. Yu, Y. Wang, and D. Ma, "Organocatalytic Michael addition of aldehydes to protected 2-amino-1-nitroethenes: the practical syntheses of oseltamivir (Tamiflu) and substituted 3-aminopyrrolidines," Angewandte Chemie International Edition, vol. 49, no. 27, pp. 4656–4660, 2010. View at Publisher · View at Google Scholar

149. J. Rehák, M. Huťka, A. Latika et al., "Thiol-free synthesis of oseltamivir and its analogues via organocatalytic Michael additions of oxyacetaldehydes to 2-acylaminonitroalkenes," Synthesis, vol. 44, pp. 2424–2430, 2012.

150. J. Weng, Y. B. Li, R. B. Wang, and G. Lu, "Organocatalytic Michael reaction of nitroenamine derivatives with aldehydes: short and efficient esymmetric eynthesis of (-)-oseltamivir," ChemCatChem, vol. 4, no. 7, pp. 1007–1012, 2012.

151. V. Hajzer, A. Latika, J. Durmis, and R. Sebesta, "Enantioselective Michael addition of the 2-(1-ethylpropoxy)acetaldehyde to N-[(1Z)-2-nitroethenyl]acetamide-optimization of the key step in the organocatalytic oseltamivir synthesis," Helvetica Chimica Acta, vol. 95, no. 12, pp. 2421–2428, 2012. View at Publisher · View at Google Scholar

152. T. Muakaiyama, H. Ishikawa, H. Koshino, and Y. Hayashi, "One-pot synthesis of (-)-oseltamivir and mechanistic insights into the organocatalyzed Michael reaction," Chemistry, vol. 19, no. 52, pp. 17789–17800, 2013. View at Publisher · View at Google Scholar

153. K. Yamatsugu, L. Yin, S. Kamijo, Y. Kimura, M. Kanai, and M. Shibasaki, "A synthesis of tamiflu by using a barium-catalyzed asymmetric diels-alder-type reaction," Angewandte Chemie International Edition, vol. 48, no. 6, pp. 1070–1076, 2009. View at Publisher · View at Google Scholar · View at Scopus

154. F. Cozzi, "Immobilization of organic catalysts: when, why, and how?" Advanced Synthesis & Catalysis, vol. 348, no. 12-13, pp. 1367–1390, 2006. View at Publisher · View at Google Scholar

155. T. E. Christensen and T. Hansen, "Polymer-supported chiral organocatalysts: synthetic strategies for the road towards affordable polymeric immobilization," European Journal of Organic Chemistry, vol. 17, pp. 3179–3204, 2010.

156. A. L. W. Demuynck, L. Peng, F. De-Clippel, J. Vanderleyden, P. A. Jacobs, and B. F. Sels, "Solid acids as heterogeneous support for primary amino acid-derived diamines in direct asymmetric aldol reactions," Advanced Synthesis and Catalysis, vol. 353, no. 5, pp. 725–732, 2011. View at Publisher · View at Google Scholar · View at Scopus

Chapter 3

MICROWAVE ASSISTED SOLVENT-FREE SYNTHESIS OF SOME IMINE DERIVATIVES

Yunus Bekdemir[1] and Kürsat Efil[1,2]

[1]Department of Molecular Biology and Genetics, Faculty of Arts and Science, Canik Basarı University, 55080 Samsun, Turkey[2]Department of Chemistry, Faculty of Arts and Sciences, Ondokuz Mayis University, 55139 Samsun, Turkey

ABSTRACT

Some imine derivatives (1a–7d) were synthesized using a rapid and an environmentally friendly method with reaction of aromatic aldehydes (a–d) and aromatic amines (1–7) (in 1 : 1 molar ratio) in the presence of □-ethoxyethanol as a wetting reagent (2 drops) under solvent-free conditions using microwave heating.

INTRODUCTION

Compounds containing the –C=N– (azomethine group) structure are known as Schiff bases, usually synthesized from the condensation of primary amines and active carbonyl groups [1]. The reaction is acid-catalyzed and is generally carried out by refluxing the carbonyl compound and amine, with an azeotroping agent if necessary, and separating the water as formed [2]. Schiff bases are well known for their biological applications as antibacterial, antifungal, anticancer, and antiviral agents; furthermore, they have been used as intermediates in medical substrates and as ligands in complex formation with some metal ions [1, 3]. The synthesis of imines was firstly reported by Hugo Schiff in 1864 and they have been known since then [4]. The imine compounds have been prepared using molecular sieves [5, 6], infrared irradiation [7], $Mg(ClO_4)_2$ [8], P_2O_5/SiO_2 [9], $ZnCl_2$ [10], CaO under microwave power [11], ethyl lactate as a tunable solvent [12], K10 clay [13], $TiCl_4$ [14], alumina [15], $CeCl_37H_2O$ [16], ultrasound irradiation [17], polymer-supported [18], nanotube TiO_2 (in sunlight) [19], and $Ti(OEt)_4$ [20, 21].

The present work reveals the comparative aspects of condensation of some aromatic amines with aldehyde derivatives using microwave and conventional methods. The amine and aldehyde compounds as starting materials, β -ethoxyethanol (β -EE) as wetting reagent and microwave power as an effective source of heating are used in this study. The corresponding imine compounds were prepared in high yields and short reaction times using this effective and environment friendly method.

RESULTS AND DISCUSSION

In this work, we synthesized quickly and efficiently a series imine derivatives (1a–7d) by condensation of some aromatic amines (1–7) and aldehyde derivatives such as salicylaldehyde (a), p-chlorobenzaldehyde (b), p-methoxybenzaldehyde (c) and cinnamaldehyde (d) under microwave-assisted solvent-free conditions using β -EE as wetting reagent. β -EE that is a polar molecule quickly absorbs microwaves and therefore heats up and heats around effectively. As a result, -EE, which increases the polarity of the reaction medium, has an active role in the heating of the reaction medium by microwaves.

The general reaction was summarized in Scheme 1. In addition, we tested the effect of different microwave power such as 180, 360, 600, 900 W and detected the microwave power of 180 W and 360 W are more appropriate choices for the reaction. Hence, the optimum microwave reaction conditions were determined using 180 and 360 W microwave power and neat and wetting with β-EE for compound 1a. The highest reaction yield (94%) and shorter reaction time (1.5 min.) were obtained at 360 W microwave power with -EE. In the absence of wetting reagent, the reaction yields are 81% for 1.5 minutes and 88% for 5 minutes. It is understood that the reaction yield was increased by wetting reagent that increases the polarity of the reaction medium. The optimization of microwave-assisted conditions was given in Table 1.

Table 1 The optimization of microwave conditions for compound 1a

Microwave method			
Watt	**Time (min)**	**Reaction conditions**	**Yield (%)**
180	5	Wetting (with β-EE)	83
360	5	Wetting (with β-EE)	93
360	**1.5**	**Wetting (with β-EE)**	**94**
360	5	Neat	88
360	1.5	Neat	81

Scheme 1

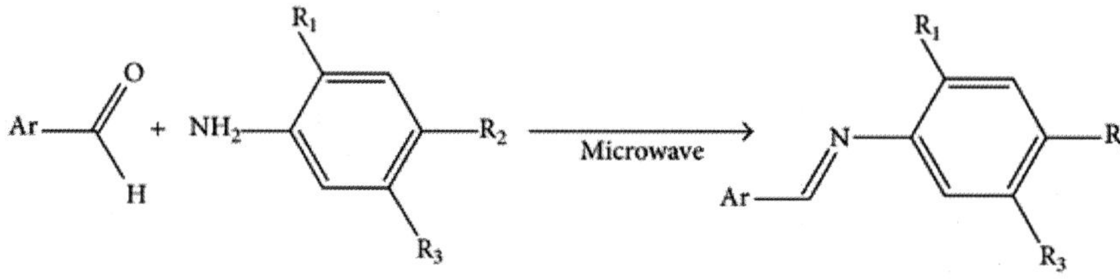

Ar–CHO + NH_2–$C_6H_2(R_1)(R_2)(R_3)$ —Microwave→ Ar–CH=N–$C_6H_2(R_1)(R_2)(R_3)$

(a–d) (1–7) (1a–7d)

In addition, we have synthesized some imine derivatives (1a, 4a, 1b, 4b, 1c, 4c, 1d, 4d) under classical conditions using a solvent (azeotropic) and traditional heating source (hot plate) for comparison with microwave conditions. Firstly, synthesis of compound 1a under the classical reaction conditions was tested at various periods of time (30, 60, 120, 240, 1300 min) to determine the optimum reaction time, which was determined as 120 minutes. Information on the optimization of the reaction time of compound1a for classic conditions was presented in Table 2.

Table 2 The optimization of classical conditions for compound 1a

	Classical method	
Time (min)	**Yield (%)**	**Reaction conditions**
30	64	Reflux
60	73	Reflux
120	**76**	**Reflux**
240	75	Reflux
1300*	74	Reflux

The compounds 1a, 4a, 1b, 4b, 1c, 4c, 1d, and 4d were obtained at 76%, 73%, 63%, 79%, 67%, 73%, 67%, and 75% yields, respectively, in 120 minutes under the classical reaction conditions, whereas in microwave conditions, the corresponding reaction yields were 94%, 97%, 95%, 98%, 91%, 97%, 95%, and 96%, respectively, in 1.0–1.5 minutes. All the results such as reactions time, yields, and melting points of the compounds were presented in Table 3. In addition, the yields of microwave and conventional reaction conditions for some imines were expressed graphically in Figure 1.

Table 3 Synthesis of imine derivatives under microwave irradiation

N	Ar	R_1	R_2	R_3	Microwave[a] Yield %	Microwave[a] t/min	Classical[b] Yield %	Classical[b] t/min	m.p °C	Lit. m.p °C
1a		OH	H	H	94	1.5	76	120	188–190	190 [22]
2a		OH	H	Cl	93	1.5	—	—	159–162	156–156.5 [23]
3a		OH	CH_3	H	93	1.5	—	—	199–202	—
4a	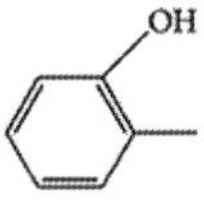	H	H	H	97	1.0	73	120	50–53	50-51 [24]
5a		H	Cl	H	94	1.5	—	—	103–105	104 [12]
6a		H	CH_3O	H	96	1.5	—	—	83–85	82–83.5 [25]
7a		H	CH_3	H	97	1.5	—	—	93-94	90–91.5 [25]
1b		OH	H	H	95	1.5	63	120	116–118	117 [26]
2b		OH	H	Cl	92	2.0	—	—	123–125	—
3b		OH	CH_3	H	91	2.0	—	—	134–137	—
4b	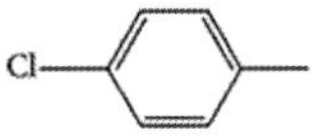	H	H	H	98	1.0	79	120	63–66	64-65 [24]
5b		H	Cl	H	94	1.5	—	—	110-111	110-111 [24]
6b		H	CH_3O	H	91	3.0	—	—	124–126	124-125 [27]
7b		H	CH_3	H	93	1.5	—	—	126–128	125 [27]
1c		OH	H	H	91	1.5	67	120	90-91	91–93 [22]
2c		OH	H	Cl	95	1.0	—	—	86–88	—
3c		OH	CH_3	H	93	1.5	—	—	86-87	—
4c	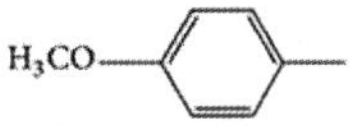	H	H	H	97	1.0	73	120	58–60	63–63.5 [24]
5c		H	Cl	H	97	1.5	—	—	90–92	92 [16]
6c		H	CH_3O	H	95	1.5	—	—	147–149	145-146 [24]
7c		H	CH_3	H	91	1.5	—	—	92-93	91-92 [24]
1d		OH	H	H	95	1.5	67	120	89–91	92 [22]
2d		OH	H	Cl	91	1.5	—	—	91–93	—
3d		OH	CH_3	H	94	1.5	—	—	97-98	—
4d	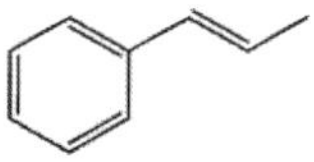	H	H	H	96	1.0	75	120	108–111	108-109 [12]
5d		H	Cl	H	96	1.5	—	—	105–107	105 [12]
6d		H	CH_3O	H	97	1.5	—	—	120-121	119-120 [12]
7d		H	CH_3	H	90	1.5	—	—	75–77	76-77 [28, 29]

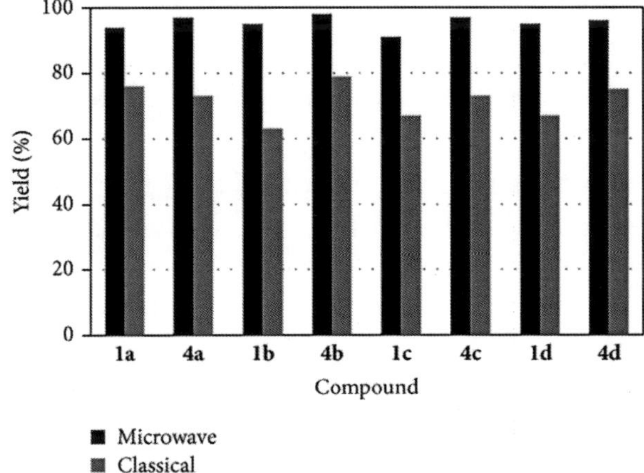

Figure 1: Comparison of yields under the microwave and classical conditions.

CONCLUSION

In conclusion, we have developed a simple microwave assisted solvent-free method for the synthesis of imines using a wetting reagent (□-ethoxyethanol). The method works well for the reaction of this type amines and aldehydes. Because, we have synthesized imine compounds for good yields and fast reaction times in our method. Our method has some advances such as higher reaction yields, shorter reaction times, and green reaction conditions than classical requirements used of a solvent and conventional heat source (like hot plate) and other some microwave methods used of catalyst and solid supports. In addition, some imines were synthesized for the first time by us with this study.

Experimental

All of the chemicals were obtained from commercial sources or prepared according to standard methods. Melting points were determined with an Electro thermal 9100 apparatus. The infrared absorption spectrum of the compounds has been recorded in the region at 4000 and 400 cm^{-1} range using a Bruker Vertex 80/80 v spectrophotometer using KBr pressed pellet technique at room temperature. The ^1H NMR and ^{13}C NMR spectra were recorded using a Bruker AC 200 NMR 200 MHz spectrometer in $CDCl_3$-and DMSO- using TMS as the internal standard. Elemental analyses were conducted at the METU Central Laboratory using a LECO, CHNS-932. All syntheses were carried out in A Bosh HMT 812 C modified microwave oven.

General Procedure for Preparation of 1a–7d. Aldehyde (1 mmoL), amine (1 mmoL) and two drops of -ethoxyethanol as a wetting reagent were mixed in a beaker. Then the beaker was placed in the microwave oven and was exposed to microwave irradiation (360 W). The product obtained by reaction was washed with cold ethanol. Then, it was recrystallized in ethanol. The physical and spectra data of the compounds3a, 2b, 3b, 2c, 3c, 2d, and 3d are as follows.

N-(Salicylidene)-2-hydroxy-4-methylaniline (3a). M.p. 199-202°C. IR: 3033, 2965, 2916, 2858, 1612 (C=N), 1592, 1527, 1486, 1422, 1278, 1221, 1141, 1113, 1012, 869, 788, 754, 720, 622, 587, 556 cm^{-1}. ^1H NMR

(DMSO-d_6): δ 8.94 (s, 1H, N=CH), 13.90 (s, 1H, OH), 9.65 (s, 1H, OH). ^{13}C NMR (DMSO-d_6): δ 160.54, 160.36, 150.82, 137.56, 132.36, 132.25, 131.92, 120.17, 119.46, 119.14, 118.46, 116.93, 116.48. Elemental Anal. Calcd. for $C_{14}H_{13}NO_2$: C, 73.99; H, 5.77; N, 6.16. Found: C, 72.81; H, 5.64; N, 6.16%.

N-(p-Chlorobenzylidene)-2-hydroxy-5-chloroaniline (2b). M.p. 123–125°C. IR: 3367, 3086, 3071, 3059, 2910, 2894, 1626 (C=N), 1588, 1569, 1478, 1424, 1370, 1274, 1234, 1194, 1154, 1087, 1011, 911, 857, 815, 799 (C–Cl), 697, 658, 608, 548, 533 cm^{-1}. ^{1}H NMR (CDCl$_3$): δ 8.66 (s, 1H, N=CH), 7.91 (d, J = 8.4 Hz, 2H, Ar), 7.54 (d, J = 8.4 Hz, 2H, Ar), 7.34 (s, 1H, Ar), 7.24 (d, J = 8.6 Hz, 1H, Ar), 7.02 (d, J = 8.6 Hz, 1H, Ar). ^{13}C NMR (CDCl$_3$): δ156.75 (N=C), 150.94, 138.21, 135.86, 133.87, 130.06, 129.28, 128.77, 125.07, 116.19, 116.16. Elemental Anal. Calcd. for $C_{13}H_9Cl_2NO$: C, 58,67; H, 3,41; N, 5,26. Found: C, 57.74; H, 3.42; N, 5.26%.

N-(p-Chlorobenzylidene)-2-hydroxy-4-methylaniline (3b). M.p. 134–137°C. IR: 3367, 3063, 3049, 3028, 2907, 2857, 1629 (C=N), 1576, 1569, 1497, 1370, 1344, 1295, 1242, 1169, 1154, 1080, 1011, 947, 874, 857, 828, 808 (C–Cl), 793, 733, 687, 610, 591, 579 cm^{-1}. ^{1}H NMR (CDCl$_3$): δ 8.72 (s, 1H, N=CH), 7.92 (d, J = 8.3 Hz, 2H, Ar), 7.53 (d, J = 8.4 Hz, 2H, Ar), 7.29 (d, J = 8.1 Hz, 1H, Ar), 6.92 (s, 1H, Ar), 6.80 (d, J = 8.0 Hz, 1H, Ar), 2.42 (s, 3H, CH$_3$). ^{13}C NMR (CDCl$_3$): δ 154.09 (N=C), 152.34, 139.81, 137.40, 134.54, 132.60, 129.72, 129.16, 120.93, 115.67, 115.36, 21.48. Elemental Anal. Calcd. for $C_{14}H_{12}ClNO$: C, 68.44; H, 4.92; N, 5.70. Found: C, 67.64; H, 4.89; N, 5.69%.

N-(p-Methoxybenzylidene)-2-hydroxy-5-chloroaniline (2c). M.p. 86–88°C. IR: 3323, 3017, 2970, 2954, 2927, 2840, 1626 (C=N), 1599, 1569, 1511, 1453, 1312, 1250, 1165, 1027, 804, 650, 594 cm^{-1}. ^{1}H NMR (CDCl$_3$): δ9.88 (s, 1H, OH), 8.55 (s, 1H, N=CH), 7.85 (d, J = 8.5 Hz, 2H, Ar), 7.23 (s, 1H, Ar), 7.11 (d, J = 8.6 Hz, 1H, Ar), 6.99 (d, J = 8.5 Hz, 2H, Ar), 6.90 (d, J = 8.6 Hz, 1H, Ar), 3.88 (s, 3H, OCH$_3$). ^{13}C NMR (CDCl$_3$): δ 162.92, 157.76 (N=C), 150.70, 136.69, 130.85, 128.49, 127.81, 124.91, 116.08, 115.77, 114.33, 55.49. Elemental Anal. Calcd. for $C_{14}H_{12}ClNO_2$: C, 64.25; H, 4.62; N, 5,35. Found: C, 62.92; H, 4.57; N, 5.45%.

N-(p-Methoxybenzylidene)-2-hydroxy-4-methylaniline (3c). M.p. 86–87°C. IR: 3344, 3066, 3018, 2974, 2934, 2907, 2840, 1622 (C=N), 1593, 1567, 1509, 1481, 1423, 1378, 1245, 1162, 1026, 969, 909, 861, 835, 809, 761, 666, 617, 559 cm^{-1}. ^{1}H NMR (CDCl$_3$): δ 8.59 (s, 1H,

N=CH), 7.83 (d, J = 8.7 Hz, 2H, Ar), 7.16 (d, J = 8.1 Hz, 1H, Ar), 6.97 (d, J = 8.7 Hz, 2H, Ar), 6.81 (s, 1H, Ar), 6.68 (d, J = 7.8 Hz, 1H, Ar), 3.86 (s, 3H, OCH_3), 2.31 (s, 3H, CH_3). ^{13}C NMR ($CDCl_3$): δ 162.52, 155.25 (N=C), 152.02, 138.71, 133.68, 130.36, 129.13, 120.74, 115.34, 115.26, 114.30, 55.45, 21.41. Elemental Anal. Calcd. for $C_{15}H_{15}NO_2$: C, 74.67; H, 6.27; N, 5.81. Found: C, 74.63; H, 6.41; N, 6.21%.

N-(Cinnamylidene)-2-hydroxy-5-chloroaniline (2d). M.p. 91–93°C. IR: 3344, 3064, 3013, 2964, 2911, 2881, 2839, 1621 (C=N), 1577, 1501, 1483, 1445, 1366, 1240, 1154, 1082, 979, 837, 804, 742, 683, 661, 593 cm^{-1} ^{1}H NMR ($CDCl_3$): δ 8.41 (d, J = 8.2 Hz, 1H, N=CH), 7.54 (d, J = 5.2, 2H, Ar), 7.43–7.00 (m, 5H, Ar & CH=CH), 7.29 (s, 1H, Ar), 7.22 (d, J = 8.1 Hz, 1H, Ar), 6.90 (d, J = 8.6 Hz, 1H, Ar). ^{13}C NMR ($CDCl_3$): δ 159.36, 150.95 (N=C), 145.23, 136.47, 135.35, 129.99, 128.99, 128.36, 128.11, 127.69, 124.94, 115.99, 115.86. Elemental Anal. Calcd, for $C_{15}H_{12}ClNO$: C, 69.91; H, 4.69; N, 5.43. Found: C, 68.80; H, 4.72; N, 5.36%.

N-(Cinnamylidene)-2-hydroxy-4-methylaniline (3d). M.p. 97-98°C. IR: 3350, 3061, 3012, 2963, 2912, 2880, 2840, 1624 (C=N), 1577, 1541, 1498, 1447, 1368, 1257, 1157, 1092, 975, 836, 804, 749, 686, 630, 552 cm^{-1} ^{1}H NMR ($CDCl_3$): δ 8.57 (d, J = 6.6 Hz, 1H, N=CH), 7.66 (dd, J = 6.0, 1.9 Hz, 2H, Ar), 7.54–7.12 (m, 5H, Ar & CH=CH), 7.30 (d, J = 7.9 Hz, 1H, Ar), 6.93 (s, 1H, Ar), 6.80 (d, J = 8.1 Hz, 1H, Ar), 2.43 (s, 3H, Ar). ^{13}C NMR ($CDCl_3$): δ 156.85, 152.28 (N=C), 143.26, 139.38, 135.71, 133.20, 129.53, 128.92, 128.73, 127.46, 120.75, 115.48, 115.04, 21.46. Elemental Anal. Calcd, for $C_{16}H_{15}NO$: C, 80.98; H, 6.37; N, 5,26. Found: C, 79.22; H, 6.18; N, 5.90%.

Conflict of Interests

The authors declare that there is no conflict of interests regarding the publication of this paper.

ACKNOWLEDGMENTS

The authors would like to thank the Middle East Technical University analysis Center (Türkiye) for elemental analyses and Özgür Özdamar (Assistant Professor) and Hasan Saral (Assistant Professor) for the analysis, NMR, and IR of some imine compounds.

REFERENCES

1. A. M. Asiri and S. A. Khan, "Synthesis and anti-bacterial activities of some novel schiff bases derived from aminophenazone," Molecules, vol. 15, no. 10, pp. 6850–6858, 2010.
2. R. W. Layer, "The chemistry of imines," Chemical Reviews, vol. 63, pp. 489–510, 1963.
3. H.-K. Fun, R. Kia, A. M. Vijesh, and A. M. Isloor, "5-Diethylamino-2-[(E)-(4-methyl-3-nitrophenyl)iminomethyl]phenol: a redetermination," Acta Crystallographica E: Structure Reports Online, vol. 65, no. 2, pp. o349–o350, 2009.
4. H. Schiff, "Sur quelques dérivés phéniques des aldéhydes," Annali Di Chimica, vol. 131, p. 118, 1864.
5. K. Taguchi and F. H. Westheimer, "Catalysis by molecular sieves in the preparation of ketimines and enamines," Journal of Organic Chemistry, vol. 36, no. 11, pp. 1570–1572, 1971.
6. M. E. Kuehne, "The application of enamines to a new synthesis of β-ketonitriles," Journal of the American Chemical Society, vol. 81, no. 20, pp. 5400–5404, 1959.
7. M. Á. Vázquez, M. Landa, L. Reyes, R. Miranda, J. Tamariz, and F. Delgado, "Infrared irradiation: effective promoter in the formation of N-benzylideneanilines in the absence of solvent," Synthetic Communications, vol. 34, no. 15, pp. 2705–2718, 2004.
8. A. K. Chakraborti, S. Bhagat, and S. Rudrawar, "Magnesium perchlorate as an efficient catalyst for the synthesis of imines and phenylhydrazones," Tetrahedron Letters, vol. 45, no. 41, pp. 7641–7644, 2004.
9. H. Naeimi, H. Sharghi, F. Salimi, and K. Rabiei, "Facile and efficient method for preparation of Schiff bases catalyzed by P_2O_5/SiO_2 under free Solvent conditions," Heteroatom Chemistry, vol. 19, no. 1, pp. 43–47, 2008.
10. J. H. Billman and K. M. Tai, "Reduction of Schiff bases. II: benzhydrylamines and structurally related compounds," Journal of Organic Chemistry, vol. 23, no. 4, pp. 535–539, 1958.
11. M. Gopalakrishnan, P. Sureshkumar, V. Kanagarajan, and J. Thanusu, "New environmentally-friendly solvent-free synthesis of imines using calcium oxide under microwave irradiation,"Research on Chemical Intermediates, vol. 33, no. 6, pp. 541–548, 2007.
12. J. S. Bennett, K. L. Charles, M. R. Miner et al., "Ethyl lactate as a tunable solvent for the synthesis of aryl aldimines," Green Chemistry, vol. 11,

no. 2, pp. 166–168, 2009.

13. R. S. Varma, R. Dahiya, and S. Kumar, "Clay catalyzed synthesis of imines and enamines under solvent-free conditions using microwave irradiation," Tetrahedron Letters, vol. 38, no. 12, pp. 2039–2042, 1997.
14. W. A. White and H. Weingarten, "A versatile new enamine synthesis," Journal of Organic Chemistry, vol. 32, no. 1, pp. 213–214, 1967.
15. F. Texier-Boullet, "A simple, convenient and mild synthesis of imines on alumina surface without solvent," Synthesis, vol. 1985, no. 6-7, pp. 679–681, 1985.
16. L. Ravishankar, S. A. Patwe, N. Gosarani, and A. Roy, "Cerium(III)-catalyzed synthesis of schiff bases: a green approach," Synthetic Communications, vol. 40, no. 21, pp. 3177–3180, 2010.
17. K. P. Guzen, A. S. Guarezemini, A. T. G. Órfão, R. Cella, C. M. P. Pereira, and H. A. Stefani, "Eco-friendly synthesis of imines by ultrasound irradiation," Tetrahedron Letters, vol. 48, no. 10, pp. 1845–1848, 2007.
18. R. Annunziata, M. Benaglia, M. Cinquini, and F. Cozzi, "Poly(ethylene glycol)-supported 4-alkylthio-substituted aniline: a useful starting material for the soluble polymer-supported synthesis of imines and 1,2,3,4-tetrahydroquinolines," European Journal of Organic Chemistry, no. 7, pp. 1184–1190, 2002.
19. M. Hosseini-Sarvari, "Nano-tube TiO_2 as a new catalyst for eco-friendly synthesis of imines in sunlight," Chinese Chemical Letters, vol. 22, no. 5, pp. 547–550, 2011.
20. G. Dutheuil, S. Couve-Bonnaire, and X. Pannecoucke, "Diastereomeric fluoroolefins as peptide bond mimics prepared by asymmetric reductive amination of α-fluoroenones," Angewandte Chemie: International Edition, vol. 46, no. 8, pp. 1290–1292, 2007.
21. J. F. Collados, E. Toledano, D. Guijarro, and M. Yus, "Microwave-assisted solvent-free synthesis of enantiomerically pure N-(tert-Butylsulfinyl)imines," Journal of Organic Chemistry, vol. 77, Article ID 300919, pp. 5744–5750, 2012.
22. E. Tauer and K. H. Grellmann, "Photochemical and thermal reactions of aromatic Schiff bases," Journal of Organic Chemistry, vol. 46, no. 21, pp. 4252–4258, 1981.
23. R. J. Argauer and C. E. White, "Effect of substituent groups on fluorescence of metal chelates," Analytical Chemistry, vol. 36, no. 11, pp. 2141–2148, 1964.
24. J. D. Margerum and J. A. Sousa, "Spectroscopic studies of substituted benzalanilines," Applied Spectroscopy, vol. 19, pp. 91–97, 1965.

25. J. E. Kuder, H. W. Gibson, and D. Wychick, "Electrochemical characterization of salicylaldehyde anils," Journal of Organic Chemistry, vol. 40, no. 7, pp. 875–879, 1975.

26. F. F. Stephens and J. D. Bower, "The preparation of benziminazoles and benzoxazoles from Schiff's bases. Part I," Journal of the Chemical Society, pp. 2971-2972, 1949.

27. J. Schmeyers, F. Toda, J. Boy, and G. Kaupp, "Quantitative solid-solid synthesis of azomethines," Journal of the Chemical Society: Perkin Transactions 2, no. 4, pp. 989–993, 1998.

28. W. Xu, S. Zhang, S. Yang et al., "Asymmetric synthesis of α-aminophosphonates using the inexpensive chiral catalyst 1,1'-binaphthol phosphate," Molecules, vol. 15, no. 8, pp. 5782–5796, 2010

29. Y. M. S. A. Al-Kahraman, H. M. F. Madkour, D. Ali, and M. Yasinzai, "Antileishmanial, antimicrobial and antifungal activities of some new aryl azomethines," Molecules, vol. 15, no. 2, pp. 660–671, 2010.

Chapter 4

CRYSTAL GROWTH, THERMAL, MECHANICAL AND OPTICAL PROPERTIES OF A NEW ORGANIC NONLINEAR OPTICAL MATERIAL: ETHYL P-DIMETHYL AMINO BENZOATE (EDMAB)

V. Natarajan, J. Kalyana Sundar, P. Selvarajan, M. Arivanandhan, K. Sankaranarayanand, S. Natarajan, and Y. Hayakawa,

a Department of Physics, Aditanar College of Arts and Science, Tiruchendur, India b School of Physics, Madurai Kamaraj University, Madurai-625021, India. C Research Institute of Electronics, Shizuoka University, Johoku 3-5-1, Hamamatsu, Japan. D School of Physics, Alagappa University, Karaikudi-630003, India.

ABSTRACT

An organic material, namely, ethyl p-dimethyl amino benzoate was crystallized for the first time by solution growth technique using pure and mixed solvents. Growth kinetics and morphology changes with solvents were investigated based on solute –solvent interactions of pure and mixed solvents. An appropriate mixed solvent for high quality crystals with well-defined morphology is reported. The absence of solvent molecules and the presence of various functional groups of the grown sample were qualitatively confirmed by FTIR spectroscopic studies. Thermal properties of the grown sample were analyzed by TG and DTA analysis. Mechanical properties of the EDMAB crystal were investigated by micro hardness studies. Moreover, the grown crystal shows high transparency in the visible and near IR regions. The material shows relatively high SHG efficiency than that of KDP.

INTRODUCTION

Benzoic acid and its derivatives are more useful in bio-medical applications especially they are good inhibitors of influenza viruses [1]. Some of the benzoic acid derivatives, such as 4- amino benzoic acid, have been extensively reported in coordination chemistry, as functional organic ligands, due to variety of their coordination modes [2]. Ethyl p-amino benzoate also known as benzocaine is one of the benzoic acid derivatives mainly used as a best local anesthesia. Its low solubility renders it unsuitable for injections but its slow absorption from mucous-lined surfaces makes it safer than other local anesthetics for ulcers and wounds [3, 4]. Recently, it was identified as an organic nonlinear optical (NLO) material which shows high second harmonic generation than that of potassium hydrogen phosphate (KDP), a well-known inorganic NLO material [5]. In this series, Ethyl p-dimethyl amino benzoate (EDMAB) is generally known as tertiary amines which are mainly used as a part of self-curing two part system for dental/medical compositions comprising degradable copolymers which are suitable for use as root canal sealants, root canal filling materials, dental restorative materials, implant materials, bone cements and pulp capping materials [6]. Moreover, since the chemical structure (Fig.1) of the material has the π – electron conjugated system attached with electron donor and acceptor groups at opposite end, it could be applicable for nonlinear optical applications [7].

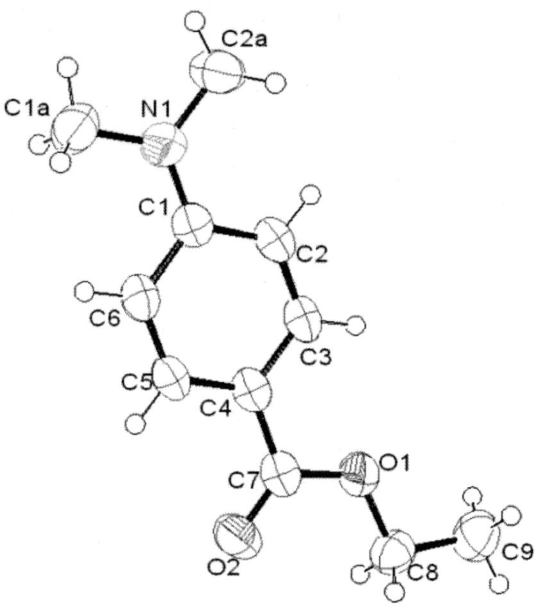

Figure 1: The molecular structure of the compound, showing the 30% probability displacement ellipsoids and the atom numbering scheme.

Recently, the crystal structure of EDMAB was solved by single crystal X-ray diffraction studies [8]. The EDMAB material crystallizes in the monoclinic system with the space group P21 a and the lattice parameters are reported as a = 12.695Å, b =6.66 Å and c =12.853 Å and β= 98.672° [8]. In the present work, large size EDMAB single crystals were grown by slow solvent evaporation method. The grown crystal was subjected to Fourier transform infrared studies to analyze the solvent incorporation during growth and ascertain various functional groups of the compound. Thermal properties of the material were investigated by means of thermo-gravimetric and differential thermal analysis.

EXPERIMENTAL METHOD

Selection of the Solvent

The EDMAB material was purchased from Aldrich and it was purified by recrystallization process using ethanol as a solvent. Organic crystal growth from solution mainly depends on the selection of a suitable solvent. Two major properties that would influence the crystal habit are assumed to be the difference in both the energy of crystallization and in the dipole moments between the crystallizing component and the solvent. If a solvent, which has similar dipole moment and the ideal solubility, is selected, large-size crystals can be expected to grow in the solution. On the other hand, a solvent has large differences in dipole moments from crystallizing substance and has poor solubility seems to be a poor solvent for the purpose of crystal growth with clear morphology. Most of the organic solvents have a dipole moment less than about 3

Debye. Unfortunately, organic dielectric materials are usually highly polar in nature so that the selection of solvent is restricted. The choice of solvent with a large dipole moment is likely to produce a crystal with a better habit growth. Since EDMAB is insoluble in water, several organic solvents such as ethanol (1.69 Debye), methanol (1.70 Debye), toluene (0.38 Debye) and acetone (2.88 Debye) have been investigated to find the suitability for the single crystal growth of EDMAB. Among the above-mentioned solvents, it was observed that EDMAB has relatively low solubility in ethanol whereas it has relatively high solubility in acetone at 32ºC. In general, high solubility leads to uncontrolled spurious nucleation which hinders bulk single crystal growth. On the other hand, low solubility leads to less availability of growth units which restricts the size and growth rate of the crystal. Based on the preliminary observations made during the crystal growth experiments, conducted using all the above-mentioned solvents, ethanol and acetone were selected for further investigation towards attainment of large size and device quality crystals.

Crystal Growth of EDMAB

Solubility of EDMAB in mixed solvent of ethanol and acetone was measured at 32°C for different volume ratio of the solvents. Figure 2 shows the solubility of EDMAB in the mixed solvent as a function of acetone measuring in volume ratio. It can be seen from the figure that the solubility of the material increases evidently while increasing the volume ratio of acetone. According to the solubility data, saturated solution of EDMAB was prepared by dissolving the purified source material in the pure and mixed solvents of ethanol and acetone at 32o C. The homogeneity of the solution was realized by continuous mild stirring of the solution using a magnetic stirrer without altering the solution temperature. Then, the near saturated solution was transferred to a crystallizer and covered by a perforated polyethylene sheet for controlled evaporation at room temperature. Transparent single crystals were harvested from the growth solution after achieved a reasonable size. Fig. 3a shows the photograph of the crystals grown from the mixed solvent of 25 % acetone and 75% ethanol. The crystallization experiments conducted using the mixed solvents which contains 50% volume ratio and higher than that of acetone resulted uncontrolled large number of nucleation. Fig. 3b shows one of the prepared specimen which was used for the following characterization studies.

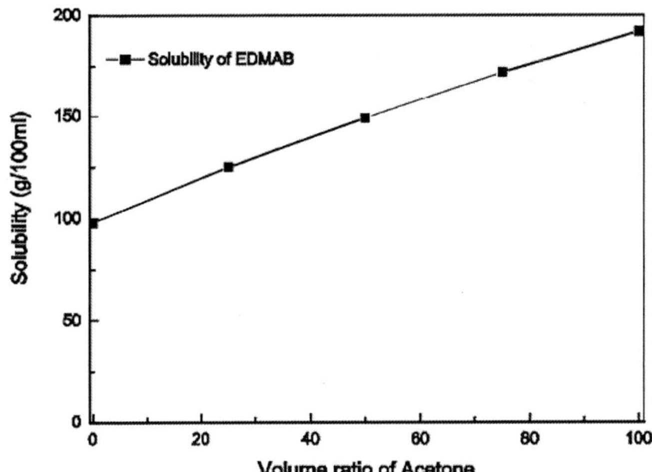

Figure 2: Solubility of EDMAB in mixed solvent of ethanol and acetone as a function of acetone volume ratio.

The physicochemical properties of the crystals grown from the mixed solvent (25 % acetone and 75% ethanol) were studied by Fourier transform infrared spectroscopy (Perkin- Elmer), TG&DTA analysis (STANTON REDCROFT), Vickers's micro hardness tester and U-3000 double beam UV-VIS-NIR spectroscopy. As a preliminary study, the second harmonic generation of the grown crystal was confirmed be powder Kurtz method [9].

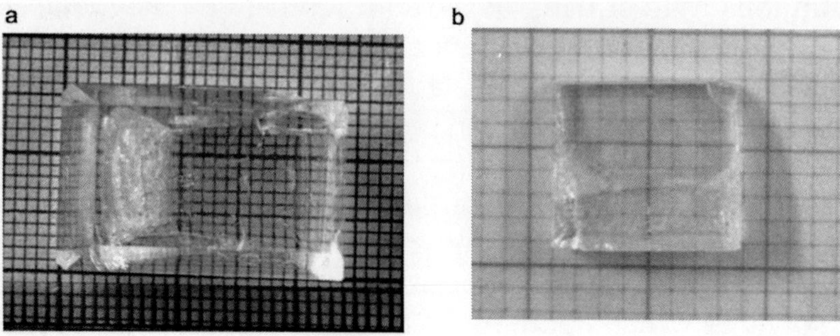

Figure 3: Photograph of the grown EDMAB crystal. (a). from the mixed solvent of ethanol and acetone of 75% and 25% volume ratio. (b). prepared specimen from the grown crystal

RESULTS AND DISCUSSION

In general, the growth rate of a growing crystal can largely be influenced by the size and rate of diffusion of molecules or growth units. In solution growth, the rate of diffusion of molecules or growth units are mainly depends on the solute-solvent interactions which can be effectively tuned by means of providing mixed solvents instead of single solvent. Due to the different vapor pressures of the constituent solvents as well as their interaction with solute molecules, the diffusion of growth units can very well be modified or controlled. Also, the vital growth parameters such as induction period and the nucleation rate can be varied by means of altering the chemical environment around the growing surface.

Based on these aspects, an attempt has been made to grow large size EDMAB crystals from mixed solvent containing ethanol and acetone, in which EDMAB has low solubility in ethanol and high solubility in acetone. Moreover, both the solvents are miscible and have different vapor pressures. This led us to investigate the controlling of super saturation thereby nucleation of EDMAB, since changing the solubility as well as the interaction between the solute and the solvent has profound effect on these vital growth parameters. It was observed that the solubility of EDMAB was increased with the increasing of acetone volume ratio in the mixed solvents (Fig. 2). The results indicates that the effective interaction between the solute and solvent is higher when the dipole moment of the solvent increases. In the case of pure ethanol, due to the low dipole moment of the solvent, weak interaction with the solute was expected which has resulted low solubility. On the other hand, since the acetone has large dipole moment, strong interaction with solute molecules was expected which led to high solubility of the material. From the growth experiments, it was found that the mixed solvent contain 25% volume ratio of acetone yielded good quality large size (25 x 15 x 8 mm3) single crystals with short period (7days) of growth. The pure ethanol solvent system yielded thin platelet crystals with prolonged growth period and the mixed solvents with ≥ 50% volume ratio of acetone resulted spurious nucleation probably due to the relatively high vapor pressure of the solvent.

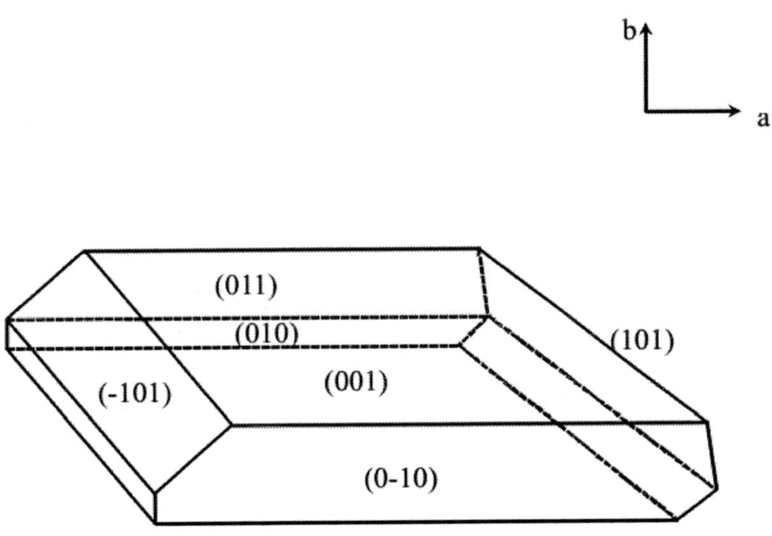

Figure 4: Morphology of the grown crystal.

The obtained large size EDMAB single crystal from the mixed solvent having 25% of acetone-volume ratio indicated an optimized growth condition. Moreover, the growth rate of EDMAB crystal along the a-axis seems to be much higher than that along b- and c-axes. Figure 4 shows the morphology of the grown crystal. Among the clearly observed facets, the crystal has two prominent flat facets such as {001} and {00-1} having large area compared to other {101}, {010} and {100} faces.

FTIR Studies

The FTIR spectrum recorded in the range of 4000–400 cm-1 for EDMAB sample is shown in Figure 5. The various functional groups of the grown material were identified by FTIR spectroscopic analysis. N-H is one of the prominent functional groups of the primary aromatic amines, and its asymmetric stretching vibration is observed at 3448 cm-1 and the symmetric stretching vibration is observed at 3392 cm-1. The aromatic C–H stretching is observed at 3088 cm-1 and the aliphatic C-H stretching is observed at 2979, 2905 and 2812 cm-1.

A strong peak observed around 1697 cm-1 is due to the out of plane bending of N-H vibration of the molecule. An absorption peak at 1605 cm-1 is due to carbonyl stretching (C=O) of the molecule. In plane bending of N-H stretching is observed at 1527 cm-1. The absorption peaks at 1482 and 1443 cm-1 are due to skeletal vibrations of aromatic ring.

Out-of-plane bending vibration modes of aromatic C–H bonds are observed at 1367, 1280, 1182, 1113 and 1012 cm-1. The absorption peaks observed below 1000 cm-1 illustrates in-plane-bending vibration modes of C–H bonds. From this spectroscopic investigation, the presence of all the fundamental functional groups of the grown sample was confirmed qualitatively. Moreover, no absorption peaks related to the solvent molecules or any other impurities were observed which confirms the absence of solvent trapping in the crystal lattice and also the purity of the grown sample.

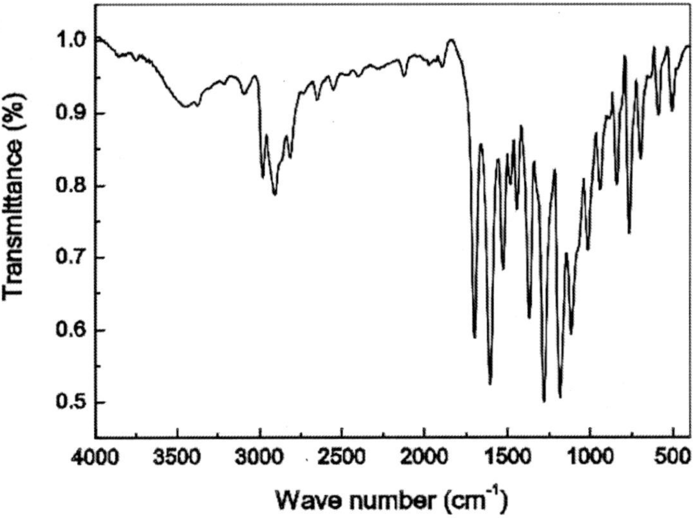

Figure 5: FTIR spectrum of the EDMAB crystal.

TG & DTA Studies

Thermal characteristics of the material were investigated by TG & DTA analysis. For the TG analysis, the known gram of material was heated from ambient temperature to 300°C at a heating rate of 20°C/min in air atmosphere. Figure 6 illustrates the TG and DTA curves for the grown EDMAB sample. The TG curve shows very small (2.5%) weight loss up to 127°C. Hence the material is thermally stable up to 127°C and above this temperature the material loses its weight gradually. As can be seen from the DTA curve in the figure, the material undergoes an endothermic transition at 50.5°C where the melting begins. The endothermic peak at 64°C represents the temperature at which the melting terminates which corresponds to melting point of the material. The TG curve shows no weight loss during the first two endothermic peaks which confirms the phase transition of the material from solid to liquid (melting) and the weight of the material starts to losses after 128°C is not due to self-degradation of EDMAB but merely its evaporation after melting. The endothermic peak at 228°C indicates a phase change from liquid to vapor state as evidenced from the huge loss of weight in TG curve

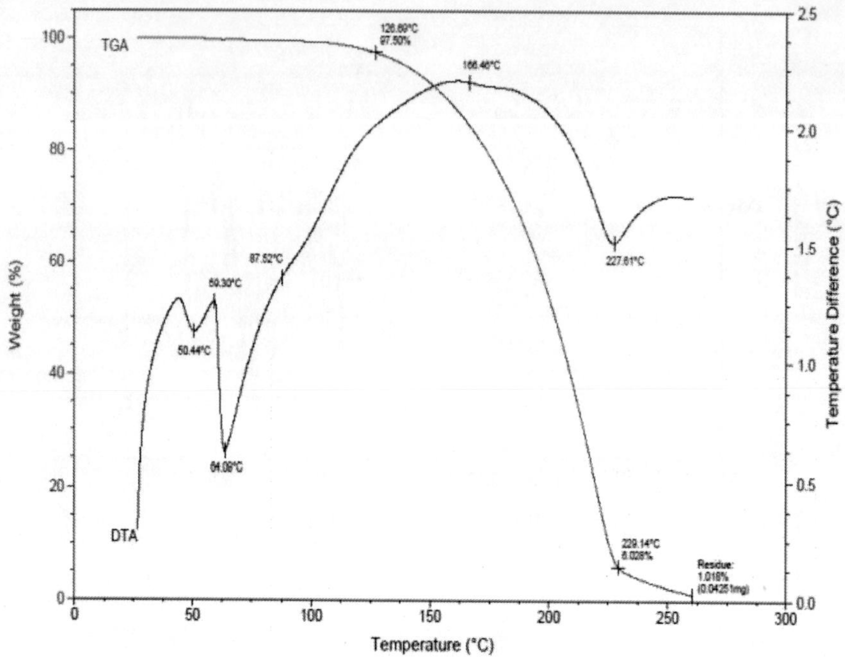

Figure 6: TG&DTA curves of the grown EDMAB sample.

Micro Hardness Studies

Mechanical hardness of a material is also one of the decisive properties especially for post growth processes and device fabrications. In order to study the mechanical properties of the as grown EDMAB crystal, micro hardness was measured from 25 to 100 gram of load using HMV Micro hardness tester. The indentation time was fixed as 10 s for all the samples. The diagonal lengths of the indented impression were measured for different loads varying from 25 to 100 g. The successive indentations were made at different sites of the sample surface. The hardness values were calculated from the formula Hv=1.8544P/d2 kg/mm2 where P is the applied load and d is the mean diagonal length of the indentation. The variation of hardness as a function of applied load is shown in Figure 7 which reveals that the hardness increases with the increase of load.

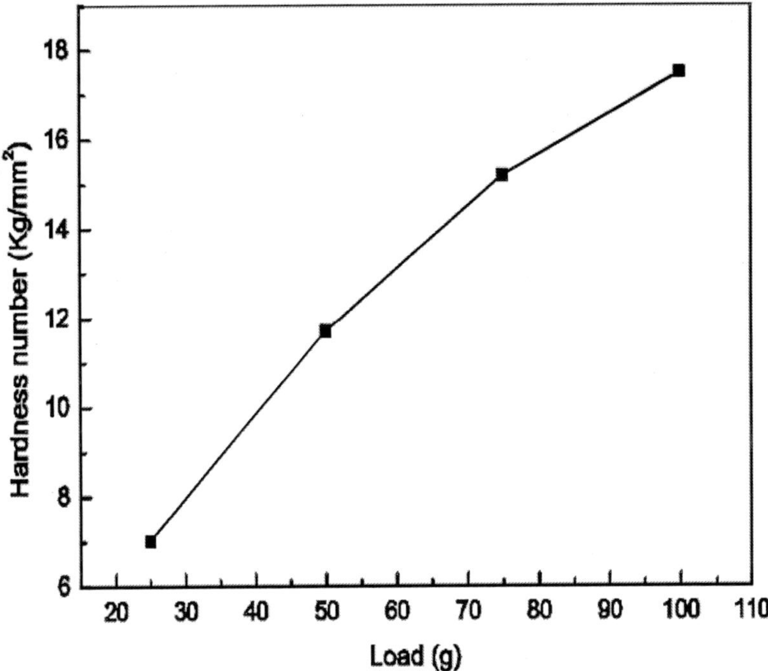

Figure 7: Load dependence of the hardness number of the EDMAB crystal

Optical Studies

The optical transmission spectrum was recorded in the wavelength range of 300–1100 nm at room temperature on the ~3 mm thick sample prepared from the grown EDMAB crystal. The recorded transmittance spectrum is shown in Figure 8. The material shows a wide transmission window in the wave length region of 400 – 1100 nm. The sharp absorption onset at 340 nm and the high transmission values of the grown EDMAB single crystal at wavelength above 400 nm exhibit the low concentration of grown-in defects and high optical quality.

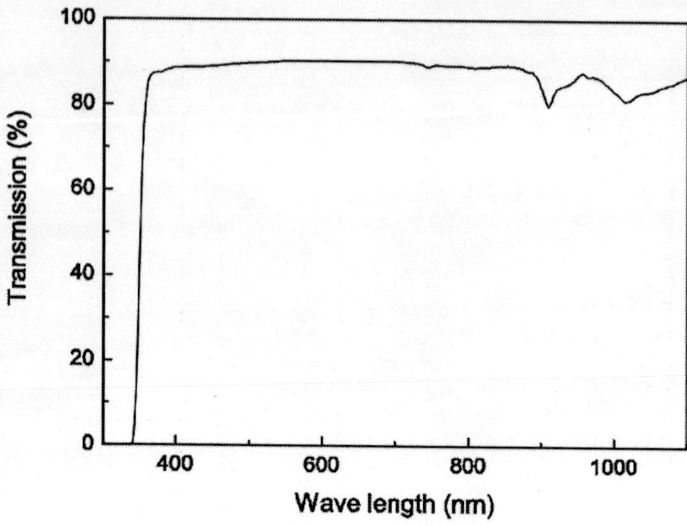

Figure 8: Optical transmission spectrum of the EDMAB sample.

SHG Studies

The SHG of the grown sample was confirmed by powder Kurtz method [9]. The powder sample prepared from the grown EDMAB crystal was packed in a triangular cell and kept in a cell holder. Nd:YAG laser of 1064 nm wavelength was used to irradiate the sample. Second harmonic signal was captured by the oscilloscope through the photomultiplier tube. Potassium dihydrogen phosphate (KDP) crystal was used as a reference material in the SHG measurement. The frequency conversion was confirmed by observing the green signal from the powdered EDMAB sample and the power of the SH signal is relatively higher (1.2 times) than that of pure KDP. Further investigations on phase matching studies of the grown crystal are under progress and the outcome will be published elsewhere.

CONCLUSIONS

EDMAB single crystals were grown by slow solvent evaporation technique using mixed solvent of ethanol and acetone. A mixed solvent having volume ratio of 75% ethanol and 25% acetone was found as a suitable solvent for bulk growth of EDMAB. Various functional groups present in the grown crystal were identified by FTIR spectroscopy. Thermal stability of the grown sample was studied by TG&DTA analysis. TG curve of this sample indicates that the crystalline sample has a melting point 64°C and chemically stable up to 150°C. Mechanical properties of the grown crystal were analyzed by micro hardness studies. The optical quality of the grown crystal was justified by optical transmission studies. The sharp absorption onset at 340 nm and the high transmission of the grown crystal at wavelength above 400 nm exhibit the optical quality, low concentration of grown-in defects and suitability for optical applications as well. The material shows high SHG efficiency than that of KDP and hence it could be applicable for nonlinear optical applications.

REFERENCES

1. M. Luo, M. J. Jedrzejas, S. Singh, C. L. White, W. J. Brouillette, G. M. Air, W. G. Laver, Acta Cryst. D51, (1995) 504-510.
2. R. Hauptmann, M. Kondo, S. Kitagawa, Z. Kristallogr. New Cryst. Struct. 215, (2000) 169- 172.
3. B. K. Sinha, V. Pattabhi, Proc. Indian Acad. Sci. Chem. Sci., 98, (1987) 229-234.
4. A. C. Schmidt, Pharm. Res. 22, (2005) 2121-2133.
5. M. Arivanandhan, K. Sankaranarayanan, P. Ramasamy, Materials Letters, 61, (2007) 4836.
6. Jia Weitao, Jin Shuhua United States Patent 6787584 (2004).
7. H. S. Nalwa, T. Watanabe, S. Miyata, in Nonlinear Optics of Organic Molecules and Polymers, Nalwa, H.S.; Miyata, S., Eds.; CRC press Inc.,: Florida, 1997; Chapter 4, pp 89- 350.
8. J. Kalyana Sundar, V. Natarajan, M. Arivanandhan, Y. Hayakawa, S. Natarajan, S, Acta Cryst. E, 66,(2010) o355.
9. S. K. Kurtz, T.T. Perry, J. Appl. Phys. 39, (1968) 3798.

Chapter 5

SYNTHESIS AND PIGMENTAL PROPERTIES OF NICKEL PHOSPHATES BY THE SUBSTITUTION WITH TETRAVALENT CERIUM CATION

Hiroaki Onoda, Takeshi Sakumura

Department of Informatics and Environmental Sciences, Faculty of Life and Environmental Sciences, Kyoto Prefectural University, Kyoto, Japan

ABSTRACT

In this paper, we report the preparation of nickel phosphate in aqueous solution and its use as inorganic pigment. Be- cause cerium phosphate is insoluble in acidic and basic solution, the addition of cerium was tried to improve the acid and base resistance of nickel phosphate pigment. The cerium substituted nickel phosphates were prepared from phosphoric acid, nickel nitrate, and ammonium cerium nitrate solution. The additional effects of tetravalent cerium cation were studied on the chemical composition, particle shape and size distribution, specific surface area, color, acid and base resistance of the precipitates and their thermal products.

INTRODUCTION

Phosphates have been used for ceramic materials, catalysts, fluorescent materials, dielectric substances, metal surface treatment, detergent, food additives, fuel cells, pigments, etc. [1-3]. Especially, as a pigment, these materials have good anticorrosion properties for oxidation reaction and suitable for coating [4-7]. However, there is a weak point that is a certain degree of solubility for acidic and basic solution.

It is well known that rare earth phosphates are insoluble for acidic and basic solution in the groups of phosphate materials. In general, the addition of rare earth elements gives higher functional properties to the material [8]. Consequently, the addition of rare earth cation had the anticipation to improve the acid and base resistance of inorganic phosphate pigments and pigmental proper- ties [9-11].

The substitution with lanthanum in nickel and cobalt phosphate materials was studied on the chemical com- position, powder condition, color, acid and base resistance in previous works [12,13]. Specific surface area of phosphates increased and particle size became larger by the substitution with lanthanum. The substitution with lanthanum was effective on acid and base resistance for de- sign of inorganic phosphate pigment. However, the color of phosphate materials was whitened by the substitution with lanthanum.

Cerium cation is also one of rare earth cation. The different phenomena are expected with lanthanum cation, because cerium cation has stable tri- and tetravalent states [14]. Generally, tetravalent cerium cation forms the yellow materials. Therefore, the substitution with tetravalent cerium cation is suitable to improve the acid and base resistance of yellow pigments. In this work, nickel- cerium (+IV) phosp hates were synthesized in aqueous solution. The obtained products were estimated from their particle shape and size distribution, specific surface area, color, acid and base resistance.

Experimental Procedure

The 0.1 mol/L of nickel nitrate, $Ni(NO_3)_2$, solution was mixed with 0.1 mol/L of phosphoric acid solution in the molar ratio of Ni/P = 3/2. This ratio is settled from the chemical composition of nickel orthophosphate, $Ni_3(PO_4)_2$. The certain part of nickel nitrate was substituted with ammonium cerium nitrate, $(NH_4)_2Ce(NO_3)_6$, in the molar ratio of Ni/Ce = 10/0, 9/1, 8/2, 5/5, 2/8, and 0/10 [12,13]. For the valence balance, two nickel cations were replaced with one cerium cation. Finally, the solutions were mixed in the molar ratio of Ce/P = 3/4. This ratio is corresponding to $Ce_3(PO_4)_4$. Then, the mixed solution was adjusted to pH 7 by ammonia solution. The precipitate was filtered off and dried in air condition. All chemicals were of guaranteed reagents from Wako Chemical Industries Ltd. (Osaka, Japan) without further purification.

A part of the precipitates was dissolved in hydro- chloric acid solution. The ratios of phosphorus and cerium in the precipitates were also calculated from ICP results of these solutions, using SPS1500VR, Seiko Instruments Inc. The thermal behavior of these materials was analyzed by TG-DTA and XRD. TG and DTA curves were measured with a Shimadzu DTG-60H at a heating rate of 10°C/min under air. XRD patterns were recorded on a Rigaku Denki RINT 1200 M X-Ray diffract meter using monochromatic Cu*Ka* radiation.

The powder properties of thermal products at 200°C, 400°C, 600°C, and 800°C were characterized by particle shape, particle size distribution, specific surface area, and their color. Particle shapes were observed by scanning electron micrographs (SEM) using JGM-5510LV, JEOL Ltd. Particle size distribution was measured with laser diffraction/scattering particle size distribution HORIBA LA-910. Specific surface areas of phosphates were calculated from the amount of nitrogen gas adsorbed at the temperature of liquid nitrogen by BET method with Belsorp mini from BEL JAPAN, INC. The color of phosphate pigments was estimated by ultraviolet—visible (UV- Vis) reflectance spectra with a Shimadzu UV2550.

Furthermore, the acid and base resistance of materials was estimated in following method. The 0.1 g of thermal products was allowed to stand in 100 ml of 0.1 wt% sulphuric acid or 0.1 wt% sodium hydroxide solutions for 1 day. Then, solid was removed off by filtration; the solution was dilute with nitric acid for ICP measurement. The concentrations of phosphorus, nickel, cerium cation were calculated by ICP results. As resistance estimation, the solubility (%) of target elements was calculated to divide by the concentration that thermal products were completely dissolved by hot hydrochloric acid.

Results and Discussion

Chemical Composition of Nickel—Cerium Phosphates

Table 1 shows ICP results of samples synthesized in Ni/Ce ratios. From the valence of nickel, cerium cations, and phosphate anion, hydrogen ratio was calculated in the following equation.

$$z(\text{hydrogen rafio}) = 3 - 2x(\text{nichel ratio}) - 4y(\text{cerium ratio}) \quad (1)$$

The negative value of z means the existence of hydroxide anion. Sample prepared in Ni/Ce = 10/0 had the near ratio with nickel hydrogen phosphate, $NiHPO_4$. On the other hand, cerium ratio was enough high at samples pre- pared in Ni/Ce = 0/10. Cerium orthophosphate was considered to form with cerium hydroxide in this condition. In the middle Ni/Ce ratios, cerium ratio was lower than that in preparation process. In previous works [12,13], rare earth ratio in precipitates was higher than that in preparation conditions. This phenomenon was caused from that rare earth phosphates have smaller solubility than transition metal phosphates. In the case of tetravalent cerium phosphate, this tendency was not appeared. How- ever, it is not clear why the ratio of cerium became lower than preparation condition.

Figure 1 shows DTA curves of nickel—cerium phosphates prepared in various Ni/Ce ratios. DTA curves of sample prepared in Ni/Ce = 10/0 had large endothermic peak at 100°C, small endothermic peak at 250°C, and small exothermic peak at 770°C (**Figure 1(a)**). These peaks were due to the volatilization of water, dehydration condensation of phosphate, and crystallization of nickel phosphate, respectively [12]. Nickel hydrogen phosphate was condensed to nickel pyrophosphate in following re- action at 250°C.

$$2NiHPO_4 \rightarrow Ni_2P_2O_7 + H_2O \quad (2)$$

The substitution with cerium produced the exothermic peaks at the range from 260°C to 330°C (**Figures 1(b)- (f)**). These peaks were related with the reduction of cerium cation [15]. **Figure 2** shows TG curves of samples prepared in various Ni/Ce ratios. Samples prepared in Ni/Ce = 10/0 had large weight loss over 50%. Nickel hydrogen phosphate had large amount of adsorbed and crystalline water. By the substitution with cerium cation, this weight loss became small.

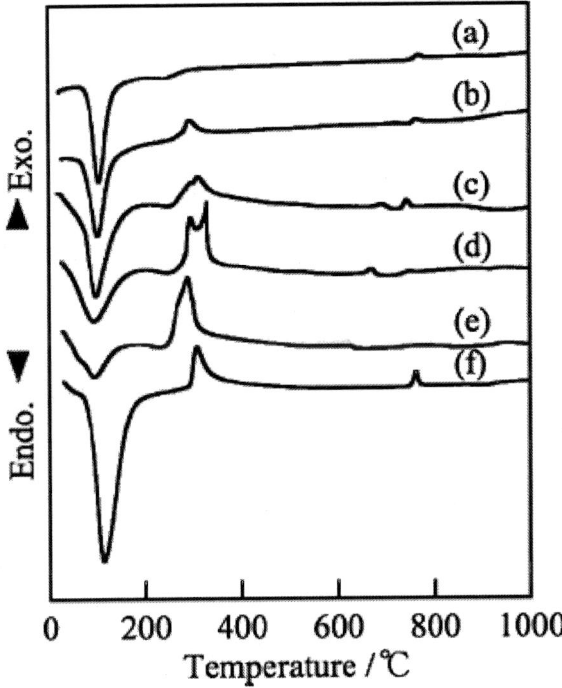

Figure 1 DTA curves of samples prepared in various Ni/Ce ratios, (a) 10/0; (b) 9/1; (c) 8/2; (d) 5/5; (e) 2/8; and (f) 0/10.

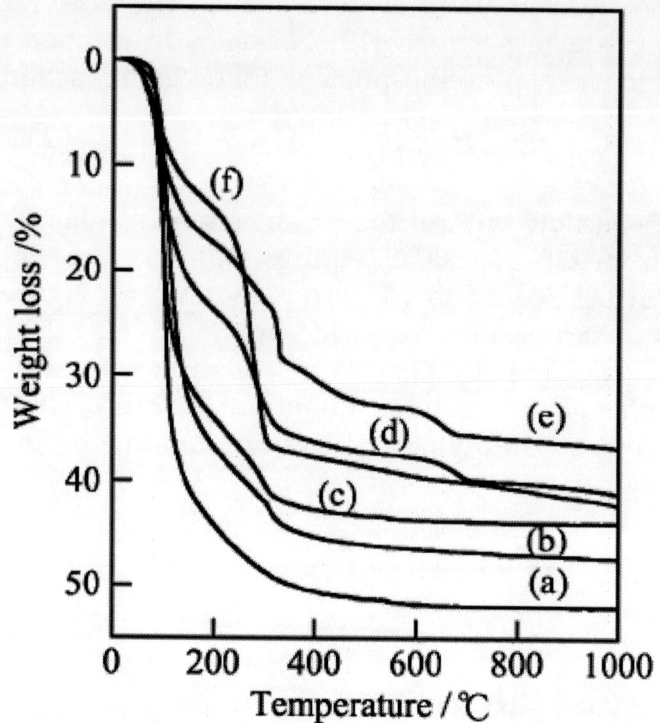

Figure 2 TG curves of samples prepared in various Ni/Ce ratios, (a) 10/0; (b) 9/1; (c) 8/2; (d) 5/5; (e) 2/8; and (f) 0/10.

Samples prepared in Ni/Ce = 10/0 - 5/5 indicated XRD peaks of NH4H2PO4. Samples heated at 200°C - 600°C were amorphous phase in XRD analyses except for sample prepared in Ni/Ce = 0/10 and then heated at 600°C. This exceptional sample had the peaks of Monazite-type $CePO_4$. Figure 3 shows XRD patterns of samples synthesized in various Ni/Ce ratios and then heated at 800°C. Sample prepared in Ni/Ce = 10/0 had strong peaks of nickel pyrophosphate, $Ni_2P_2O_7$. The peaks of Monazite- type $CePO_4$ were observed in XRD patterns of cerium- substituted samples. Tetravalent cerium phosphate changed to trivalent cerium compound by heating.

We considered that the following reaction occurred in sample prepared at Ni/Ce = 0/10.

$$Ce_3(PO_4)_4 + Ce(OH)_4 \rightarrow 4CePO_4 + H_2O + O_2 \quad (3)$$

The starting materials, the mixture of cerium phosphate and hydroxide, was suited to the Ce/P ratio (=0.986) in Table 1. Sample heated at 800°C indicated XRD peaks of CePO4 (Figure 3(f)). TG curves of cerium- substituted samples had weight loss at the range from 260°C to 330°C corresponding with exothermic peak in DTA curves (Figures 1, 2). These weight losses were caused from the volatilization of water and oxygen. The color of sample prepared in Ni/Ce = 0/10 changed from yellow to white by heating at 400°C. Tetravalent cerium phosphate was yellow powder and trivalent one was white powder

Powder Properties of Nickel—Cerium Phosphates

Figure 4 shows SEM images of samples synthesized in various Ni/ Ce ratios. Sample prepared in Ni/Ce = 10/0 consisted of small particles. On the other hand, samples prepared at high cerium ratios had large particles.

Table 1 Chemical composition, NixCeyHzPO4, of precipitates from ICP measurements.

Ni/Ce*	x	y	z	Ni/Ce**
10/0	1.064	0	0.872	10/0
9/1	0.786	0.027	1.320	9/0.31
8/2	0.915	0.179	0.454	8/1.57
5/5	0.668	0.401	0.060	5/3.00
2/8	0.267	0.687	−0.015	2/5.15
0/10	0	0.986	−0.944	0/10

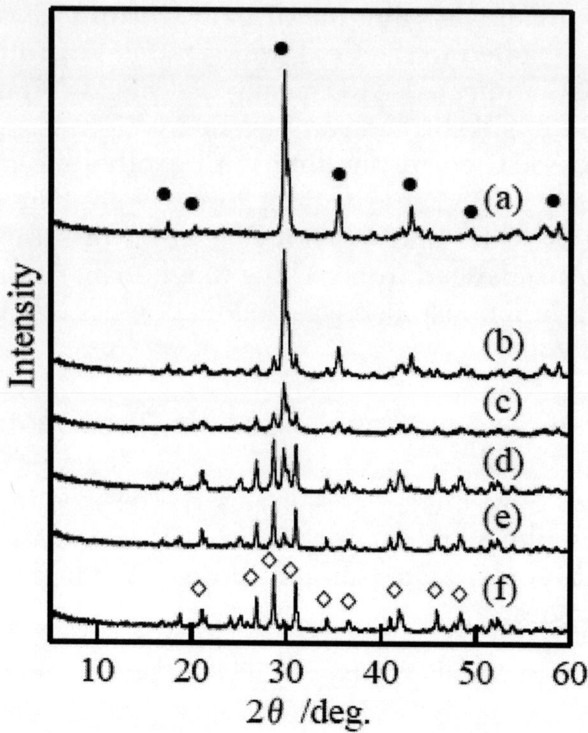

Figure 3. XRD patterns of samples prepared in various Ni/Ce ratios and then heated at 800°C, (a) 10/0; (b) 9/1; (c) 8/2; (d) 5/5; (e) 2/8; and (f) 0/10, •; $Ni_2P_2O_7$ and ◇; $CePO_4$.

The specified shape like pillar was observed in samples prepared at Ni/Ce = 9/1 and 5/5. Samples prepared in Ni/Ce = 2/8 and 0/10 had large particles.

Figure 5 shows the particle size distribution of samples synthesized in various Ni/Ce ratios. The main part of particle of sample prepared in Ni/Ce = 10/0 was from 100 to 10 μm in size. Samples prepared in Ni/Ce = 9/1 and 8/2 had smaller particle size than sample prepared in Ni/Ce = 10/0. In contrast, samples prepared in Ni/Ce = 2/8 and 0/10 had much larger particles. Particle size distribution of samples prepared in this work was less affected by heating temperature.

Table 2 shows the specific surface area of samples synthesized in various Ni/Ce ratios. Specific surface area as well as particle size has influence on color, and solubility of phosphate materials [16-19].

Sample prepared in Ni/Ce = 8/2 had large specific surface area. On the other hand, sample prepared in Ni/Ce = 2/8 had small specific surface area.

Figure 4 SEM images of samples prepared in various Ni/Ce ratios, (a) 10/0; (b) 9/1; (c) 8/2; (d) 5/5; (e) 2/8; and (f) 0/10.

Table 2 Specific surface area of nickel–cerium phosphates / $m^2 \cdot g^{-1}$

Temp.	Ni/Ce					
/°C	10/0	9/1	8/2	5/5	2/8	0/10
200	42.83	43.63	62.43	22.43	1.13	29.99
400	34.58	34.38	35.64	15.88	14.80	13.33
600	26.54	24.64	28.40	11.33	21.34	38.12
800	3.11	4.80	5.72	2.10	4.65	6.08

The change of specific surface area was not proportional to the Ni/Ce ratio. Samples pre- pared in all Ni/Ce ratios had smaller specific surface area by heating at 800°C.

Pigmental Properties of Phosphates

The color of samples without heating changed from light green to yellow with the increase of cerium ratio. By heating, light green of nickel hydrogen phosphate trans- formed to dark yellow powder of nickel pyrophosphate.

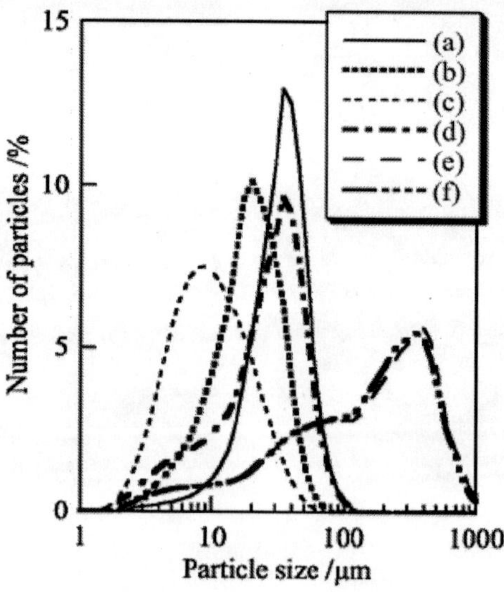

Figure 5 Particle size distribution of samples prepared in various Ni/Ce ratios, (a) 10/0; (b) 9/1; (c) 8/2; (d) 5/5; (e) 2/8; and (f) 0/10.

The yellow powder of tetravalent cerium phosphate changed to the white powder of trivalent cerium phosphate by heating over 400°C. Figure 6 shows UV-Vis reflectance

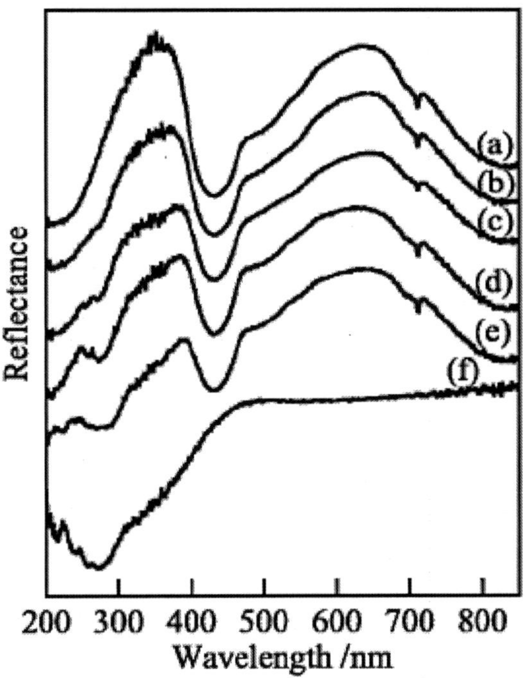

Figure 6. UV-Vis reflectance spectra of samples prepared in various Ni/Ce ratios and then heated at 800°C, (a) 10/0; (b) 9/1; (c) 8/2; (d) 5/5; (e) 2/8; and (f) 0/10.

spectra of samples synthesized in various Ni/Ce ratios and then heated at 800°C. Sample in Ni/Ce = 10/0 had strong reflectance at 350 and 640 nm and weak reflectance at 430 nm. By the substitution with cerium, the adsorption at 430 nm became smaller. The optical band gap energy of thermal products at 800 decreased from 5.17 eV to 4.43 eV [20]. The color of materials changed from yellow to white (Figure 7) [21].

Figure 8 shows acid resistance of samples synthesized in various Ni/Ce ratios. The small number of solubility means high acid resistance. The eluted ratio of phos- phorus changed from 100 to 28% with the increase of cerium ratio [22]. Formation of cerium phosphate inhi- bited the elution of phosphate materials. The elution ratio of nickel cation changed a little smaller by the sub- stitution with cerium to Ni/Ce = 5/5. Sample prepared in

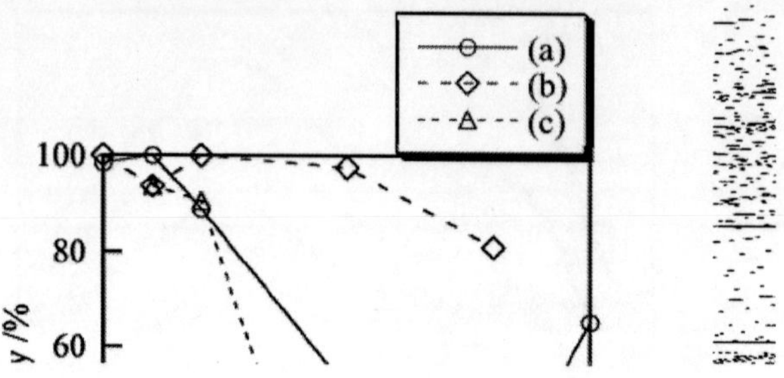

Figure 7 Photographs of samples prepared in various Ni/Ce ratios.

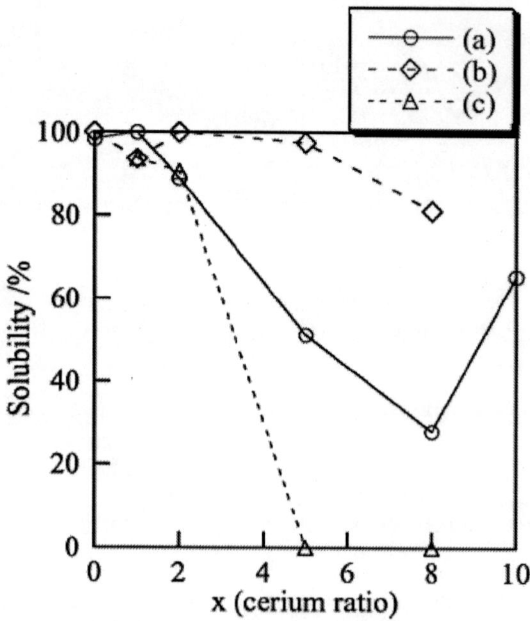

Figure 8 Acid resistance of samples prepared in various Ni/Ce ratios, (a) phosphorus; (b) nickel; and (c) cerium.

Table 3 Elution ratio of phosphorus in basic solution/ %

Temp.	Ni/Ce					
/°C	10/0	9/1	8/2	5/5	2/8	0/10
R.T.	87.47	93.67	90.54	75.00	48.43	40.23
200	67.96	83.20	67.71	64.49	4.14	22.12
400	62.70	52.26	47.50	5.13	2.05	16.16
600	46.59	38.85	40.65	8.41	4.84	15.71
800	0	1.01	0.05	0	8.28	0

Ni/Ce = 5/5 indicated small elution ratio of cerium cation.

In basic solution, all samples indicated no elution of nickel and cerium cations, because of the formation of their hydroxides. **Table 3** shows the solubility of phosphorus in sodium hydroxide solution. The eluted ratio of phosphorus had the tendency to decrease with the increase of cerium ratio. Samples had high base resistance by heating at high temperature.

CONCLUSIONS

In summary, nickel-cerium phosphates were prepared in aqueous solution. The obtained products were estimated for design of functional materials. The cerium ratio in precipitates was lower than that in preparation process. By the substitution of cerium cation, samples changed the mixture of nickel phosphate and cerium phosphate. The color of nickel phosphate was changed from light green to dark yellow by the substitution with cerium. The cerium substituted nickel phosphate changed to yellow powder by heating. The acid and base resistance of nickel phosphate materials improved by the substitution with cerium.

REFERENCES

1. H. Onoda, H. Nariai, A. Moriwaki, H. Maki and I. Mo- tooka, "Formation and Catalytic Characterization of Va- rious Rare Earth

Phosphates," *Journal of Materials Che- mistry*, Vol. 12, No. 6, 2002, pp. 1754-1760. doi:10.1039/b110121h

2. H. Onoda, T. Ohta, J. Tamaki and K. Kojima, "Decom- position of Trifluoromethane over Nickel Pyrophosphate Catalysts Containing Metal Cation," *Applied Catalysis A: General*, Vol. 288, No. 1-2, 2005, pp. 98-103. doi:10.1016/j.apcata.2005.04.028
3. H. Onoda, K. Yokouchi, K. Kojima and H. Nariai, "Ad- dition of Rare Earth Cation on Formation and Properties of Various Cobalt Phosphates," *Materials Science and En- gineering B*, Vol. 116, No. 2, 2005, pp. 189-195. doi:10.1016/j.mseb.2004.10.002
4. D. M. Lenz, M. Delamar and C. A. Ferreira, "Improve- ment of the Anticorrosion Properties of Polypyrrole by Zinc Phosphate Pigment Incorporation," *Progress in Or- ganic Coating*, Vol. 58, No. 1, 2007, pp. 64-69.
5. A. M. Mahdavian and M. M. Attar, "Investigation on Zinc Phosphate Effectiveness at Different Pigment Vo- lume Concentrations via Electrochemical Impedance Spec- troscopy," *Eletrochimica Acta*, Vol. 50, No. 24, 2005, pp. 4645-4648. doi:10.1016/j.electacta.2005.02.015
6. M. A. Hernandez, F. Galliano and D. Landolt, "Mecha- nism of Cathodic Delamination Control of Zinc-Alumi- num Phosphate Pigment in Waterborne Coatings," *Co- rrosion Science*, Vol. 46, No. 9, 2004, pp. 2281-2300. doi:10.1016/j.corsci.2004.01.009
7. M. C. Deya, G. Blustein, R. Romagnoli and B. Amo, "The Influence of the Anion Type on the Anticorrosive Behaviour of Inorganic Phosphates," *Surface and Coating Technology*, Vol. 150, No. 2-3, 2002, pp. 133-142. doi:10.1016/S0257-8972(01)01522-5
8. G. Li, Y. Shi, H. Hao, Z. Xia, Y. Lei, F. Guo and X. Li, "Effect of Rare Earth Addition on Shear Strength of SnAgCu Lead-Free Solder Joints," *Journal of Materials Science: Materials in Electronics*, Vol. 20, No. 2, 2009, pp. 186-192. doi:10.1007/s10854-008-9696-z
9. V. S. Vishnu, G. George and M. L. P. Reddy, "Effect of Molybdenum and Praseodymium Dopants on the Optical Properties of Sm2Ce2O7: Tuning of Band Gaps to Realize Various Color Hues," *Dyes and Pigments*, Vol. 85, No. 3, 2010, pp. 117-123. doi:10.1016/j.dyepig.2009.10.012
10. M. G. B. Nunes, L. S. Cavalcante, V. Santos, J. C. Sczancoski, M. R. M. Santos-Junior and E. Longo, "Sol- Gel Synethsis and Characterization of Fe2O3 ·CeO2 Doped with Pr Ceramic Pigments," *Journal of Sol-Gel Science and Technology*, Vol. 47, No. 1, 2008, pp. 38-43. doi:10.1007/s10971-008-1751-y
11. V. S. Vishnu, G. George and M. L. P. Reddy, "Synthesis and Characterization of New Environmental Benign Tan- talum-Doped

Ce0.8Zr0.2O2 Yellow Pigments: Applications in Coloring of Particles," *Dyes and Pigments,* Vol. 82, No. 1, 2009, pp. 53-57. doi:10.1016/j.dyepig.2008.11.001

12. H. Onoda, H. Matsui and I. Tanaka, "Improvement of Acid and Base Resistance of Nickel Phosphate Pigment by the Addition of Lanthanum Cation," *Materials Science and Engineering B,* Vo. 141, No. 1-2, 2007, pp. 28-33.
13. H. Onoda H., K. Tange and I. Tanaka, "Influence of Lan- thanum Addition on Preparation and Powder Properties of Cobalt Phosphates," *Journal of Materials Science,* Vol. 43, No. 16, 2008, pp. 5483-5488. doi:10.1007/s10853-008-2831-7
14. D. Bregiroux, O. Terra, F. Audubert, N. Dacheux, V. Se- rin, R. Podor and D. Bernache-Assollant, "Solid-State Synthesis of Monazite-Type Compounds Containing Te- travalent Elements," *Inorganic Chemistry,* Vol. 46, No. 24, 2007, pp. 10372-10382. doi:10.1021/ic7012123
15. A. A. Hanna, S. M. Mousa, G. M. Elkomy and M. A. Sherief, "Synthesis and Microstructure Studies of Nano- sized Cerium Phosphates," *European Journal of Che- mistry,* Vol. 1, No. 3, 2010, pp. 211-215. doi:10.5155/eurjchem.1.3.211-215.69
16. F. Rohner, F. O. Ernst, M. Arnold, M. Hilde, R. Biebinger, F. Ehrensperger, S. E. Pratsinis, W. Langhans, R. F. Hurrell and M. B. Zimmermann, "Synthesis, Characteri- zation, and Bioavailability in Rats of Ferric Phosphate Nanoparticles," *The Journal of Nutrition,* Vol. 137, No. 3, 2007, pp. 614-619.
17. M. Badsar and M. Edrissi, "Synthesis and Characteriza- tion of Different Nanostructures of Cobalt Phosphate," *Materials Research Bulletin,* Vol. 45, No. 9, 2010, pp. 1080-1084. doi:10.1016/j.materresbull.2010.06.022
18. N. H. M. Kamel, W. S. Hegazy and J. D. Navratil, "So- lubility and Sorption Properties of Some Phosphate Fer- tilizer Components on Soils," *Journal of Radio-analy- tical and Nuclear Chemistry,* Vol. 284, No. 3, 2010, pp. 653-658. doi:10.1007/s10967-010-0535-3
19. T. J. Brunner, M. Bohner, C. Dora, C. Gerber and W. J. Stark, "Comparison of Amorphous TCP Nanoparticles to Micron-Sized *a*-TCP as Starting Materials for Calcium Phosphate Cements," *Journal of Biomedical Materials Research,* Vol. 83B, No. 2, 2007, pp. 400-407. doi:10.1002/jbm.b.30809
20. J. C. Sczamcoski, L. S. Cavalcante, N. L. Marana, R. O. da. Silve, R. L. Tranquilin, M. R. Joya, P. S. Pizani, J. A. Sambrano, M. S. Li, E. Longo and J. Andres, "Electro- nic Structure and Optical Properties of BaMoO4 Pow- ders," *Current Applied Physics,* Vol. 10, No. 2, 2010, pp. 614-624. doi:10.1016/j.cap.2009.08.006

21. E. P. Lokshin, O. A. Tareeva and T. G. Kashulina, "Ef- fect of Sulfuric Acid and Sodium Cation on the Solubility of Lanthanides in Phosphoric Acid," *Russian Journal of Applied Chemistry*, Vol. 81, No. 1, 2008, pp. 1-7. doi:10.1134/S1070427208010011
22. E. P. Lokshin, O. A. Tareeva and T. G. Kashulina, "Ef- fect of Sulfuric Acid and Sodium Cation on the Solubility of Lanthanides in Phosphoric Acid," *Russian Journal of Applied Chemistry*, Vol. 81, No. 1, 2008, pp. 1-7. doi:10.1134/S1070427208010011

Chapter 6

SYNTHESIS, CRYSTAL GROWTH AND CHARACTERIZATION OF ORGANIC NLO MATERIAL: M-NITROACETANILIDE

Ramesh Rajendran[1], Thangammal Harris Freeda[1], Udaya Lakshmi Kalasekar[2], Rajesh Narayana Peruma[3]

[1]Physics Research Center, S. T. Hindu College, Nagercoil, India

[2]Department of Physics, College of Engineering and Technology, Saveetha University, Chennai, India

[3]Center for Crystal Growth, SSN College of Engineering, Kalavakkam, Chennai, India

ABSTRACT

Single crystals of m-Nitroacetanilide (mNAa) were successfully grown by slow evaporation method at a constant temperature 40°C from methanol solution. The solubility studies for mNAa were estimated. The cell dimensions were obtained by single crystal X-ray diffraction (XRD) study. The functional groups have been confirmed using Fourier transform infrared (FTIR) analysis. The placement of protons was identified from Nuclear Magnetic Resonance Spectroscopy (NMR) spectral analysis. UV-visible and fluorescence spectral analyses were carried out for the grown crystals. Thermo gravimetric analysis and differential thermal analy-sis were carried out to determine the thermal properties of the as grown crystal. The Second Harmonic Gen-eration (SHG) efficiency of mNAa was also determined.

INTRODUCTION

Nonlinear optical materials (NLO) have proven to be an interesting candidate for a number of applications such as second harmonic generation, frequency mixing, electrooptic modulation, etc. In recent years, organic NLO materials are attracting a great deal of attention for possible use in optical devices because of their large optical nonlinearity, low cut-off wavelengths, short response time and high laser damage thresholds [1]. Considerable work has been done in order to understand the microscopic origin of nonlinear behavior of organic materials [2-5]. The NLO properties of large organic molecules and polymers have been the subject of extensive theoretical and experimental investigations during the past two decades and they have been investigated widely due to their high nonlinear optical properties, rapid response in electrooptic effect and large second- or third-order hyperpolarizibilities compared to inorganic NLO materials [6]. Thus, there is much impetus to design and understand organic compounds for SHG applications.

To possess NLO property organic materials should contain highly conjugated π electron system affected by electron donor and acceptor groups. Hence in this class one such acetanilide derivatives, mNAa was taken under study which showed efficient NLO property. Some of the acetanilide derivatives such as Acetoacetanilide [7,8] and *p*-aminoacetanilide [9] were found to exhibit NLO prop-erties. mNAa is a meta substituted aromatic compound with molecular formula $C_8H_8N_2O_3$. The molecular struc-ture of mNAa, given in **Figure 1**, shows the charge transfer between electron acceptor (NO_2) and electron donor (NHCOR where R = CH_3) groups. This compound crystallizes in the monoclinic system in the chiral space group P21 with four independent molecules in the asymmetric unit.

In this paper, we report the material synthesis, solubility, crystal growth, single crystal X-ray diffraction (XRD), Fourier Transform Infrared Spec-troscopy (FTIR), optical, Fluorescence, thermal and NLO studies of mNAa.

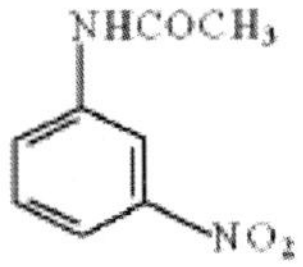

$NHCOCH_3$

NO_2

Figure 1. Molecular structure of mNAa.

EXPERIMENTAL

Material Synthesis

The title compound was synthesized from analytical re- agent (AR) mNA and acetic anhydride following the procedure given by Mahalakshmi et al. [10]. Required quantity of m-nitroaniline was dissolved in acetic anhy- dride at room temperature. The direct reaction between them as shown in the **Scheme 1** immediately yielded yellow colour compound. The precipitated product was filtered and dried using vacuum filtration. The material was repurified by recrystallization processes.

Solubility and Crystal Growth

The solubility of mNAa was determined using methanol, since methanol is found to be a suitable solvent to grow considerable size crystals. Recrystallized salt was dis- solved in methanol and the solution was maintained at 30°C in a constant temperature bath and stirred continu- ously to ensure homogenization of the solution. On re- aching the saturation, the amount of the salt in the solu- tion was analyzed gravimetrically. The same procedure was repeated for the temperatures 35°C, 40°C, 45°C and 50°C and results are shown in **Figure 2**. The mNAa ex- hibits good solubility and a positive solubility-tempera- ture gradient in methanol. From the figure we understand that the solubility of mNAa is going saturated at higher temperatures.

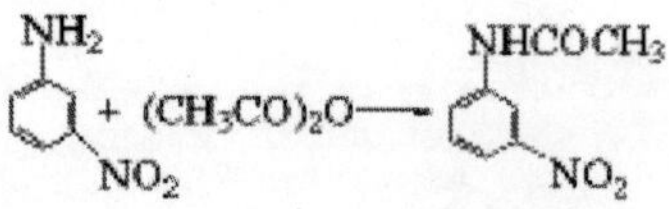

NH_2 + $(CH_3CO)_2O$ → $NHCOCH_3$
NO_2 NO_2

Scheme 1. Reaction mechanism of m-Nitroacetanilide.

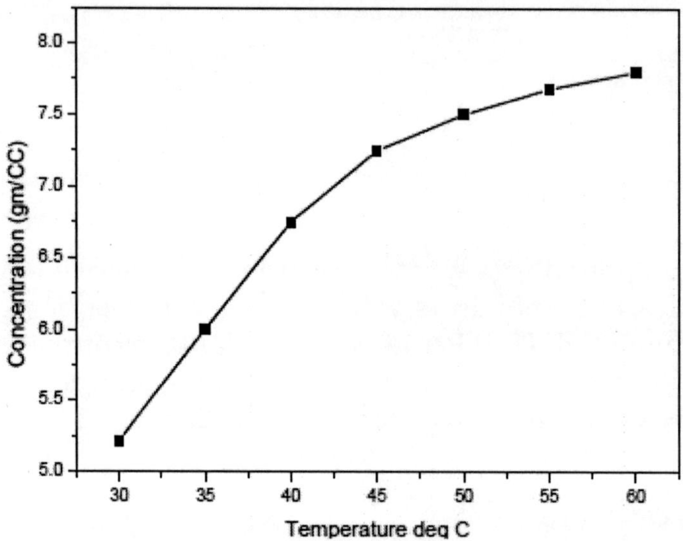

Figure 2. Solubility curve of mNAa.

Single crystals of mNAa were grown by slow evapo- ration growth technique using methanol as the solvent. About 250 ml of saturated solution was prepared at 40°C and it was carefully filtered at the same temperature using Whatman filter paper of pore size 11 μm. The filtered solution was taken in a beaker and placed in a constant temperature bath maintained at 40°C having an accuracy of ±0.01°C. Optically good quality seed crystal of di- mension 7 mm × 2 mm × 1 mm, obtained from slow evaporation method, was introduced into this solution. Crystal of dimension 10 mm × 4 mm ×3 mm was har- vested in a growth period of two days using solvent evaporation method. The morphology of the harvested crystal was tetragonal bipyramid as shown in **Figure 3**.

RESULTS AND DISCUSSIONS

Single Crystal X-Ray Diffraction

Single crystal XRD studies were carried out on the as grown mNAa crystal using Enraf-Nonius CAD-4 single crystal XRD reveals that mNAa belongs to monoclinic system. The unit cell parameters obtained are a = 9.7609 Å (9.767 Å), b = 13.3084 Å (13.298 Å), c = 13.3124 Å (13.272 Å), β = 103.15° (102.99°) and cell volume is 1683.8 $Å^3$ (1679.8 $Å^3$). These values are in close agreement with the corresponding values given in parentheses reported by Mahalakshmi *et al.* [10].

Fourier Transform Infrared Spectroscopy

FTIR spectrum of the as grown crystals was recorded in the range 400 cm^{-1} - 4000 cm^{-1} at room temperature using JASCO 460 plus FTIR spectrometer. The sample was prepared following the pressed KBr pellet technique. The presence of functional groups of the sample, were identified from the spectrum as shown in **Figure 4**. The absorption at 3263 cm^{-1} is due to N-H stretching. The peak at 1674.10 cm^{-1} corresponds to C = O stretching vibration of carbonyl group. The presence of nitro group is confirmed by the peaks at 1556.88 cm^{-1} and 1599 cm^{-1}. The peaks at 1380 cm^{-1}, 1472.81 cm^{-1} and 1426.40 cm^{-1} are due to C = C stretching [11-13].

Nuclear Magnetic Resonance Spectroscopy (NMR)

NMR spectrum of mNAa was recorded using JEOL GS × 400 model FT-NMR spectrometer. mNAa crystal was powdered and dissolved in deuterated Dimethyl Sulfoxide (DMSO). FT-NMR spectrum recorded for mNAa is shown in **Figure 5**. A triplet at 7.5 ppm is due to aromatic proton. The singlet at 2.07 ppm is assigned to

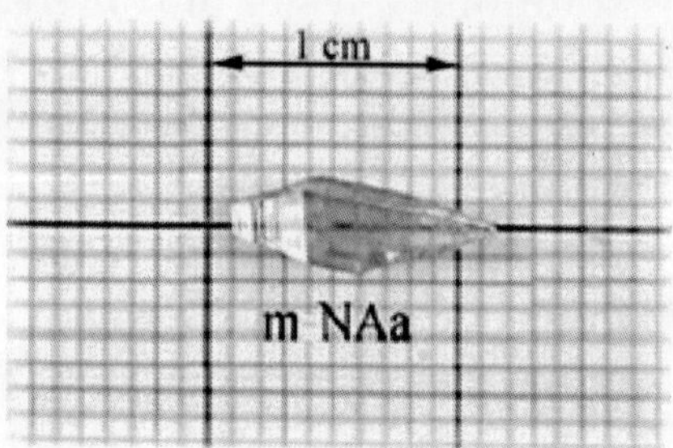

Figure 3. Photograph of as grown crystal

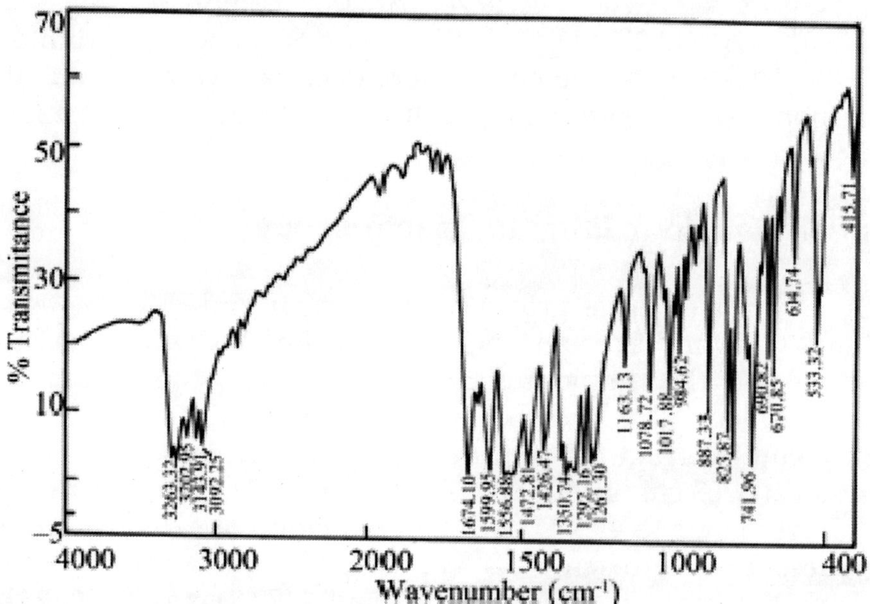

Figure 4. FTIR spectrum of mNAa

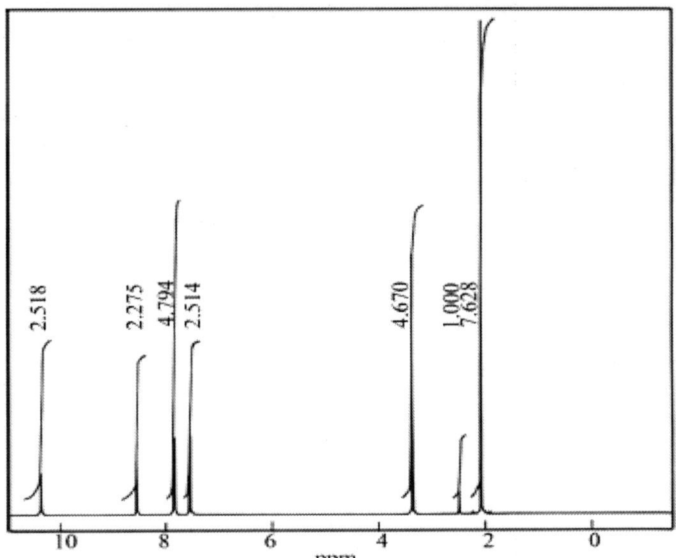

Figure 5. NMR spectrum of mNAa.

CH_3 proton. Singlet at 3.37 ppm is due to NH. A singlet at 10.39 ppm is due to NH proton [12].

UV-Vis Studies

Optical transmittance spectrum of mNAa single crystal was recorded in the region 200 nm - 1100 nm using SHIMADZU 1601 UV-Vis spectrophotometer. The ma- ximum transmittance is about 61% for mNAa crystal of 3 mm thickness. UV-Vis transmission spectrum presented in **Figure 6** shows that the crystal has good transparency in the range 340 nm - 1100 nm, which indicates that this crystal can be employed in the NLO applications in the entire visible and IR region. The absence of the absorp- tion in the visible region is the necessity for this com- pound as it is to be exploited for NLO applications in the room temperature

Fluorescence Studies

Fluorescence may be expected generally in molecules that are aromatic or contain multiple conjugated double bonds with a high degree of resonance stability [14]. Fluorescence finds wide application in the branches of biochemistry and medicine. It is also used as lighting in fluorescent lamps, Light Emitting Diode (LED) lamps etc. The excitation and emission spectra for mNAa recorded using FP-6500 Spectrofluorometer shown in **Figure 7**. The emission spectrum was measured in the range 350 nm - 600 nm. It is observed that the compound was excited at 340 nm and the corresponding emission was observed at 409 nm. The compound mNAa fluoresces due to the carbonyl chromophore [15].

NLo Studies

A preliminary study of the powder SHG conversion efficiency was carried using Kurtz and Perry powder technique [16]. Q-switched Nd:YAG laser (QUANTA RAY ICR 11) of wavelength 1064 nm with an input power of 5 mJ and pulses of 8 ns with the repetition rate of 10 Hz was used. The crystalline sample of mNAa was pow-

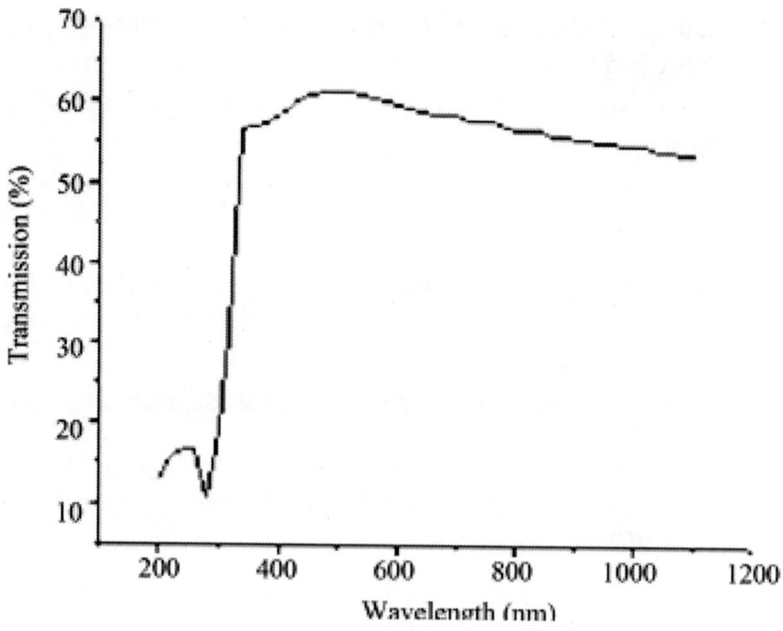

Figure 6. UV-Vis spectrum of mNAa.

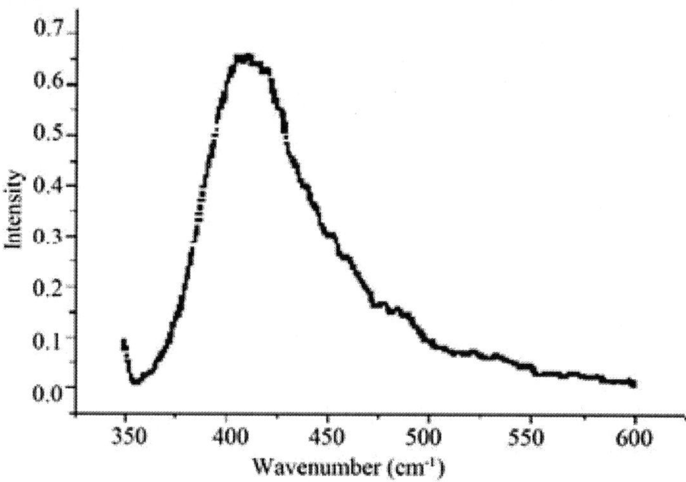

Figure 7. Emission spectrum of mNAa.

dered to a particle size of ≈ 125 μm. When the powder sample of mNAa was illuminated with this laser source emission of green light was observed. In order to deter- mine the efficiency of mNAa, a sample of parent compound mNA, which is also found to be an important material in the research field of nonlinear optics [17] was powdered to the same particle size and hence was used as reference material. The SHG conversion efficiency of mNAa is found to be 0.1 times that of mNA.

Thermo Gravimetric Analysis

Thermo Gravimetric Analysis (TGA) and Differential Thermal Analysis (DTA) were carried out for mNAa and spectra are shown in **Figure 8**. They were recorded using a simultaneous thermal analyzer PL-STA 1500 in nitro- gen atmosphere for temperature range 20°C to 800°C at a heating rate of 20°C/min. The sharp endothermic peak in DTA at 148°C indicates the melting point of the crystal. The melting point measured directly using TEMPO melting point apparatus was 149°C. There is no exother- mic or endothermic peak below this endotherm. This illustrates the absence of any absorbed water in the crystal sample. It also shows the absence of any isomorphic transition. The material exhibits single sharp weight loss starting at 215°C and below this temperature no significant weight loss is observed. The sharpness of the peaks indicates a good degree of crystallinity of the sample.

CONCLUSIONS

A single crystal of mNAa, an organic NLO material, was grown by solvent evaporation method from methanol solution. The single crystal X-ray analysis revealed that the crystal belongs to monoclinic system. The functional groups were identified using FT-IR spectroscopic technique. NMR spectral analysis were carried out to identify

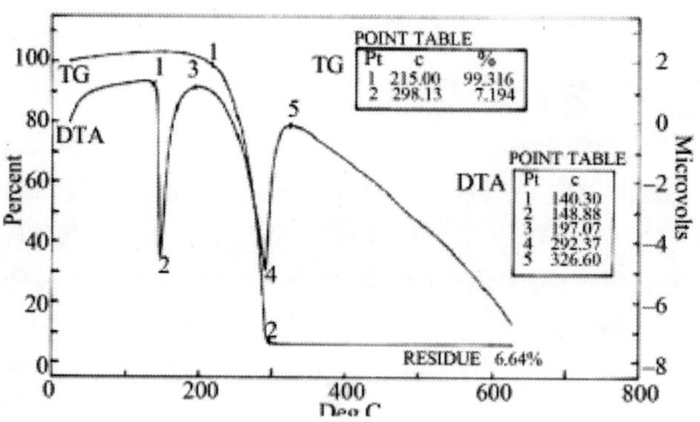

Figure 8. TGA-DTA curves of mNAa.

the position of protons. The optical properties such as UV-Vis in transmittance mode and second harmonic generation (SHG) conversion efficiency were investigated to explore the nonlinear optical characteristics of the above crystal. In addition, the thermal properties of the mNAa crystal were studied with TG analysis.

ACKNOWLEDGEMENTS

This work supported by the Department of Science and Technology, New Delhi, India under the grant of project ref-SR/FTP/PS-20/2005, is hereby gratefully acknowledged. The authors thank Prof. K. Panchanatheswaran, School of Chemistry, Bharathidasan University, Tiruchirappalli for fruitful discussions. The authors also thank Regional Sophisticated Instrumentation Centre, IIT, Chennai for recording NMR, FTIR and single crystal data collection. The authors acknowledge Prof. P. K. Das and Sampa Ghosh, IISc, Bangalore for having extended the laser facilities for SHG measurements and Centre for Electrochemical Research Institute (CECRI) for having extended the TGA-DTA measurements

REFERENCES

1. M. Narayan Bhat and S. M. Dharmaprakash, J. Cryst Growth, Vol. 236, 2002, pp. 376-380. doi:10.1016/S0022-0248(01)02094-2
2. T. Suthan and N. P. Rajesh, J. Cryst. Growth, Vol. 312, 2010, pp. 3156-3160. doi:10.1016/j.jcrysgro.2010.08.002
3. Huaihong Zhang, Yu Sun, Xiaodan Chen, Xin Yan and Baiwang Sun, J. Cryst. Growth, Vol. 324, 2011, pp. 196-200. doi:10.1016/j.jcrysgro.2011.03.044
4. Natalia Zaitseva, Leslie Carman, Andrew Glenn, Jason Newby, Michelle Faust, Sebastien Hamel, Nerine Cherepy and Stephen Payne, J. Cryst. Growth, Vol. 314, 2011, pp. 163-170. doi:10.1016/j.jcrysgro.2010.10.139
5. D. S. Chemla and J. Zyss, "NonlinearOptical Properties of Organic Molecule and Crystals," Academic press, NewYork, 1987.
6. N. Bloembergen and J. Nonlinear, *Opt. Phys. Mater*, Vol. 15, 1996, pp. 1-8. doi:10.1142/S0218863596000027
7. S. G. Prabhu and P. Mohan Rao, *J. Cryst Growth*, Vol. 210, 2000, pp. 824-827. doi:10.1016/S0022-0248(99)00843-X
8. N. Vijayan, R. Ramesh Babu, R. Gopalakrishnan and P. Ramasamy, *J. Cryst. Growth*, Vol. 267, 2004, pp. 646-653. doi:10.1016/j.jcrysgro.2004.04.008
9. D. Sajan, I. Hubert Joe and V. S. Jayakumar, *J. Physics: Conference Series*, Vol. 28, 2006, pp. 123-126.
10. L. Mahalakshmi, V. Upadhyaya and T. N. Guru Row, *Acta Cryst*, Vol. E58, 2002, pp. 983-984.
11. J. R. Dyer, "Applications of Absorption Spectroscopy of Organic Compounds," Prentice-Hall of India, New Delhi, 1994.
12. R. M. Silverstein, G. Clayton Bassler and T. C. Morrill, "Spectroscopic

Identification of Organic Compounds," 4th edition, John Wiley & Sons, New York, 1981.

13. Hobart H. Willard, Lynne L. Merritt jr., John A. Dean and Frank A. Settle jr., "Instrumental Methods of Analysis, Sixth Edition," Wadsworth Publishing Company, Florence, 1986, p. 609.
14. N. J. Turro, "Molecular Photochemistry," Benjamin, New York, 1965.
15. K. Biemann, "Tables of Spectral Data for Structure Determination of Organic Compounds," SpringerVerlag, Berlin Heidelberg, 1989.
16. S. K. Kurtz and T. T. Perry, J. Appl. Phys., Vol. 39, 1968, pp. 3798-3813. doi:10.1063/1.1656857
17. H. -X. Cang, W. -D. Huang and Y. -H. Zhou, J. Cryst. Growth, Vol. 192, 1998, pp. 236-242. doi:10.1016/S0022-0248(98)00408-4

Chapter 7

OXALYL RETRO-PEPTIDE GELATORS. SYNTHESIS, GELATION PROPERTIES AND STEREOCHEMICAL EFFECTS

Janja Makarevic, Milan Jokic, Leo Frkanec, Vesna Caplar, Nataša ŠijakoviC Vujicic and Mladen Žinic

Laboratory for Supramolecular and Nucleoside Chemistry, Ruđer Bošković Institute, P.O.B. 180, HR-10002 Zagreb, Croatia

ABSTRACT

In this work we report on gelation properties, self-assembly motifs, chirality effects and morphological characteristics of gels formed by chiral retro-dipeptidic gelators in the form of terminal diacids (**1a–5a**) and their dimethyl ester (**1b–5b**) and dicarboxamide (**1c–5c**) derivatives. Terminal free acid retro-dipeptides (*S,S*)-bis(LeuLeu) **1a**, (*S,S*)-bis(PhgPhg) **3a** and (*S,S*)-bis(PhePhe) **5a** showed moderate to excellent gelation of highly polar water/DMSO and water/DMF solvent mixtures. Retro-peptides incorporating different amino acids (*S,S*)-(LeuPhg) **2a** and (*S,S*)-(PhgLeu) **4a** showed no or very weak gelation. Different gelation effectiveness was found for racemic and single enantiomer gelators. The heterochiral (*S,R*)-**1c** diastereoisomer is capable of immobilizing up to 10 and 4 times larger volumes of dichloromethane/DMSO and toluene/DMSO solvent mixtures compared to homochiral (*S,S*)-**1c.** Based on the results of

^{1}H NMR, FTIR, CD investigations, molecular modeling and XRPD studies of diasteroisomeric diesters (*S,S*)-**1b**/(*S,R*)-**1b** and diacids (*S,S*)-**1b**/(*S,R*)-**1a**, a basic packing model in their gel aggregates is proposed. The intermolecular hydrogen bonding between extended gelator molecules utilizing both, the oxalamide and peptidic units and layered organization were identified as the most likely motifs appearing in the gel aggregates. Molecular modeling studies of (*S,S*)-**1a**/(*S,R*)-**1a** and (*S,S*)-**1b**/(*S,R*)-**1b**diasteroisomeric pairs revealed a decisive stereochemical influence yielding distinctly different low energy conformations: those of (*S,R*)-diastereoisomers with lipophilic *i*-Bu groups and polar carboxylic acid or ester groups located on the opposite sides of the oxalamide plane resembling bola amphiphilic structures and those of (*S,S*)-diasteroisomers possessing the same groups located at both sides of the oxalamide plane. Such conformational characteristics were found to strongly influence both, gelator effectiveness and morphological characteristics of gel aggregates.

INTRODUCTION

Reversible processes of peptide, protein and nucleic acids self-assembly are of paramount importance in biotic systems and are central to vital biological functions. On the other hand, some pathological changes leading to diseases such as Alzheimer's, Parkinson's and prion diseases, type II diabetes, etc. are associated with anomalous self-assembly of smaller peptides into amyloid fibrils which finally result in the formation of amyloid plaques [1-4]. During the last two decades there has been a growing interest in self-organization of small peptide models capable of self-assembling into highly organized supramolecular structures with potential use as novel bio- or nano-materials possessing advanced properties and functions [5-10]. It has been shown that even short peptides, such as dipeptides, tripeptides or tetrapeptides, themselves or incorporated into more complex structures, are capable of self-assembling into fibers or fibrils [11-15]. In a number of cases such organization results in formation of gels consisting of self-assembled fibrous aggregates usually containing a large volume of solvent [16]. In gels, the fibers are heavily entangled into 3-dimensional networks which immobilize the solvent and prevent fluidity in the system

[17-23]. In the last 15 years many low molecular weight gelling molecules of wide structural diversity, including a variety of amino acid and small peptide derivatives, have been prepared and studied [24-41]. These investigations revealed that gelator assemblies of various morphologies, including fibers and fiber bundles of diverse diameters, helical fibers or ribbons, tapes and nano-tubules, sometimes simultaneously present with micelles or vesicles, could be found [17-23]. For many gel systems evidence for hierarchical organisation was provided which determined the final morphological appearance of the aggregates[26]. It appears that gelation induced by aggregation of small abiotic or bio-inspired organic molecules represents an advantageous experimental system allowing in depth studies of the self-assembly as a general phenomenon. Such studies should ultimately result in revealing the relationship between gelator structures, self-assembly motifs and apparent morphologies of final assemblies as well as assisting in the elucidation of the role of solvent, which has been largely neglected in the majority of studies carried out to date. However, such an understanding of gelation is still out of reach; it is still hardly possible to predict gelation capability on the basis of the structure of a candidate molecule and it is even more difficult to predict which solvents and how effectively they would be gelled[42,43]. Hence, systematic studies of gels formed by structurally diverse small gelator molecules comprising elucidation of their self-assembly motifs, gelation effectiveness toward solvents of different structure and physical characteristics, estimation of solvation and stereochemical effects and their influence on the morphological characteristics of final gel assemblies may be rewarding, and should provide a much better understanding of the self-assembly processes involved in gelation.

In this work we report on gelation properties, self-assembly motifs, chirality effects and morphological characteristics of gels formed by chiral bis(dipeptide)oxalamides. Structurally, such gelators belong to the group of retro-peptides, which have been intensively studied as peptidomimetics due to their higher proteolytic stability and bioavailability compared to natural counterparts [44-47]. Despite very promising biomedicinal properties, very little is known about the self-assembly potential of this class of compounds in solution. Computer simulations of some malonamide retro-peptides have shown that the extended conformations are less stable than the helical ones[48,49]. Nevertheless, the crystal structure of the retro–inverso peptide Bz–S–gAla–R–mAla–NHPh revealed it's unidirectional self-assembly by intermolecular β-sheet type of hydrogen bonding so that malonamide retro-peptides could be considered as potential candidates for development of new gelator molecules [50]. The oxalamide based retro-peptides are relatively rare and much less studied than the more flexible malonamide retro-peptides [51-54]. In contrast to the malonamide group, the planar and much more rigid oxalamide fragment is self-complementary and exhibit a strong tendency for intermolecular hydrogen bonding both in the solid state and in the solution [55-60]. Hence, the oxalyl retro-peptides are expected to preferably form extended conformations capable of intermolecular oxalamide–oxalamide hydrogen bonding and the formation of unidirectional assemblies the latter being a necessary condition for gelation [24-31,58-60]. Herein, we provide experimental and molecular modeling evidence that the oxalamide retro-peptides indeed tend to form unidirectional hydrogen bonded assemblies of gelator molecules that adopt fully extended conformations. We also present the evidence that for the *(S,R)*-bis(LeuLeu) **1a** and *(S,S)*-bis(LeuLeu) **1a** retro-peptidic gelators, the stereochemistry has a decisive impact on their gelation effectiveness and final gel morphology in their water/DMSO gels.

RESULTS AND DISCUSSION

Synthesis of Oxalyl Retro-dipeptidic Gelators

A series of chiral bis(dipeptide)oxalamides was prepared as outlined in Scheme 1. The synthesis and analytical characterization of the prepared compounds are collected in the Supporting Information. Two sets of bis(dipeptide)oxalamides were prepared: the first incorporating a single amino acid and variable terminal groups such as carboxylic acid, methyl ester and carboxamide, namely (*S*,*S*)-bis(LeuLeu) **1a, b, c**; (*S*,*S*)-bis(PhgPhg) **3a,b, c** and (*S*,*S*)-bis(PhePhe) **5a, b, c** and, the second containing two different amino acids, (*S*,*S*)-bis(LeuPhg) **2a, b,c** and (*S*,*S*)-bis(PhgLeu) **4a, b, c** (configurations of only two of the four stereogenic centers are denoted corresponding to that of oxalamide α- and β-amino acid, respectively, as depicted in Scheme 1).

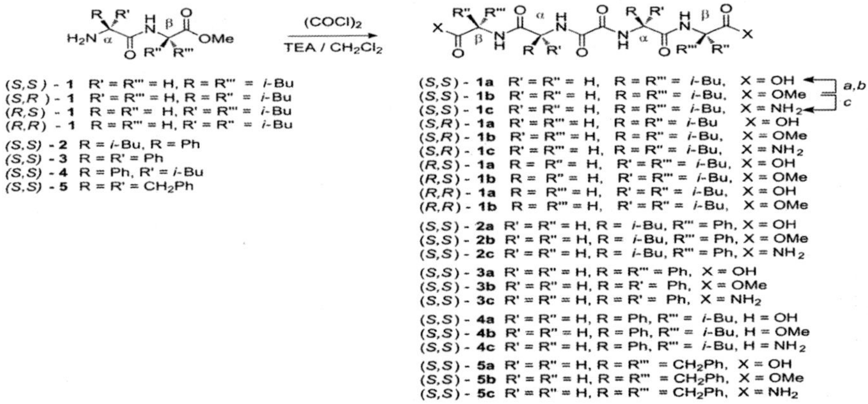

(COCl)$_2$
TEA / CH$_2$Cl$_2$

(*S,S*) - **1** R' = R''' = H, R = R''' = *i*-Bu
(*S,R*) - **1** R' = R''' = H, R = R'' = *i*-Bu
(*R,S*) - **1** R = R'' = H, R' = R''' = *i*-Bu
(*R,R*) - **1** R = R''' = H, R' = R'' = *i*-Bu

(*S,S*) - **2** R = *i*-Bu, R = Ph
(*S,S*) - **3** R = R' = Ph
(*S,S*) - **4** R = Ph, R' = *i*-Bu
(*S,S*) - **5** R = R' = CH$_2$Ph

(*S,S*) - **1a** R' = R'' = H, R = R''' = *i*-Bu, X = OH
(*S,S*) - **1b** R' = R'' = H, R = R''' = *i*-Bu, X = OMe
(*S,S*) - **1c** R' = R'' = H, R = R''' = *i*-Bu, X = NH$_2$
(*S,R*) - **1a** R' = R''' = H, R = R'' = *i*-Bu X = OH
(*S,R*) - **1b** R' = R''' = H, R = R'' = *i*-Bu, X = OMe
(*S,R*) - **1c** R' = R''' = H, R = R'' = *i*-Bu, X = NH$_2$
(*R,S*) - **1a** R = R'' = H, R' = R''' = *i*-Bu, X = OH
(*R,S*) - **1b** R = R'' = H, R' = R''' = *i*-Bu, X = OMe
(*R,R*) - **1a** R = R''' = H, R' = R'' = *i*-Bu, X = OH
(*R,R*) - **1b** R = R''' = H, R' = R'' = *i*-Bu, X = OMe

(*S,S*) - **2a** R' = R'' = H, R = *i*-Bu, R''' = Ph, X = OH
(*S,S*) - **2b** R' = R'' = H, R = *i*-Bu, R''' = Ph, X = OMe
(*S,S*) - **2c** R' = R'' = H, R = *i*-Bu, R''' = Ph, X = NH$_2$

(*S,S*) - **3a** R' = R'' = H, R = R''' = Ph, X = OH
(*S,S*) - **3b** R' = R'' = H, R = R' = Ph, X = OMe
(*S,S*) - **3c** R' = R'' = H, R = R' = Ph, X = NH$_2$

(*S,S*) - **4a** R' = R'' = H, R = Ph, R''' = *i*-Bu, H = OH
(*S,S*) - **4b** R' = R'' = H, R = Ph, R''' = *i*-Bu, H = OMe
(*S,S*) - **4c** R' = R'' = H, R = Ph, R''' = *i*-Bu, H = NH$_2$

(*S,S*) - **5a** R' = R'' = H, R = R''' = CH$_2$Ph, X = OH
(*S,S*) - **5b** R' = R'' = H, R = R''' = CH$_2$Ph, X = OMe
(*S,S*) - **5c** R' = R'' = H, R = R''' = CH$_2$Ph, X = NH$_2$

a,b
c

Scheme 1: Oxalyl retro-dipetide gelators; each **b** to **a**, (*a*) LiOH/MeOH, H_2O; (*b*) H^+; each **b** to **c**: (*c*) NH_3/MeOH.

Compared to the previously studied bis(amino acid)-oxalamide gelators (Figure 1), the retro-dipeptidic gelators, in addition to the oxalamide hydrogen bonding unit, also contain two peptidic units with specifically oriented hydrogen bond donor and acceptor sites and amino acid lipophilic substituents. Such structural characteristics enable multiple structural and stereochemical variations of the basic gelator structure and subsequent studies of structural and stereochemical influences on gelation properties, self-assembly motifs and gel morphology.

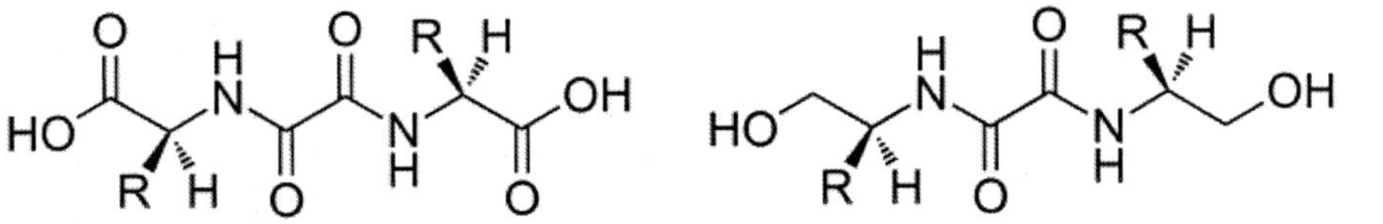

I

II

Figure 1: Chiral bis(amino acid)-(**I**) and bis(amino alcohol)-(**II**)-oxalamide gelators.

The influence of stereochemistry on self-aggregation and morphology was studied with **1a–c** combining different configurations of Leu: (*S*,*R*)-**1a**, **b**, **c** and (*R*,*S*)-**1a**, **b**, **c**. Gelation properties of pure enantiomers (*S*,*S*)-**1b** and (*S*,*R*)-**1b** are compared with those of (*S*,*S*)-**1b**/(R,R)-**1b** and (*S*,*R*)-**1b**/(*R*,*S*)-**1b** racemic mixtures (Scheme 1).

GELATION PROPERTIES

Terminal diacid Retro-dipeptides

Gelation observed for selected gelator–solvent pairs is expressed by gelator effectiveness (G_{eff}, mL) corresponding to the maximal volume of solvent that could be immobilized by 10 mg of the gelator (Table 1). The oxalamides **1a**, **3a** and **5a** were insoluble in water but showed moderate to excellent gelation of water/DMSO and water/DMF solvent mixtures. The Leu containing gelator **1a** appeared more than 2 times more effective in gelation of water/DMSO or DMF mixtures than the aromatic acid containing gelators **3a** and **5a**. However, **3a** and **5a** also gelled small to moderate volumes of EtOH and *rac*-2-octanol, whilst Leu incorporating **1a** formed gels with the more lipophilic solvents, decalin and tetralin. The retro-dipeptides containing two different amino acids showed no or only weak gelation; (*S*,*S*)-(LeuPhg) **2a** lacked any gelation ability toward the tested solvents, while (*S*,*S*)-(PhgLeu)**4a** showed only weak gelation of water/DMSO, dichloromethane and toluene. Apparently, the retro-peptides incorporating aliphatic and aromatic amino acids are less versatile gelators compared to retro-peptides containing identical amino acid fragments. The latter points to the importance of intermolecular lipophilic interactions for the stabilization of gel assemblies being stronger in the case of identical amino acids either lipophilic or aromatic, and weaker for mixed aromatic-lipophilic amino acid fragments present in the gelator molecule.

Interestingly, the (*S*,*R*)-bis(LeuLeu) retro-peptide **1a**, the diastereomer of (*S*,*S*)-**1a**, exhibited an increased G_{eff} for the water/DMSO mixture and decalin which however is absent for water/DMF solvent mixture (Table 1). The latter exemplifies the strong stereochemical influence on gelator effectiveness in certain solvents.

Table 1: Gelator effectivenesses (G_{eff}, mL) of retro-dipeptides **1a–5a** and bis(Leu)oxalamide **I** in gelation of various solvents and solvent mixtures (sol.: soluble; ins.: insoluble; cr.: crystallization; [A] gel/sol mixture).

Solvent	I	(*S,S*)-1a	(*S,R*)-1a	(*S,R*)-1a/ (*R,S*)-1a	(*S,S*)-1a/ (*R,R*)-1a	2a	3a	4a	5a
H_2O	0.4	ins.	ins.	ins.	ins.	ins.	ins.	ins.	ins.
H_2O/DMSO	0.8+0.4	5.09+2.98	15.0+4.5	5.3+5.3	cr.	cr.	2.3+1.6	1.0+1.0	1.1+0.5
H_2O/DMF	cr.	5.84+3.25	7.5+2.0	2.8+1.7	cr.	cr.	2.5+1.1	cr.	2.8+0.7
EtOH	1.5	sol.	cr.	cr.	cr.	sol.	cr.	sol.	1.1
±2-octanol	10.95	ins.	sol.		cr.	sol.	1.25	cr.	5.75
THF	0.4	sol.	sol.	sol.	sol.	sol.	sol.	sol.	sol.
CH_2Cl_2	1.5+0.05[a]	ins.	cr.	ins.	ins.	ins.	ins.	0.25	ins.
CH_3CN	0.95	cr.	cr.	cr.	cr.	sol.	cr.	ins.	7.0
toluene	1.95	cr.	cr.	0.25	cr.	ins.	ins.	0.5	sol.
p-xylene	2.45	ins.	ins.		0.25	ins.	ins.	[A]	cr.
decalin	0.2	0.2	0.8	1.7	jelly	ins.	ins.	ins.	ins.
tetralin	3.0	1.8	jelly	0.5	0.2	sol.	ins.	jelly	4.0

[a]DMSO.

In many cases of chiral gelators, pure enantiomers were found more effective gelators than the racemates, although several exceptions were observed showing that the racemic form could be a more effective gelator of certain solvents than the corresponding pure enantiomer [60-68]. Therefore, we also compared gelation properties of selected enantiomers and racemates and found that the (*S,R*)-**1a**/(*R,S*)-**1a** racemate was considerably less effective in gelation of both, water/DMSO and water/DMF solvent mixtures compared to the pure enantiomer (*S,R*)-**1a**, while the racemate (*S,S*)-**1a**/(*R,R*)-**1a** lacked any gelation ability and tended to crystallize from both solvent mixtures.

Generally, it can be concluded that the retro-dipeptides are less effective and less versatile gelators compared to the previously studied bis(amino acid)oxalamides. Table 1 shows that bis(Leu)oxalamide **I** is much more versatile and a more efficient gelator compared to (*S,S*)-**1a** and (*S,R*)-**1a**, and is capable of gelling water and various solvents of medium and low polarity. However, **I** is considerably less efficient in gelation of highly polar water/DMSO and water/DMF solvent mixtures compared to both **1a** diasteroisomers. Hence, the presence of more hydrogen bonding sites and lipophilic groups in the retro-dipeptides appears less favorable for gelation of water and solvents of medium and low polarity presumably due to decreased solubility and increased crystallization tendency compared to bis(amino acid) oxalamides. However, their efficient gelation of water/DMSO and water/DMF solvent mixtures presents a striking difference where DMSO and DMF co-solvents could sufficiently increase their solubility up to the point necessary for aggregation into sufficiently long fibers capable of networking.

Terminal dimethyl ester retro-dipeptides

In the previously studied series of bis(amino acid)oxalamide gelators transformation of terminal carboxylic acid groups into methyl esters resulted in the complete loss of gelation ability [59]. In the retrodipeptide series the gelation properties of methyl ester derivatives **1b–5b**, were not significantly different from those of the respective diacid derivatives **1a–5a** except that the diester derivatives appear slightly more versatile exhibiting gelation also with some lipophilic solvents (Table 1 and Table 2).

This could be explained by the increased lipophilicity of the diester derivatives and, consequently increased solubility in more lipophilic solvents compared to the diacid gelators. It should be noted that the diester racemates (*S,R*)-**1b**/(*R,S*)-**1b** and (*S,S*)-**1b**/(*R,R*)-**1b** showed significantly increased effectiveness in gelation of water/DMSO and water/DMF solvent mixtures compared to the respective free acid racemates (*S,R*)-**1a**/(*R,S*)-**1a** and (*S,S*)-**1a**/(*R,R*)-**1a** (Table 1), respectively.

Also, in contrast to the free acid gelators, the diester racemates were up to two times more efficient in the gelation of water/DMF and water/DMSO mixtures than their pure enantiomer counterparts (*S,R*)-**1b** and (*S,S*)-**1b** (Table 2). The latter provides additional examples that in some cases racemates could be more effective gelators than the pure enantiomers. Hence, in the search for highly effective gelators for targeted solvents, the racemic form of a chiral gelator must be tested.

Table 2: Gelator effectiveness (G_{eff}, mL) of bis(dipeptide)oxalamide dimethyl esters **1b–5b** in gelation of various solvents and solvent mixtures (sol.: soluble; ins.: insoluble; cr.: crystallization; [A] gel/sol mixture; (F): cotton-like fiber aggregates; ** the mixture of crystals and gel.).

Solvent	(S,S)-1b	(S,R)-1b	(S,R)-1b/ (R,S)-1b	(S,S)-1b/ (R,R)-1b	2b	3b	4b	5b
H_2O	ins.	ins.	ins.	ins.	ins.	ins.	ins.	ins.
H_2O/DMSO	3.95+6.75	9.7+10.0	13.2+9.65	13.8+9.1	0.15+0.5	1.1+2.7	0.2+0.5	cr.
H_2O/DMF	5.1+4.8	11.5+11.7	13.1+7.8	5.4+3.4	cr.	cr.	cr.	cr.
EtOH	cr.	cr.	cr.	cr.	cr.	2.00	(F1.3)	0.9
±2-octanol	cr.	cr.	cr.	cr.	1.85	cr.	(F0.5)	2.05
THF	sol.	cr.	sol.	sol.	sol.	sol.	sol.	cr.
CH_2Cl_2	sol.	sol.	sol.	sol.	sol.	ins.	sol.	cr.
CH_3CN	cr.	cr.	cr.	cr.	cr.	cr.	cr	cr.
toluene	[A]	0.15	**	0.55	1.1	ins.	1.25	2.0
p-xylene	[A]	0.5	**	1.3	1.1	ins.	1.1	2.5
decalin	2.3	0.8	1.75	5.6	2.0	ins.	4.8	1.7
tetralin	cr.	0.2	sol.	sol.	0.6	0.4	0.9	0.5

Terminal Dicarboxamide Retro-dipeptides

The diamide derivatives bis(LeuLeuNH_2) **1c**, bis(PhgPhgNH_2) **3c** and bis(PhePheNH_2) **5c** appeared more versatile being capable of gelling a larger set of tested solvents compared to the respective dicarboxylic acid (**1a**, **3a**, and**5a**) and dimethyl ester derivatives (**1b**, **3b** and **5b**) (Tables 1–3). The influence of stereochemistry on gelator versatility and effectiveness can be illustrated by the considerably improved gelation properties of the heterochiral (*S*,*R*)-**1c** diastereoisomer compared to homochiral (*S*,*S*)-**1c**; the former is capable of immobilizing up to 10 and 4 times larger volumes of dichloromethane and toluene solvent mixtures containing a little DMSO, respectively (Table 3). Also the bis(LeuPhgNH_2) **2c** and bis(PhgLeuNH_2) **4c** incorporating different amino acids appeared more versatile than the corresponding diacids (**2a**, **4a**) and diesters (**3b**, **4b**). It appears that the increased hydrogen bonding potential of terminal diamide derivatives provides somewhat more versatile gelators capable of gelating solvents of medium and low polarity where intermolecular hydrogen bonding is favored.

Table 3: Gelation effectiveness (G_{eff}, mL) of bis(amino acid and dipeptide-$CONH_2$)oxalamides **1c–5c** in gelation of various solvents and solvent mixtures (sol.: soluble, ins.: insoluble; cr.: crystalline; [A] gel/sol mixture).

Solvent	(*S,S*)-1c	(*S,R*)-1c	2c	3c	4c	5c
H_2O	ins.	ins.	ins.	ins.	ins.	ins.
H_2O/DMSO	0.55+1.4	0.8+1.2	1.8+2.9[a]	cr.	4.7+5.0	1.3+1.6
H_2O/DMF	0.45+0.75	0.8+0.6	1.05+11[a]	cr.	1.7+1.8	2.7+3.0
EtOH	cr.	[A]	ins.	ins.	3.4	ins.
±2-octanol	ins.	3.0	5.0+0.1[a]	ins.	cr.	ins.
THF	ins.	ins.	1.1+0.05[a]	ins.	ins.	ins.
CH_2Cl_2	1.05+0.2[a]	11.0+0.84[a]	2.0+0.1[a]	ins.	1.5+0.4	ins.
CH_3CN	ins.	0.5+0.04[a]	ins.	ins.	ins.	ins.
toluene	1.05+0.2[a]	4.0+0.4[a]	2.2+0.2[a]	ins.	0.75+0.1[a]	ins.
p-xylene	ins.	7.5+0.4[a]	6.9+0.2[a]	ins.	0.95+0.08[a]	ins.
decalin	ins.	ins.	0.8	ins.	ins.	0.5(ins.)
tetralin	7.3	sol.	2.0	ins.	1.0	2.0

[a]DMSO.

TEM and DSC investigations

As reported previously, TEM investigations of bis(amino acid) oxalamide gels revealed in most cases formation of very dense networks consisting of heavily entangled tiny fibers with diameters in the range of 10–20 nm [58-60]. A similar morphology was observed for the bis(PhePhe)-**5a**-EtOH gel (fiber *d*'s 6–20 nm) and bis(PhgLeu) **4a**water/DMSO gel (fiber *d*'s 5–15 nm) (see Supporting Information File 1, Figure S1a,b). TEM images of diastereomeric (*S,S*)-**1a** and (*S,R*)-**1a** water/DMSO gels (Figure 2 and Figure 3) show highly distinct morphology of gel networks. In the first gel rather straight fibers and fiber bundles with diameters in the range of 40–100 nm could be observed. However, the (*S,R*)-**1a** network showed a lower bundling tendency (Figure 3) and contained mostly fibers with diameters between 20–40 nm. In contrast to water/DMSO gels, the TEM image of the methyl ester derivative (*S,R*)-**1b** gel with toluene had a totally different morphology characterized by the presence of short and very wide tapes (Figure 4). As observed earlier for other gel systems, gelator effectiveness G_{eff} depends not only solubility but also depends on the thickness of fibers constituting the network [43,59,60]. Since solvent is entrapped by capillary forces, the formation of a dense network composed of thin fibers should possess smaller compartments and hence a higher solvent immobilization capacity compared to those less dense formed by thick fibers. The TEM observed thicknesses of gel aggregates existing in water/DMSO and toluene gels could be also correlated with gelator effectiveness (G_{eff}, mL, see Table 1 and Table 2). It appears that (*S*,*R*)-**1b**(G_{eff} 19.5 mL, water/DMSO gel) organized in thinner fibers is more than twice effective a gelator than its diasteroisomer (*S*,*S*)-**1a** (G_{eff} 8.0 mL, water/DMSO gel) which forms thicker fiber bundles. In the toluene gel, (*S*,*R*)-**1b** organizes into wide and short tapes with low networking capacity which is reflected in a very low (G_{eff}0.15 mL) gelator effectiveness.

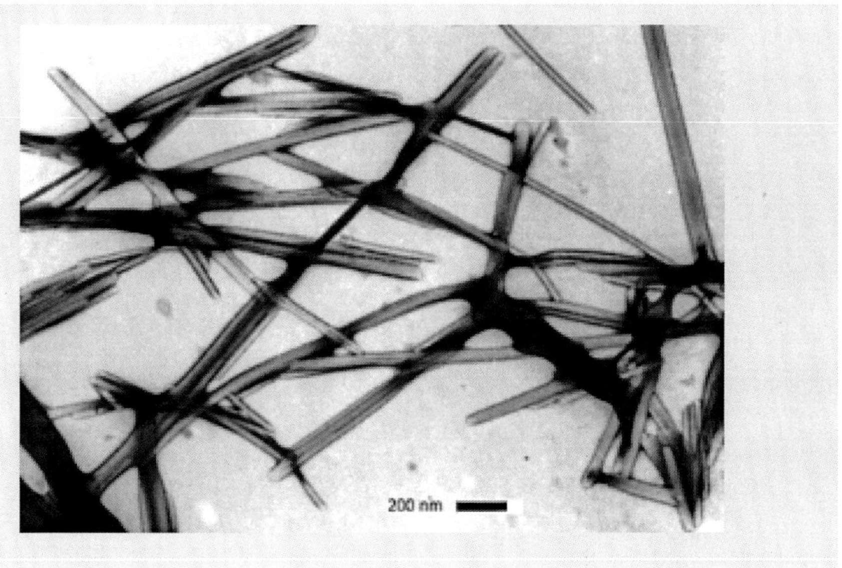

Figure 2: TEM images (PWK staining) of: (*S*,*S*)-**1a** H_2O/DMSO gel.

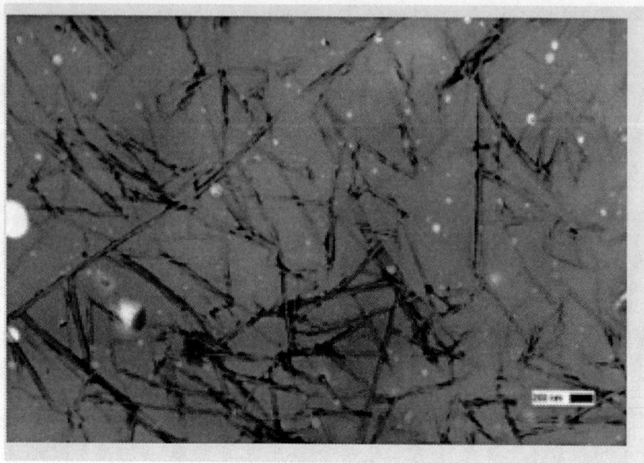

Figure 3: TEM images (PWK staining) of: (*S,R*)-**1a** H_2O/DMSO gel.

Figure 4: TEM images (PWK staining) of: (*S,R*)-**1b** toluene gel showing the presence of short tape like aggregates.

DSC investigation of the highly efficient (*S,R*)-**1a** gelator of water/DMSO solvent mixture showed only one transition in the heating (T_m) and cooling (T_c) cycle with gelation enthalpy changes of 37.70 and −38.20 kJ/mol, respectively (Table 4). The (*S,R*)-**1a**/(*R,S*)-**1a** racemic mixture being almost two times less effective in gelation of the same solvent mixture compared to (*S,R*)-**1a**, showed two transitions in the DCS heating and cooling cycle neither of which corresponded to those observed with (*S*,*R*)-**1a**. Moreover, the racemate showed considerably lower enthalpy changes compared to pure enantiomer gel (Table 4). The latter observations for the racemate gel indicate higher complexity of such systems and suggest possible interactions of enantiomers that lead to diasteromeric assemblies with a certain level of organization.

Table 4: ΔH and transition temperatures for selected retropeptide DMSO/water obtained from DSC heating and cooling cycles.

Gelator/solvent	T_m	ΔH_m	T_c	ΔH_c
	°C	kJ/mol	°C	kJ/mol
(*S*,*R*)-**1a**/water/DMSO	93.2	37.70	82.6	-38.20
(*S*,*R*)-**1a**/(*R*,*S*)-**1a**/ water/DMSO	97.6 137.0	20.36 2.32	78.2 123.2	-23.95 -2.80

FTIR, ^{1}H NMR and CD investigations

To identify supramolecular interactions that stabilize gel assemblies, the selected gels were studied by ^{1}H NMR, FTIR and CD spectroscopy. Valuable information on the self-assembly of gelator molecules in the pre-gelation state and in the gel could be obtained by analysis of the concentration and temperature dependent ^{1}H NMR and FTIR spectra. It was previously reported that the planar and self-complementary oxalamide unit persistently forms intermolecular hydrogen bonds and represents the major organizational element in the gel assemblies of both, bis(amino acid)- and bis(amino alcohol) oxalamides, and also has the major influence on their organization in the solid state [55-60]. In addition, the latter gelators tend to exhibit layered organization in their gel assemblies due to their structural resemblance to bola-amphiphiles.

In the FTIR spectra of (*S*,*S*)-**1b** and (*S*,*R*)-**1b** toluene gels one wide band or two poorly resolved NH bands, respectively, appear in the region of 3260–3320 cm^{-1} corresponding to a hydrogen bonded NH. In addition, the ester carbonyl and amide I bands are located at 1750 and 1653 cm^{-1}, respectively, the position of the latter being in accord with its participation in hydrogen bonding.

^{1}H NMR investigation showed significant concentration dependence of N-H and C*H proton shifts in the (*S*,*R*)-**1b**(Figure 5),

(*S,S*)-**1b** and its racemate (*S,S*)-**1b**/(*R,R*)-**1b** (Figure 6) toluene-d_8 gel samples. In the first case the oxalamide NH and Leu-NH protons were downfield shifted by 1.6 and 1.8 ppm, respectively, for a gelator concentration increase from 0.001–0.1 mol dm^{-3}. The oxalamide-α-Leu methine protons (C*H) were also significantly downfield shifted ($\Delta\delta_{C*H}$ = 0.566 ppm) while the β-Leu-C*H proton shifts were less significant. Strong downfield shifts of the oxalamide- and Leu-NH protons as well as the α-Leu-C*H proton closest to the oxalamide unit suggest simultaneous participation of both the oxalamide and Leu-NH protons in intermolecular hydrogen bonding. A comparison of the magnitudes of concentration induced shifts for diastereomeric gelators (*S,R*)-**1b** (Figure 5; oxalamide-NH protons Δδ 0.25 ppm; Leu-NH protons Δδ 0.17 ppm; α-Leu-C*H Δδ 0.13 ppm) and (*S,S*)-**1b** (Figure 6, oxalamide-NH protons Δδ 1.50 ppm; Leu-NH protons Δδ 1.55 ppm; α-Leu-C*H Δδ 0.53 ppm) shows large differences. The monitored protons of (*S,S*)-**1b** are more strongly downfield shifted than the corresponding protons of (*S,R*)-**1b** for the same concentration range. Similar trends of NH and C*H concentration induced shifts are observed for (*S,S*)-**1b** and its racemate (*S,S*)-**1b**/(*R,R*)-**1b** toluene-d_8 gels (Figure 6). Again the magnitudes of the NH and C*H concentration induced shifts are higher for the (*S,S*)-**1b** than for the racemate gel. It should be noted that the higher concentration induced shifts are observed for (*S,S*)-**1b** which forms sol–gel mixture in toluene compared to both (*S,R*)-**1b** and the racemate (*S,S*)-**1b**/(*R,R*)-**1b** forming stable toluene gels (Table 2).

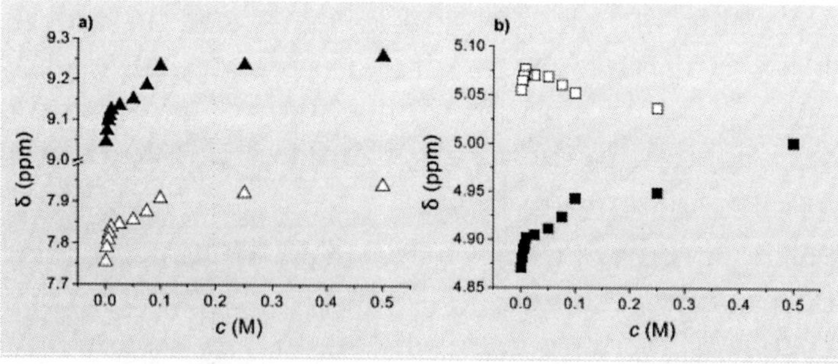

Figure 5: The concentration dependence of NH and C*H chemical shifts in (*S,R*) **1b** toluene-d_8 gel samples (concentration range 0.001–0.1 M): a) oxalamide-NH protons (▲); Leu-NH's (Δ) and b) oxalamide-α-Leu-C*H (■) and oxalamide-β-Leu-C*H (□) protons.

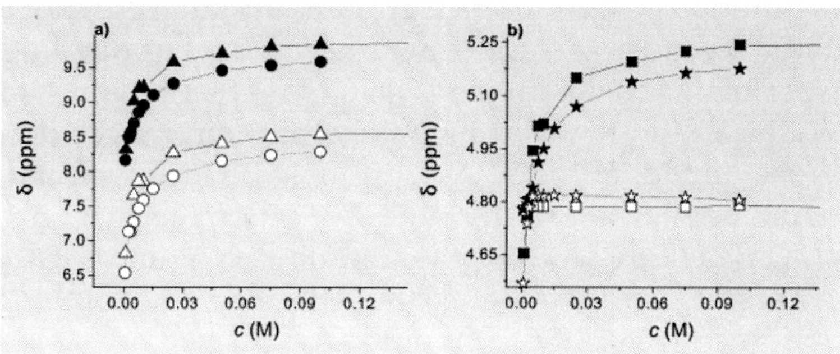

Figure 6: The concentration dependence of NH and C*H chemical shifts in (*S*,*S*)-**1b** and its racemate (*S*,*S*)-**1b**/(*R*,*R*)-**1b** toluene-d_8 gels (concentration range 0.001–0.1 mol dm^{-3}): a) (*S*,*S*)-**1b** oxalamide-NH protons (▲) and Leu-NH protons (Δ); the racemate oxalamide-NH protons (•) and Leu-NH protons (O); b) (*S*,*S*)-**1b**oxalamide-α-Leu-C*H (■);(*S*,*S*)-**1b** oxalamide-β-Leu-C*H (□); the racemate oxalamide-α-Leu-C*H (★); racemate oxalamide-β-Leu-C*H (☆).

The concentration induced shift curves show that for the examined gelators, the self-assembly equilibrium is reached at different gelator concentrations; for (*S*,*S*)-**1b** and its racemate (*S*,*S*)-**1b**/(*R*,*R*)-**1b**, the saturation point is reached at the same concentration of 0.03 mol dm^{-3} (Figure 6a,b) which corresponds to the experimentally determined minimal gelation concentration (*MGC*) for the racemate of 0.034 mol dm^{-3}. If the racemate were organized in the conglomerate as separate enantiomeric (*S*,*S*)-**1b** and (*R*,*R*)-**1b** assemblies, the magnitudes of the concentration induced shifts should be similar to those measured the for (*S*,*S*)-**1b** assemblies. Since this was not observed (Figure 6a, b) the results suggest formation of racemic gel assemblies composed of both enantiomers. The latter also points to the lack of any resolution at the supramolecular level which was found to occur for some racemic gelators and specific solvents [60]. The observation that (*S*,*S*)-**1b** with toluene gives a sol/gel mixture while the racemate gives a stable gel implies that the enantiomer forms insufficiently long assemblies incapable of efficient networking and of forming of self-supported gel, while the opposite holds for the racemic assemblies which are capable of forming the gel network

The discontinuous concentration induced shift curves obtained for (*S*,*R*)-**1b** diasteroisomer may indicate the presence of different assemblies at lower and higher gelator concentrations. It appears that the first saturation point is reached at a concentration around 0.03 mol dm^{-3} and the second at 0.12 mol dm^{-3}, the latter corresponding nicely to the experimentally determined *MGC* of 0.116 mol dm^{-3}. It should be noted that in the low and high concentration ranges downfield shifts of oxalamide- and Leu-NH protons are observed indicating that both assemblies are formed by intermolecular hydrogen bonding. The determined higher saturation point and *MGC* of 0.12 mol dm^{-3} for (*S*,*R*)-**1b** compared to the saturation point of the (*S*,*S*)-**1b** diastereoisomer (0.03 mol dm^{-3}) could be explained by the increased solubility of the (*S*,*R*)-**1b** assemblies in toluene compared to those formed by the second diastereoisomer [69]. However, despite of the lower saturation point and lower solubility, the (*S*,*S*)-**1b**assemblies cannot form the gel which points toward possible solvation effects taking a decisive role in the self-assembly of the diastereoisomers. Recently, Meijer et al. [70] presented convincing evidence that co-organization of solvent at the periphery of the gel aggregates plays a direct role in the assembly processes evident, even during the formation of the pre-aggregates. The influence of solvent structure on the length of the aggregates was clearly demonstrated. Hence, different solvation effects of toluene operating in the self-assembly of (*S*,*S*)-**1b** and (*S*,*R*)-**1b** may be responsible for the formation of insufficiently long aggregates of the first diastereoisomer resulting in the formation of the sol–gel mixture, and sufficiently long assemblies of the second one being capable of networking and the formation of a self-supported gel.

The variations of oxalamide NH, Leu-NH, α- and β-Leu-C*H proton chemical shifts with increasing temperature in the toluene-d_8 gel samples of the diastereomeric (*S*,*R*)-**1b** and (*S*,*S*)-**1b** (concentrations of 0.5 mol dm^{-3}) are shown in Figure 7. For the (*S*,*R*)-**1b** gel, a temperature increase from 20–50 °C induced only slight downfield shifts of both the oxalamide- and Leu-NH protons as well as the α- and β-Leu-C*H protons; in the higher temperature interval (50–90 °C) all protons were downfield shifted in accord with the breaking of intermolecular hydrogen bonds involving both the oxalamide and Leu NH protons. In contrast, the respective protons of the (*S*,*S*)-**1b** are continuously shifted downfield with increasing temperature (Figure 7c). Hence, a clear difference in the thermal behavior of (*S*,*S*)-**1b** weak gel and (*S*,*R*)-**1b** gels was observed. Similar discontinuous temperature variation curves to those observed for (*S*,*R*)-**1b** were also found for bis(amino acid)oxalamide gelators which were shown to exhibit the layered type of organization in their gel assemblies [59,60]. Small downfield shifts of the oxalamide- and LeuNH protons observable in the low temperature regime (Figure 7a) were explained by the less energy demanding disassembly that occurred at lipophilic sites of the interacting bilayers resulting in small deshielding of these protons. In the higher temperature regime the downfield shifts of the same protons indicate the breaking of intermolecular hydrogen bonds. This conclusion is supported by molecular modeling (see the respective paragraph) which showed that the low energy conformation of (*S*,*R*)-**1b** is similar to those found for bis(amino acid)oxalamides and that both show a strong resemblance to bola-amphiphiles which are known to organize into bilayers [71].

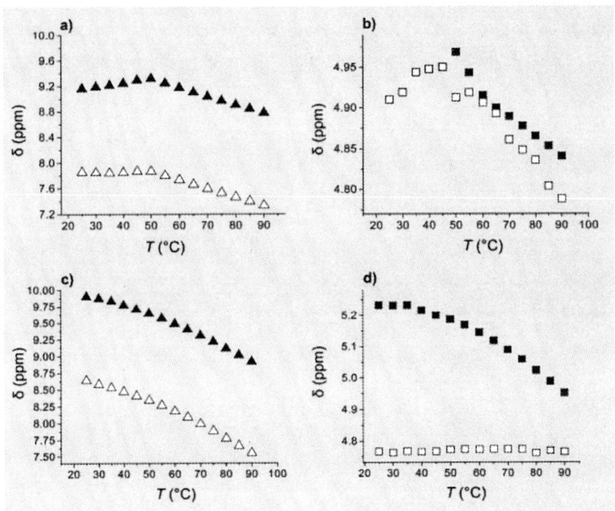

Figure 7: Temperature dependence of: a) oxalamide NH protons (▲), Leu-NH protons (Δ) and b) oxalamide-α-Leu-C*H (■) and oxalamide-β-Leu-C*H (□) chemical shifts in 0.5 mol dm^{-3} (*S*,*R*)-**1b** toluene-d_8 gel sample; c) and d) induced shifts of the respective protons in (*S*,*S*)-**1b** toluene-d_8 gel.

The temperature dependence of CD spectra of decalin gels formed by diastereoisomeric methyl esters (*S*,*R*)-**1b**and (*S*,*S*)-**1b** (Figure 8a, b, respectively) was also investigated. At room temperature (*S*,*R*)-**1b** shows negative Cotton effect at λ = 245 nm of moderate intensity which decreases on increasing temperature from 20 to 50 °C. Further temperature increase of the gel sample resulted in the appearance of a new negative CD peak at λ = 234 nm corresponding to the shoulder band in the gelator UV spectrum (Figure 8e); the intensity of the band increased with increasing temperature (Figure 8a). As reported for the self-assembled alanine based gelators, the CD signal at around 232 nm can be ascribed to the n,π*-transition of the amide carbonyl [72-74]. Hence, the λ = 234 nm band that appeared at 60 °C could be ascribed to the intrinsic chirality of disassembled gelator molecules. Although the origin of the λ = 245 nm CD band is not clear, it could be the consequence of circular differential scattering which was shown to contribute to the CD spectra of large aggregated biomolecules [75].

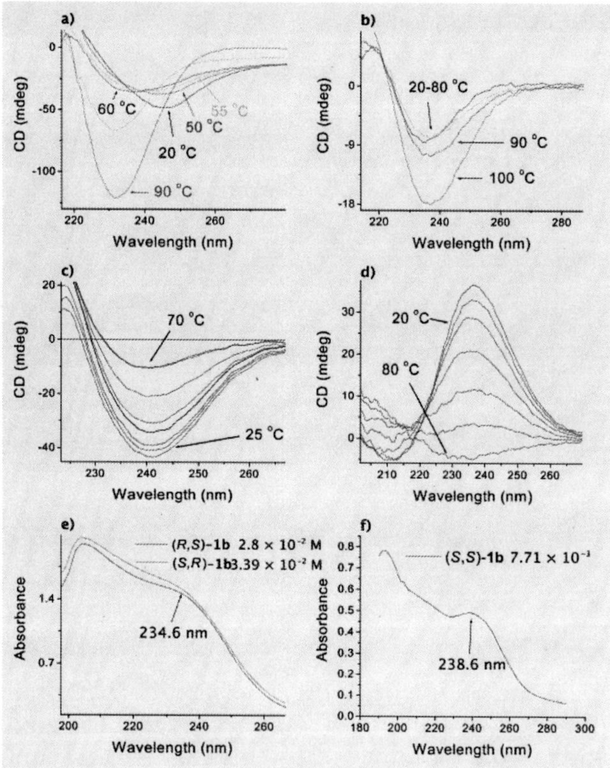

Figure 8: Temperature dependent CD spectra of: a) (*S*,*R*)-**1b** decalin gel (c = 3.4·10^{-2} M); b) (*S*,*S*)-**1b** decalin gel (c =7.6·10^{-3} M); c) **5a** ethanol gel (c =1·10-2 M) and d) (*S*,*S*)-bis(Leu)oxalamide **I** 1-butanol gel (c = 2.8·10^{-2} M); e), f) UV spectra of (*S*,*R*)-**1b** (red curve), (*R*,*S*)-**1b** and (*S*,*S*)-**1b** taken in decalin at specified concentrations.

By contrast, the CD spectra of the (*S*,*S*)-**1b** gel showed a negative CD band at λ_{min} 238.6 nm (Figure 8b) corresponding to its electronic absorption band (Figure 8f), but similarly to (*S*,*R*)-**1b** the intensity of CD band increased with increasing temperature.

It should be noted that the temperature induced changes in the CD spectra of both, (*S*,*R*)-**1b** and (*S*,*S*)-**1b** decalin gels are different to those obtained for **5a** ethanol and bis(Leu)oxalamide 1-butanol gels (Figure 8c, d). With these latter gels a decrease of CD peak intensities with increasing temperature was observed in accord with the disassembly of the chiral gel aggregates which has also been observed for some other chiral gels [76]. Consequently, the CD results described for the (*S*,*R*)-**1b** and (*S*,*S*)-**1b** gels, characterized by the increase of CD signals with increasing gel temperature, indicate that in these systems there is no aggregation increased chirality as observed for some other gels of the chiral gelators.

XRPD, MOLECULAR MODELING AND PACKING MODEL

The X-ray powder diffraction (XRPD) pattern of (*S*,*S*)-**1b** xerogel showed strong peaks corresponding to periodic distance *d* of 16.1 and 13.4 Å and a weaker peaks corresponding to *d's* of 15.1 and 8.6 Å (Figure 9b).

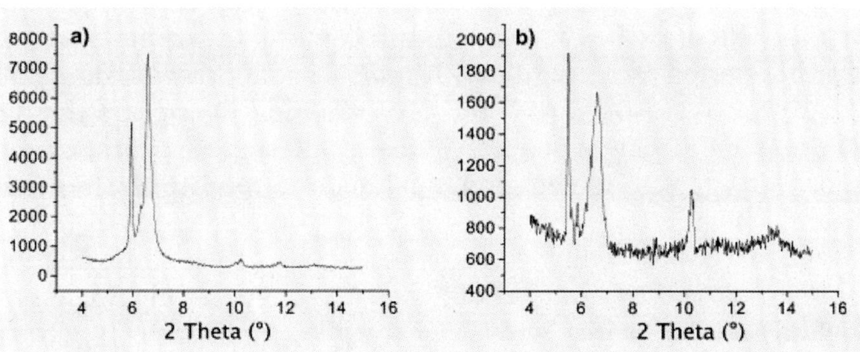

Figure 9: X-ray powder diffractograms of (a) (*S*,*R*)-**1b** and (b) (*S*,*S*)-**1b** xerogels prepared from their toluene gels.

For the (*S,R*)-**1b** xerogel, strong peaks corresponding to *d*'s of 14.9 and 13.4 Å and smaller peaks to *d*'s of 13.98, 8.6 and 7.5 Å could be observed (Figure 9a). Molecular modeling of (*S,R*)-**1b** and (*S,S*)-**1b** yields low energy extended conformations with lengths of 15.1 and 15.9 Å, respectively (Figure 10a) [77].

The measured extended conformation lengths correspond nicely to the largest periodic distances of 14.9 and 16.1 Å obtained by XRP diffraction of (*S,R*)-**1b** and (*S,S*)-**1b** xerogels indicative of the formation of assemblies of the extended gelator molecules. However, the low energy conformations of (*S,R*)-**1b** and (*S,S*)-**1b** are distinctly different with all *i*-Bu groups *cis*-oriented in (*S,R*)-**1b**, while the *i*-Bu groups in α- and β-Leu of (*S,S*)-**1b** have the*trans*-arrangement with respect to the plane containing the amide and oxalamide groups (Figure 10a). Conformational analysis reveals why such arrangements of *i*-Bu groups occur (Figure 10b, Newman projections of two stereogenic centers only). Our earlier results based on single crystal X-ray analysis of bis(amino acid)oxalamides showed that their most stable conformations are characterized by vicinal positioning of the methine proton at the stereogenic centre and oxalamide carbonyl oxygen atom which produces the lowest steric repulsion. Similarly, in the conformation A of (*S,R*)-**1b** with *cis*-arrangement of the *i*-Bu groups, the smallest group (H) of the β-Leu chiral centre is located close to amide carbonyl; the conformations with *trans*-arrangement of *i*-Bu groups should be less stable due to increased steric repulsion between the amide carbonyl oxygen and either the*i*-Bu or carboxymethyl group. Among the conformations of (*S,S*)-**1b** denoted B, C and D with *trans-*, *cis-* and*trans*-arrangement of *i*-Bu groups, respectively, the conformation D appears the most stable due to the vicinal position of the smallest group (H) and amide carbonyl oxygen atom. These conclusions are supported by molecular modeling (Figure 10a); the lowest energy conformations of (*S,R*)-**1b** and (*S,S*)-**1b** generated by systematic search of their conformational space correspond to A and D of Figure 10b, respectively. In support, the values of the vicinal NH-Cα-H coupling constants $J_{NH\text{-}CH}$ for the (*S,R*)-**1b** oxalamide NH-Leu Cα-H and Leu NH-Leu Cα-H (8.63 and 8.48 Hz) and (*S,S*)-**1b** (8.63 and 8.33 Hz) obtained from their ^{1}H NMR spectra taken in $CDCl_3$ correspond to dihedral angles close to *trans*-coplanar positioning of NH and Cα-H protons in both groups

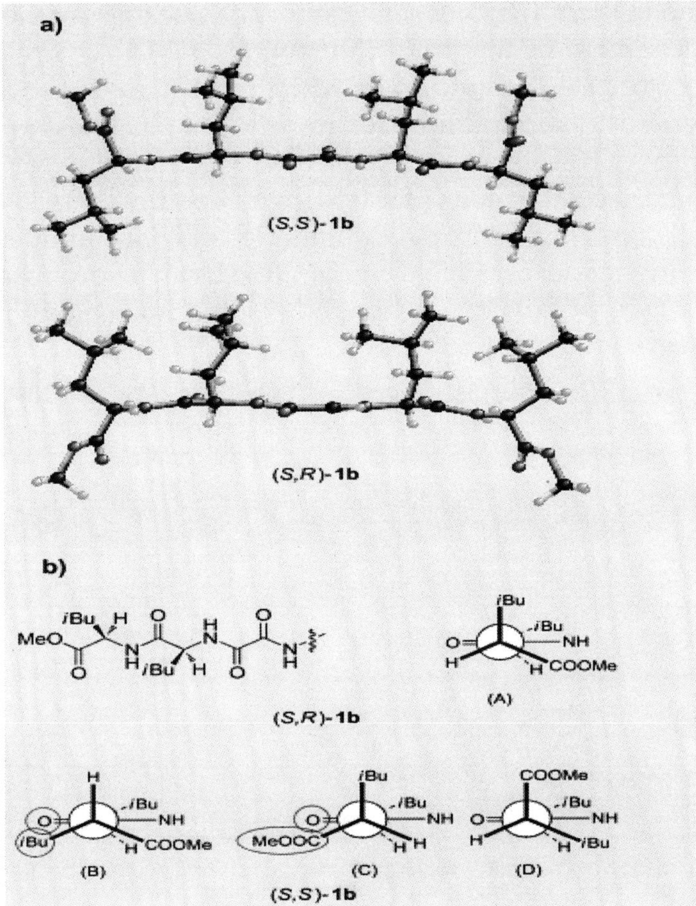

Figure 10: (a) Fully minimized the lowest energy conformations of (*S*,*S*)-**1b** (top) and (*S*,*R*)-**1b** generated by systematic conformational search (SYBYL package; second graphic); (b) Partial Newman projections of two stereogenic centers of (*S*,*R*)-**1b** and (*S*,*S*)-**1b** showing conformations with *cis*-arrangement of *i*-Bu groups in the former (A) and *trans*- (B, D) and *cis*- (C)-arrangements of *i*-Bu groups in the latter.

The low energy conformations of (*S*,*R*)-**1b** and (*S*,*S*)-**1b** were used for docking calculations to generate the hydrogen bonded dimmers of extended gelator molecules involving both, the oxalamide and Leu-NH protons (Supporting Information File 1, Figure S3). The thicknesses of such dimmers estimated from models are between 7.5 and 8.6 Å (Figure 11) which correspond well to periodic distance *d* of

8.6 Å found in their XRP diffractograms. The thickness of the (*S*,*R*)-**1b** dimmer generated by lipophilic interactions is 13.4 Å corresponding exactly to *d* of 13.4 Å found in its XRPD. The (*S*,*S*)-**1b** model of the dimer formed by lipophilic packing gives a thickness of 13.9 Å. Based on these results and those of the FTIR and ^{1}H NMR studies, which suggested intermolecular hydrogen bonding between gelator molecules involving both the oxalamide and Leu amide units, a basic packing model for (*S*,*S*)-**1b**and (*S*,*R*)-**1b** can be proposed which consists of layers of hydrogen bonded gelator molecules (Figure 11).

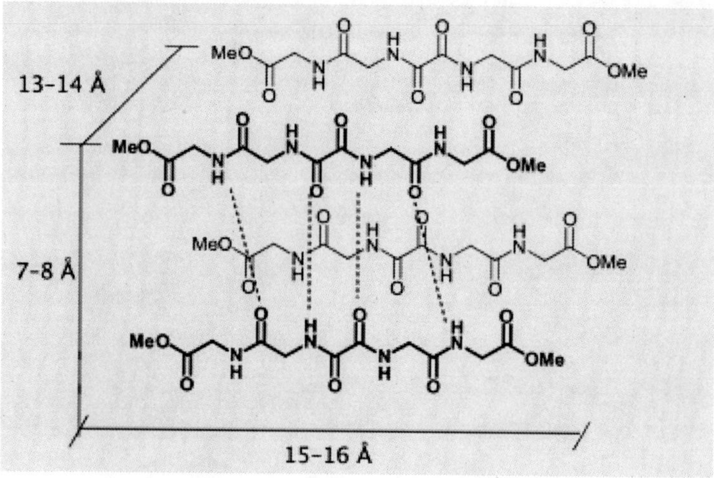

Figure 11: Schematic presentation of the proposed (*S*,*S*)-**1b** and (*S*,*R*)-**1b** basic packing model based on XRPD,^{1}H NMR, FTIR and molecular modeling results.

In contrast to the diester gelators (*S*,*R*)-**1b** and (*S*,*S*)-**1b** a detailed spectroscopic investigation of (*S*,*S*)-**1a** and (*S*,*R*)-**1a** organization in their water/DMSO gel assemblies was not possible due to solvent unsuitability.

Nevertheless, the FTIR spectrum of the (*S*,*R*)-**1b** xerogel prepared from its water/DMSO gel was found to differ from that of the crystalline sample; the positions of NH stretching, carboxylic acid and amide I carbonyl stretching, and NH bending amide II bands in the spectra of crystalline and xerogel samples appear at 3281.2 1724.4 1655.6 1543.6 1510.4 cm^{-1} and 3303.4 3273.5 1728.5 1651.6 1534.2 1510.5 cm^{-1}, respectively. The positions of the xerogel bands are similar to those found in the spectrum of previously studied bis(Leu)oxalamide water/DMSO gel assemblies (3300 1729 1658 1515 cm^{-1}) and which was shown to organize by intermolecular hydrogen bonding between oxalamide units and lateral carboxylic acid hydrogen bonding [59]. The appearance of two NH stretching and amide II bands in the (*S*,*R*)-**1b** xerogel spectrum (3303.4 3273.5 1534.2 1510 cm^{-1}) can be attributed to the intermolecular hydrogen bonds formed by Leu amide units. The XRPD of (*S*,*R*)-**1a** water/ DMSO gel (Figure 12) showed two diffraction peaks at 2θ = 5.509 and 10.501 corresponding to periodic distances *d* of 16.04 and 8.42 Å which also suggests formation of hydrogen bonded assemblies between extended forms of gelator molecules as in the cases of the diester derivatives (*S*,*S*)-**1b** and (*S*,*R*)-**1b** (Figure 11).

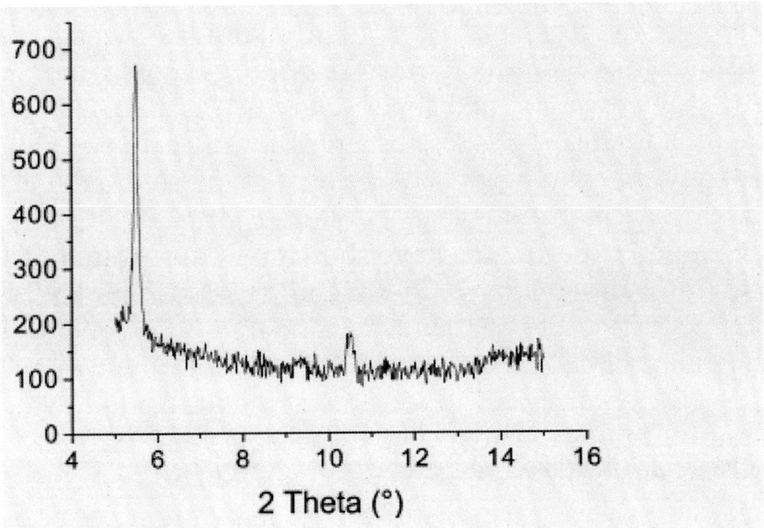

Figure 12: X-ray powder diffraction (XRPD) diagram of (*S*,*R*)-**1a** water/DMSO xerogel.

These results indicate that the dicarboxylic retro-dipeptides (*S*,*S*)-**1a** and (*S*,*R*)-**1a** also show similar basic organization as their dimethyl ester counterparts (*S*,*S*)-**1b** and (*S*,*R*)-**1b** (Figure 11). Also molecular modeling of (*S*,*S*)-**1a** and (*S*,*R*)-**1a** generated very similar low energy conformations to those of (*S*,*S*)-**1b** and (*S*,*R*)-**1b** shown in Figure 10a, b. In both cases, the major organizational driving force is provided by extensive intermolecular hydrogen bonding. In lipophilic solvents, where such types of intermolecular interactions are highly favored, formation of wide and relatively short tapes could be observed (TEM, (*S*,*R*)-**1b** toluene gel, Figure 4) possibly due to the enhanced self-assembling in the direction of intermolecular hydrogen bonds. In contrast, gelling of the highly polar and hydrogen bond competitive water/DMSO solvent mixture (TEM, (*S*,*S*)-**1a** and (*S*,*R*)-**1a** water/DMSO gels,Figure 2 and Figure 3) results in the formation of tiny fibers or fiber bundles due to less favored self-assembly in the direction of intermolecular hydrogen bonding and more pronounced intermolecular lipophilic interactions.

CONCLUSION

A series of chiral bis(dipeptide)oxalamides was prepared representing a novel family of retro-peptidic gelators. Their gelation properties towards a defined set of solvents was assessed and, their conformational characteristics, organization in gel assemblies and thermal and morphological characteristics of selected gels were studied by molecular modeling, ^{1}H NMR, FTIR, CD, DCS, TEM and XRPD. Gelation experiments have shown that the group of terminal free acid retro-dipeptides (*S*,*S*)-bis(LeuLeu) **1a**, (*S*,*S*)-bis(PhgPhg) **3a** and (*S*,*S*)-bis(PhePhe) **5a** showed moderate to excellent gelation of polar water/DMSO and water/DMF solvent mixtures and were much less efficient in gelating solvents of medium and low polarity. Interestingly, the free acid gelators incorporating different amino acids (*S*,*S*)-(LeuPhg) **2a** and (*S*,*S*)-(PhgLeu) **4a** had no or very weak gelating ability. The observed difference in gelation between retro-peptides incorporating identical or two different amino acids is intriguing. It seems that the intermolecular lipophilic interactions that stabilize gel assemblies in polar solvents are more favorable for identical amino acid fragments (lipophilic or aromatic) and less favorable when aromatic–lipophilic amino acid fragments are present in the gelator molecule. Gelation properties of methyl ester derivatives**1b–5b**, were not significantly different to those of the respective diacid derivatives **1a–5a** except that the former appear slightly more versatile and were capable of gelating some lipophilic solvents presumably due to increased solubility (Table 1 and Table 2). The diamide derivatives bis(LeuLeuNH_2) **1c**, bis(PhgPhgNH_2)

3c and bis(PhePheNH$_2$) **5c** were even more versatile and were capable of gelating a larger set of tested solvents compared to the respective dicarboxylic acid (**1a**, **3a**, and **5a**) and dimethyl ester derivatives (**1b**, **3b** and **5b**) (Tables 1–3). It appears that the increased hydrogen bonding potential of terminal diamide derivatives gives somewhat more versatile gelators which could also gel solvents of medium and low polarity where intermolecular hydrogen bonding is favored. Stereochemical influences on gelation properties are exemplified by the following observations: (i) the racemate (*S*,*R*)-**1a**/(*R*,*S*)-**1a** exhibited considerably lower gelation effectiveness than the pure enantiomer (*S*,*R*)-**1a** while the (*S*,*S*)-**1a**/(*R*,*R*)-**1a** racemate had no gelation ability and tended to crystallize, (ii) terminal diester racemates (*S*,*R*)-**1b**/(*R*,*S*)-**1b** and (*S*,*S*)-**1b**/(*R*,*R*)-**1b** were two times more efficient in the gelation of water/DMF and water/DMSO mixtures, respectively, than the pure enantiomers (*S*,*R*)-**1b** and (*S*,*S*)-**1b**; the latter provides additional examples that some racemates could be more effective gelators of specific solvents than the pure enantiomers; (iii) among the terminal carboxamide gelators the heterochiral (*S*,*R*)-**1c** diastereoisomer is capable of immobilizing up to 10 and 4 times larger volumes of dichloromethane and toluene solvent mixtures containing a little DMSO, respectively, compared to homochiral (*S*,*S*)-**1c**. The combined results of ^{1}H NMR, FTIR, XRPD and molecular modeling studies of terminal diester (*S*,*S*)-**1b** and (*S*,*R*)-**1b** and terminal free acid (*S*,*S*)-**1a** and (*S*,*R*)-**1a** derivatives gave a consistent picture of their basic organization in gel assemblies. In lipophilic solvents and also in the highly polar water/DMSO mixture, the intermolecular hydrogen bonding between extended gelator molecules utilizing both, the oxalamide and Leu amide hydrogen bonding functionalities represents the major organizational driving force for aggregation. The TEM investigations have shown that in the highly lipophilic solvents the extensive intermolecular hydrogen bonding may lead to the formation of wide and relatively short tapes giving a gel network of low solvent immobilization capacity. Molecular modeling studies revealed that the homochiral (*S*,*S*)-**1a** and (*S*,*S*)-**1b** form the low energy conformations with *cis-trans*-arrangement of *i*-Bu groups in contrast to the heterochiral (*S*,*R*)-**1a** and (*S*,*R*)-**1b** conformations possessing the all-*cis*-arrangement of *i*-Bu groups with respect to the oxalamide plane. Such conformational differences were found to strongly influence both, gelation effectiveness and the morphology characteristics of gel network.

ACKNOWLEDGEMENTS

The authors thank L. Horvat, B.sc and Dr. N. Ljubešić for TEM investigations. The financial support from the Croatian Ministry of Science, Education and Sports (Projects No. 098-0982904-2912) and EU COST Action D31 is gratefully acknowledged.

REFERENCES

1. Barta-Szalai, G.; Borza, I.; Bozó, É.; Kiss, C.; Ágai, B.; Proszenyák, Á.; Keserű, G. M.; Gere, A.; Kolok, S.; Galgóczy, K.; Horváth, C.; Farkasa, S.; Domány, G. *Bioorg. Med. Chem. Lett.* **2004,** *14,* 3953-3956. doi:10.1016/j.bmcl.2004.05.053 Return to citation in text: [1]
2. Di Stefano, A.; Mosciatti, B.; Cingolani, G. M.; Giorgioni, G.; Ricciutelli, M.; Cacciatore, I.; Sozioa, P.; Claudi, F.*Bioorg. Med. Chem. Lett.* **2001,** *11,* 1085–1088. doi:10.1016/S0960-894X(01)00140-8 Return to citation in text: [1]
3. Medou, M.; Priem, G.; Quélever, G.; Camplo, M.; Kraus, J. K. *Tetrahedron Lett.* **1998,** *39,* 4021–4024.doi:10.1016/S0040 4039(98)00680-7 Return to citation in text: [1]
4. Jadhav, P. K.; Man, H.-W. *Tetrahedron Lett.* **1996,** *37,* 1153–1156. doi:10.1016/0040-4039(96)00013-5 Return to citation in text: [1]
5. Zhang, S.; Marini, D. M.; Hwang, W.; Santoso, S. *Curr. Opin. Chem. Biol.* **2002,** *6,* 865–871.doi:10.1016/S1367 5931(02)00391-5 Return to citation in text: [1]
6. Zhao, X.; Zhang, S. *Trends Biotechnol.* **2004,** *22,* 470–476. doi:10.1016/j.tibtech.2004.07.011 Return to citation in text: [1]
7. Zhang, S. *Biotechnol. Adv.* **2002,** *20,* 321–339. doi:10.1016/S0734-9750(02)00026-5 Return to citation in text: [1]
8. Llusar, M.; Sanchez, C. *Chem. Mater.* **2008,** *20,* 782–820. doi:10.1021/cm702141e Return to citation in text: [1]
9. Vintiloiu, A.; Leroux, J. C. *J. Controlled Release* **2008,** *125,* 179 192. doi:10.1016/j.jconrel.2007.09.014 Return to citation in text: [1]
10. Yang, Y.; Khoe, U.; Wang, X.; Horii, A.; Yokoi, H.; Zhang, S. *Nano Today* **2009,** *4,* 193–210.doi:10.1016/j.nantod.2009.05.001 Return to citation in text: [1]

11. Zhang, S. *Nat. Biotechnol.* **2003,** *21,* 1171–1178. doi:10.1038/nbt874 Return to citation in text: [1]
12. Gao, X. Y.; Matsui, H. *Adv. Mater.* **2005,** *17,* 2037–2050. doi:10.1002/adma.200401849 Return to citation in text: [1]
13. Bong, D. T.; Clark, T. D.; Granja, J. R.; Ghadiri, M. R. *Angew. Chem., Int. Ed.* **2001,** *40,* 988–1011.doi:10.1002/1521-3773(20010316)40:6<988::AID-ANIE9880>3.0.CO;2-N Return to citation in text: [1]
14. Valery, C.; Paternostre, M.; Robert, B.; Gulik-Krzywicki, T.; Narayanan, T.; Dedieu, J. C.; Keller, G.; Torres, M. L.; Cherif-Cheikh, R.; Calvo, P.; Artzner, F. *Proc. Natl. Acad. Sci. U. S. A.* **2003,** *100,* 10258 10262. doi:10.1073/pnas.1730609100 Return to citation in text: [1]
15. Aggeli, A.; Nyrkova, I. A.; Bell, M.; Harding, R.; Carrick, L.; McLeish, T. C. B.; Semenov, A. N.; Boden, N.*Proc. Natl. Acad. Sci. U. S. A.* **2001,** *98,* 11857 11862. doi:10.1073/pnas.191250198 Return to citation in text: [1]
16. Fages, F.; Vögtle, F.; Žinić, M. *Top. Curr. Chem.* **2005,** *256,* 77–131. Return to citation in text: [1]
17. Terech, P.; Weiss, R. G., Eds. *Molecular Gels: Materials with Self-Assembled Fibrillar Networks;* Springer: Dordrecht, 2006. Return to citation in text: [1] [2]
18. Smith, D. K. Chapter 5. In *Organic Nanostructures;* Steed, J. W.; Atwood, J. L., Eds.; Wiley-VCH: Weinheim, 2008. Return to citation in text: [1] [2]
19. Fages, F. *Top. Curr. Chem.* **2005,** *256,* 1–273. Return to citation in text: [1] [2]
20. Terech, P.; Weiss, R. G. *Chem. Rev.* **1997,** *97,* 3133–3160. doi:10.1021/cr9700282 Return to citation in text: [1] [2]
21. van Esch, J. H.; Feringa, B. L. *Angew. Chem., Int. Ed.* **2000,** *39,* 2263–2266. doi:10.1002/1521-3773(20000703)39:13<2263::AID-ANIE2263>3.0.CO;2-V Return to citation in text: [1] [2]
22. de Loos, M.; Feringa, B. L.; van Esch, J. H. *Eur. J. Org. Chem.* **2005,** 3615–3631. doi:10.1002/ejoc.200400723 Return to citation in text: [1] [2]
23. Estroff, L. A.; Hamilton, A. D. *Chem. Rev.* **2004,** *104,* 1201 1218. doi:10.1021/cr0302049 Return to citation in text: [1] [2]
24. Percec, V.; Ungar, G.; Peterca, M. *Science* **2006,** *313,* 55–56. doi:10.1126/science.1129512 Return to citation in text: [1] [2]
25. Huang, X.; Terech, P.; Raghavan, S. R.; Weiss, R. G. *J. Am. Chem. Soc.* **2005,** *127,* 4336–4344.doi:10.1021/ja0426544 Return to citation in text: [1] [2]

26. Aggeli, A.; Nyrkova, I. A.; Bell, M.; Harding, R.; Carrick, L.; McLeish, T. C. B.; Semenov, A. N.; Boden, N.*Proc. Natl. Acad. Sci. U. S. A.* **2001,** *98,* 11857 11862. doi:10.1073/pnas.191250198 Return to citation in text: [1] [2] [3]

27. de Loos, M.; van Esch, J.; Kellogg, R. M.; Feringa, B. L. *Angew. Chem., Int. Ed.* **2001,** *40,* 613–616.doi:10.1002/1521 3773(20010202)40:3<613::AID-ANIE613>3.0.CO;2-K Return to citation in text: [1] [2]

28. Arnaud, A.; Bouteiller, L. *Langmuir* **2004,** *20,* 6858–6863. doi:10.1021/la049365d Return to citation in text: [1] [2]

29. Simic, V.; Bouteiller, L.; Jalabert, M. *J. Am. Chem. Soc.* **2003,** *125,* 13148–13154. doi:10.1021/ja037589x Return to citation in text: [1] [2]

30. Bouteiller, L.; Colombani, O.; Lortie, F.; Terech, P. *J. Am. Chem. Soc.* **2005,** *127,* 8893–8898.doi:10.1021/ja0511016 Return to citation in text: [1] [2]

31. Webb, J. E. A.; Crossley, M. J.; Turner, P.; Thordarson, P. *J. Am. Chem. Soc.* **2007,** *129,* 7155–7162.doi:10.1021/ja0713781 Return to citation in text: [1] [2]

32. Ellis-Behnke, R. G.; Liang, Y.-X.; You, S.-W.; Tay, D. K. C.; Zhang, S.; So, K.-F.; Schneider, G. E.*Proc. Natl. Acad. Sci. U. S. A.* **2006,** *103,* 5054 5059. doi:10.1073/pnas.0600559103 Return to citation in text: [1]

33. Silva,G.A.;Czeisler,C.;Niece,K.L.;Beniash,E.;Harrington,D.A.;Kessler, J. A.; Stupp, S. I. *Science* **2004,***303,* 1352 1355. doi:10.1126/science.1093783 Return to citation in text: [1]

34. Liang, Z.; Yang, G.; Ma, M.; Abbah, A. S.; Lu, W. W.; Xu, B. *Chem. Commun.* **2007,** 843–845. Return to citation in text: [1]

35. Sada, K.; Takeuchi, M.; Fujita, N.; Numata, M.; Shinkai, S. *Chem. Soc. Rev.* **2007,** *36,* 415–436.doi:10.1039/b603555h Return to citation in text: [1]

36. Fukushima, T.; Asaka, K.; Kosaka, A.; Aida, T. *Angew. Chem., Int. Ed.* **2005,** *44,* 2410–2413.doi:10.1002/anie.200462318 Return to citation in text: [1]

37. Puigmarti-Luis, J.; Laukhin, V.; Perez del Pino, A.; Vidal-Gancedo, J.; Rovira, C.; Laukhina, E.; Amabilino, D. B.*Angew. Chem., Int. Ed.* **2007,** *46,* 238 241. doi:10.1002/anie.200602483 Return to citation in text: [1]

38. Vemula, P. K.; John, G. *Chem. Commun.* **2006,** 2218–2220. doi:10.1039/b518289a Return to citation in text: [1]

39. Yang, Z.; Xu, B. *Chem. Commun.* **2004,** 2424–2425. doi:10.1039/b408897b Return to citation in text: [1]

40. Tiller, J. C. *Angew. Chem., Int. Ed.* **2003,** *42,* 3072–3075. doi:10.1002/

anie.200301647 Return to citation in text: [1]

41. Vinogradov, S. V.; Bronich, T. K.; Kabanov, A. V. *Adv. Drug Delivery Rev.* **2002,** *54,* 135–147.doi:10.1016/S0169-409X(01)00245-9 Return to citation in text: [1]
42. van Esch, J. H. *Langmuir* **2009,** *25,* 8392–8394. doi:10.1021/la901720a Return to citation in text: [1]
43. Frkanec, L.; Žinić, M. *Chem. Commun.* **2010,** *46,* 522–537. doi:10.1039/b920353m Return to citation in text: [1] [2]
44. Fletcher, M. D.; Campbell, M. M. *Chem. Rev.* **1998,** *98,* 763–795. doi:10.1021/cr970468t Return to citation in text: [1]
45. Chorev, M.; Goodman, M. *Trends Biotechnol.* **1995,** *13,* 438–445. doi:10.1016/S0167-7799(00)88999-4 Return to citation in text: [1]
46. Chorev, M.; Goodman, M. *Acc. Chem. Res.* **1993,** *26,* 266–273. doi:10.1021/ar00029a007 Return to citation in text: [1]
47. Ceretti, S.; Luppi, G.; De Pol, S.; Formaggio, F.; Crisma, M.; Toniolo, C.; Tomasini, C. *Eur. J. Org. Chem.* **2004,**4188–4196. doi:10.1002/ejoc.200400242 Return to citation in text: [1]
48. Stern, P. S.; Chorev, M.; Goodman, M.; Hagler, A. T. *Biopolymers* **1983,** *22,* 1885–1900.doi:10.1002/bip.360220806 Return to citation in text: [1]
49. Stern, P. S.; Chorev, M.; Goodman, M.; Hagler, A. T. *Biopolymers* **1983,** *22,* 1901–1917.doi:10.1002/bip.360220807 Return to citation in text: [1]
50. Carotti, A.; Carrieri, A.; Cellamare, S.; Fanizzi, F. P.; Gavuzzo, E.; Mazza, F. *Biopolymers* **2001,** *60,* 322–332.doi:10.1002/10970282(2001)60:4<322::AID-BIP9993>3.0.CO;2-Y Return to citation in text: [1]
51. Ranganathan, D.; Vaish, N. K.; Shah, K.; Roy, R.; Madhusudanan, K. P. *J. Chem. Soc., Chem. Commun.* **1993,**92–94. Return to citation in text: [1]
52. Karle, I.; Rangahathan, D.; Shah, K.; Vaish, N. K. *Int. J. Pept. Protein Res.* **1994,** *43,* 160 165.doi10.1111/j.1399 3011.1994.tb00517.x Return to citation in text: [1]
53. Karle, I.; Rangahathan, D. *Int. J. Pept. Protein Res.* **1995,** *46,* 18 23. doi:10.1111/j.1399-3011.1995.tb00577.x Return to citation in text: [1]
54. Karle, I.; Rangahathan, D. *Biopolymers* **1995,** *36,* 323–331. doi:10.1002/bip.360360307 Return to citation in text: [1]
55. Coe, S.; Kane, J. J.; Nguyen, T. L.; Toledo, L. M.; Wininger, E.; Fowler, W.; Lauher, J. W. *J. Am. Chem. Soc.***1997,** *119,* 86–93. doi:10.1021/ja961958q Return to citation in text: [1] [2]
56. Nguyen, T. L.; Fowler, F. W.; Lauher, J. W. *J. Am. Chem. Soc.* **2001,** *123,*

11057–11064.doi:10.1021/ja016635v Return to citation in text: [1] [2]

57. Nguyen, T. L.; Scott, A.; Dinkelmeyer, B.; Fowler, F. W.; Lauher, J. W. *New J. Chem.* **1998,** 22, 129–135.doi:10.1039/a707642h Return to citation in text: [1] [2]
58. Jokić, M.; Makarević, J.; Žinić, M. *J. Chem Soc., Chem. Commun.* **1995,** 1723–1724. Return to citation in text: [1] [2] [3] [4]
59. Makarević, J.; Jokić, M.; Perić, B.; Tomišić, V.; Kojić-Prodić, B.; Žinić, M. *Chem.–Eur. J.* **2001,** *7,* 3328–3341.doi:10.1002/1521-3765(20010803)7:15<3328::AID-CHEM3328>3.0.CO;2-C Return to citation in text: [1] [2] [3] [4] [5] [6] [7] [8]
60. Makarević, J.; Jokić, M.; Raza, Z.; Štefanić, Z.; Kojić-Prodić, B.; Žinić, M. *Chem.–Eur. J.* **2003,** *9,* 5567–5580.doi:10.1002/chem.200304573 Return to citation in text: [1] [2] [3] [4] [5] [6] [7] [8]
61. D'Aléo, A.; Pozzo, J.-L.; Fages, F.; Schmutz, M.; Mieden-Gundert, G.; Vögtle, F.; Čaplar, V.; Žinić, M.*Chem. Commun.* **2004,** 190–191. doi:10.1039/b307846a Return to citation in text: [1]
62. Čaplar, V.; Žinić, M.; Pozzo, J.-L.; Fages, F.; Mieden-Gundert, G.; Vögtle, F. *Eur. J. Org. Chem.* **2004,** 4048–4059. doi:10.1002/ejoc.200400105 Return to citation in text: [1]
63. Luo, X.; Liu, B.; Liang, Y. *J. Chem. Soc., Chem. Commun.* **2001,** 1556–1557. doi:10.1039/b104428c Return to citation in text: [1]
64. Hanabusa, K.; Kobayashi, H.; Suzuki, M.; Kimura, M.; Shirai, H. *Colloid Polym. Sci.* **1998,** *276,* 252–259.doi:10.1007/s003960050236 Return to citation in text: [1]
65. Hanabusa, K.; Okui, K.; Karaki, K.; Kimura, M.; Shirai, H. *J. Colloid Interface Sci.* **1997,** *195,* 86–93.doi:10.1006/jcis.1997.5139 Return to citation in text: [1]
66. Bhattacharya, S.; Acharya, S. N. G.; Raju, A. R. *Chem. Commun.* **1996,** 2101–2102.doi:10.1039/cc9960002101 Return to citation in text: [1]
67. Fuhrhop, J.-H.; Schneider, P.; Rosenberg, J.; Boekema, E. *J. Am. Chem. Soc.* **1987,** *109,* 3387–3390.doi:10.1021/ja00245a032 Return to citation in text: [1]
68. Terech, P.; Rodriguez, V.; Barnes, J. D.; McKenna, G. B. *Langmuir* **1994,** *10,* 3406–3418.doi:10.1021/la00022a009 Return to citation in text: [1]
69. Hirst, A. R.; Coates, I. A.; Boucheteau, T. R.; Miravet, J. F.; Escuder, B.; Castelletto, V. I.; Hamley, W.; Smith, D. K. *J. Am. Chem. Soc.* **2008,** *130,* 9113–9121. doi:10.1021/ja801804c Return to citation in text: [1]
70. Jonkheijm, P.; van der Schoot, P.; Schenning, A. P. H. J.; Meijer, E. W. *Science* **2006,** *313,* 80–83.doi:10.1126/science.1127884 Return to citation in text: [1]

71. Fuhrhop, J.-H.; Wang, T. *Chem. Rev.* **2004,** *104,* 2901–2937. doi:10.1021/cr030602b Return to citation in text: [1]

72. Lee, S. J.; Kim, E.; Seo, L. M.; Do, Y.; Lee, Y.-A.; Lee, S. S.; Jung, J. H.; Kogiso, M.; Shimizu, T. *Tetrahedron***2008,** *64,* 1301–1308. doi:10.1016/j.tet.2007.11.062 Return to citation in text: [1]

73. Jung, J. H.; John, G.; Yoshida, K.; Shimizu, T. *J. Am. Chem. Soc.* **2002,** *124,* 10674–10675.doi:10.1021/ja020752o Return to citation in text: [1]

74. Sreerama, N.; Woody, R. W. *Circular Dichroism: Principles and Applications*. 2nd ed.; Berova, N.; Nakanishi, K.; Woody, R. W., Eds.; John Wiley & Sons: New York, NY, 2000; pp 601–620. Return to citation in text: [1]

75. Bustamante, C.; Tinoco, I., Jr.; Maestre, M. F. *Proc. Natl. Acad. Sci. U. S. A.* **1983,** *80,* 3568–3572.doi:10.1073/pnas.80.12.3568 Return to citation in text: [1]

76. Cai, W.; Wang, G.-T.; Du, P.; Wang, R.-X.; Jiang, X.-K.; Li, Z.-T. *J. Am. Chem. Soc.* **2008,** *130,* 13450–13459.doi:10.1021/ja8043322 Return to citation in text: [1]

77. *SYBYL®,* Version 7.3; TRIPOS Inc.: St. Louis, MO, 2007. Return to citation in text: [1]

Chapter 8

EXPANDING THE GELATION PROPERTIES OF VALINE-BASED 3,5-DIAMINO-BENZOATE ORGANOGELA-TORS WITH N-ALKYLUREA FUNCTIONALITIES

Hak-Fun Chow[1,2,3] and Chin-Ho Cheng[1]

[1]Department of Chemistry, The Chinese University of Hong Kong, Shatin, NT, Hong Kong SAR

[2]The Center of Novel Functional Molecules, The Chinese University of Hong Kong, Shatin, NT, Hong Kong SAR

[3]Institute of Molecular Functional Materials, Areas of Excellence Scheme, University Grants Committee, Hong Kong SAR

ABSTRACT

A new family of valine-containing 3,5-diaminobenzoate derivatives **2** with *N*-alkylurea moieties attached to the valine moieties was prepared. By appending these two new *N*-alkylurea chains to the molecular structure, their organogelating properties were extended from only aromatic solvents, to a wide range of other types of solvents such as alicyclic hydrocarbons, alcohols and polar solvents such as DMSO and DMF. It was also found that a longer *N*-alkylurea chain conferred improved gelation power and higher thermal stability as compared to those of the shorter ones.

INTRODUCTION

Organic gelators are an interesting group of molecules that are able to form a non-covalent three dimensional network with a particular solvent system. The ultimate result is the immobilization of solvent molecules. These molecules possess many potential applications in biomedical science, environmental and separation technology[1-6].

One key problem in the design and synthesis of organogelators lies in the difficulty in predicting their gelation properties beforehand. Most often a subtle change of the substituents of the organgelator can lead to substantial modification in its gelating properties. This is because gelation is the result of a delicate balance of various driving forces. If the binding interactions between the organogelator molecules are too strong, precipitation or crystallization will take place. On the other hand, if the interactions are too weak, dissolution of the organogelator will result. In addition, non-specific interactions such as hydrophobic and van der Waals interactions are difficult to quantify, yet they can play very important roles on a cumulative scale when such interactions actually involve a large ensemble of solvent and gelator molecules.

Our group has been interested in the synthesis and self-assembled gelating properties of amino acid-containing non-dendritic [7,8] and dendritic molecules [9,10]. We recently showed that the organogelation strength in aromatic solvents of a series of α-amino acid-based low molecular weight organogelators **1** (Figure 1) could be enhanced by appending additional aromatic-containing substituents [7]. Both the Cbz protecting group and the benzyl ester functionality were responsible for the enhanced gelating power in aromatic solvents. However, they are poor organogelators in non-aromatic solvents such as alkanes, alcohols, acetone, acetonitrile and DMSO. In order to further expand their gelating power in other solvents, we decided to further modify the appending Cbz groups with other functionalities. One particular interesting moiety is the urea group, which can act simultaneously as a donor and an acceptor of hydrogen bonds and is known to be a key structural element in many organogelators [11-14]. Herein we report that the organogelating properties of **1** could be expanded to include alicyclic hydrocarbon, alcohols and even polar aprotic solvents such as DMSO and DMF by replacing the Cbz group with an *N*-alkylurea functionality (e.g., **2**).

In addition, the length of the alkyl chain $-(CH_2)_nMe$ also exhibited some interesting effects on their gelation ability [15].

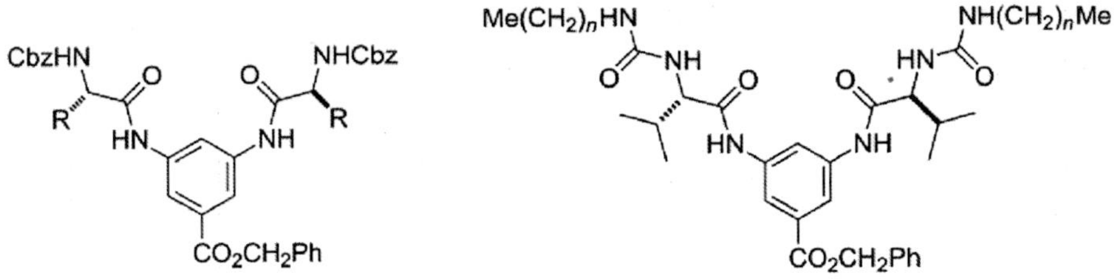

Figure 1: Amino acid based organogelators 1 and 2.

RESULTS AND DISCUSSION

Synthesis

In our earlier report it was shown that, amongst the many α-amino acids used, valine-based **1** (R = $CHMe_2$) possessed much better organogelating properties. Hence in the present study we focused only on valine derivatives **2**. A homologous series of *N*-alkylurea side chain derivatives **2** (n = 3–6, 9, 10, 12, 15, 18 and 20) was prepared in order to evaluate effect of the length of the alkyl side chain on the resulting gelation properties.

The target organogelators were prepared according to Scheme 1. The known Boc protected benzyl ester **3** [7]was converted into the corresponding diamino compound **4** in 98% yield by treatment with trifluoroacetic acid (TFA) followed by neutralization with $NaHCO_3$. The *N*-alkylurea derivatives **2** were then obtained in 88–94% as white solids by coupling compound **4** with various *O*-succinimidyl alkylcarbamates **6**, prepared from the corresponding alkanoic acids **5** according to a literature procedure [16], in the presence of *N*-diisopropylethylamine.

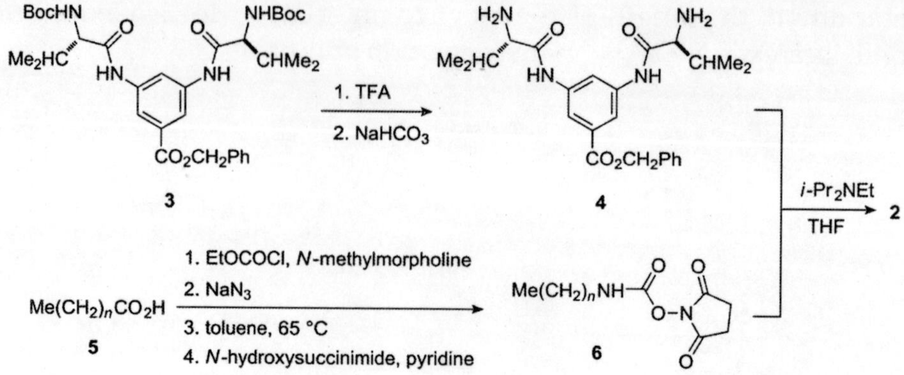

Figure: Synthesis of organogelators **2**.

Structural Characterization

All synthesized compounds were characterized by ^{1}H and ^{13}C nuclear magnetic resonance spectroscopy, mass spectrometry, elemental analysis, and optical polarimetry. Due to the poor solubilities of the target organogelators **2**in chloroform and acetone, their ^{1}H and ^{13}C NMR spectra were recorded in DMSO-d_6. In addition, the ^{1}H and ^{13}C NMR spectra of analogues with thirteen or more carbon atoms in the aliphatic hydrocarbon chains were recorded at 100 °C to avoid gelation of the solvent. The ^{1}H NMR spectra of compounds **2** showed the presence of the aliphatic hydrocarbon chains and the urea moieties. The methyl group of the aliphatic chains appeared as a triplet at ~ δ 0.86, while the methylenes of the aliphatic chains were located in close proximity to the signals of the valine side chain. On the other hand, the signals from the isopropyl group of the valine residue overlapped with those of the methylenes of the aliphatic chains. The two NH protons of the urea groups were different and appeared as a doublet at δ 6.04 and a triplet at δ 6.07. The benzyl ester group appeared as a singlet at δ 5.24 and a set of multiplets at δ 7.35–7.47. Finally, the aromatic protons of the 3,5-diaminobenzoate moiety appeared as two singlets at δ 7.98 and δ 8.26.

The ^{13}C NMR signals of the target molecules **2** could be grouped into three separate regions. First, in the downfield region, the ^{13}C signals located at δ 157, 165 and 171 were attributed to the urea, ester and anilide C=O groups, respectively. Second, the ^{13}C signals of the central 3,5-diamino-substituted aromatic core were scattered between δ 110–140, while the aromatic signals due to the benzyl ester were found at ~ δ 127–135. Finally, the ^{13}C signals of the valine side chain and the aliphatic hydrocarbon side chain were located in a range of δ 12–66. Specifically, the ^{13}C signal situated at δ 58 corresponded to the valine α-carbon. The benzylic carbon appeared at δ 66. The remaining aliphatic ^{13}C signals due to the aliphatic carbons and some of the resonance peaks overlapped with each other. Hence, the number of the observed signals was sometimes less than theoretically predicted. One ^{13}C signal (CH_2NH) was located at δ 39 and was often buried within the residual solvent signals of DMSO-d_6. For organogelators with longer aliphatic chains, this signal was too weak to be observed and assignment could only be made based on the chemical shift value of other homologues where this signal could be observed.

High-resolution mass spectra were recorded for all synthesized compounds. The experimental determined M^+ results were in accordance with the theoretical values. Generally, the abundance of the M^+ decreased with increasing length of the alkyl hydrocarbon side chain.

GELATION PROPERTIES

The gelation behavior of bis(*N*-alkylurea valine) benzyl esters **2** were examined at a concentration of 2% w/v in various solvents (Table 1). Most organogelators **2** were found to be insoluble in common non-aromatic organic solvents at room temperature. In contrast, they were soluble in aromatic solvents and organic solvents of high polarity such as 1,4-dioxane, DMF and DMSO upon warming. Interestingly, homologues having shorter aliphatic chain ($n < 10$) were poor organogelators, while the longer aliphatic chain analogues ($n \geq 10$) formed transparent gels in aromatic solvents and translucent gels in 1,4-dioxane, DMF and DMSO. Most interestingly, it was observed that organogelator **2** ($n = 20$) with the longest hydrocarbon chain also formed opaque gels in alcoholic solvents and translucent gels in alicyclic hydrocarbon solvents, respectively. Hence, the range of the solvents that these new compounds can gel was significantly expanded by replacing the Cbz group with an *N*-alkylurea functionality.

Table 1: Gelation behavior of bis(*N*-alkylurea valine) compounds **2** at 2% w/v [a].

Solvent	**2 (*n* = 3–6)**	**2 (*n* = 9)**	**2 (*n* = 10)**	**2 (*n* = 12)**	**2 (*n* = 15)**	**2 (*n* = 18)**	**2 (*n* = 20)**
THF	I	P	P	P	P	P	P
acetone	I	I	I	I	I	I	I
chloroform	I	I	I	I	I	I	I
dichloromethane	I	I	I	I	I	I	I
ethyl acetate	I	I	I	I	I	I	I
ethylene glycol	I	I	I	I	I	I	I
1-butanol	–	–	–	–	–	–	OG
1-heptanol	–	–	–	–	–	–	OG
1-dodecanol	–	–	–	–	–	–	OG
cyclooctanol	–	–	–	–	–	–	TG
hexane	I	I	I	I	I	I	I
cyclohexene	–	–	–	–	–	–	TG
cyclooctene	–	–	–	–	–	–	TG
1,4-dioxane	S	S	S	TG	TG	TG	TG
DMF	S	S	S	TG	TG	TG	TG
DMSO	S	S	TG	TG	TG	TG	TG
anisole	S	S	CG	CG	CG	CG	CG
benzene	S	S	S	CG	CG	CG	CG
benzyl alcohol	S	CG	CG	CG	CG	CG	CG
o-dichlorobenzene	S	CG	CG	CG	CG	CG	CG
nitrobenzene	S	S	TG	TG	TG	TG	TG
toluene	S	CG	CG	CG	CG	CG	CG
o-xylene	S	CG	CG	CG	CG	CG	CG

[a]I = insoluble; P = precipitation; S = soluble; CG = clear transparent gel; OG = opaque gel; TG = translucent gel.

Based on these experiments, organogelators **2** with longer aliphatic hydrocarbon chains ($n \geq 10$) were found to exhibit better gelation behavior in aromatic solvents. However, it was interesting to note that compounds with less than or equal to eight carbon atoms in the side chain failed to induce gelation. The effect of aliphatic hydrocarbon chain length (n = 10, 12, 15, 18 and 20) on their gelation power was then compared by the determination of their minimum gelation concentration (MGC) and gel-sol transition temperature (T_g) in different solvents (Table 2). It was found that the longer the alkyl chain, the lower the MGC but the higher the T_g value. Hence, the cumulative hydrophobic interaction between organogelator molecules with longer aliphatic chains must be stronger than that of the shorter ones, and this factor should contribute to the higher thermal stability of the longer chain organogels. In addition, the presence of the long aliphatic chain also prevented the organogelators from crystallizing by imposing a higher degree of local disorder.

Table 2: Minimum gelation concentration and gel-to-sol transition temperature of compounds 2 (n = 10, 12, 15, 18 and 20).

Solvent	2 (n = 10)		2 (n = 12)		2 (n = 15)		2 (n = 18)		2 (n = 20)	
	MGC[a]	T_g[b]	MGC	T_g	MGC	T_g	MGC	T_g	MGC	T_g
benzyl alcohol	1.8	–	1.8	70	1.5	79	1.2	88	1.0	94
o-dichlorobenzene	1.5	–	1.2	109	1.3	113	1.0	116	0.8	120
o-xylene	1.6	–	1.5	75	1.5	80	1.0	83	1.0	86

[a]In w/v%; [b]In °C as determined by inverted tube method.

Fourier transform infrared spectroscopy (FT-IR) was employed to elucidate the extent of intermolecular hydrogen bonding in the different macroscopic phases of organogelator **2** (n = 20) in o-xylene (Table 3). In 1% w/v hot o-xylene solution (100 °C), two peaks at 3342 and 3247 cm^{-1} in the $\nu_{N\text{-}H}$ regions were identified. These two signals could be assigned to the stretching bands of urea N–H and anilide N–H, respectively. The two peaks at 1729 and 1629 cm^{-1} in the $\nu_{C=O}$ region were found and they were attributed to the anilide C=O and urea C=O, respectively. Signal assignments of the urea C=O and N–H were based on the spectral data of a model compound, namely 1,3-didodecylurea. In the FT-IR spectrum of **2** (n = 20) (1% w/v) in *o*-xylene gel at 25 °C, the corresponding peaks were identified at 3335, 3268, 1728 and 1628 cm^{-1}, respectively. Hence, there was little difference in terms of the stretching frequencies both in solution and in the gel state. Furthermore, both sets of values are very similar to those of **2** (n = 20) in solid KBr, where the C=O and N–H moieties are known to form extensive intermolecular hydrogen bonds. Hence, it is very likely that compound **2** exists as aggregates via intermolecular hydrogen bonding in solution state due to the extremely strong hydrogen bonding property of the urea moiety. In addition, broadening of absorption signals in the gel state was observed which suggests that further intermolecular hydrogen bonding occurred during gel formation.

Table 3: FT-IR data of compound 2 (n = 20) and 1,3-di(dodecyl)ureaa.

Samples	**urea N–H**	**anilide N–H**	**anilide C=O**	**urea C=O**
2 (n = 20) in *o*-xylene solution at 100 °C	3342	3247	1729	1629
2 (n = 20) in *o*-xylene gel at 25 °C	3335	3268	1728	1628
2 (n = 20) in solid in KBr pellet	3348	3248	1730	1629
1,3-di(dodecyl)urea in solid KBr pellet	3340	—	—	1622
[a]in cm^{-1}.				

CONCLUSION

We have reported the synthesis of novel valine-containing 3,5-diaminobenzoate derivatives **2** with additional *N*-alkylurea functionality at the *N*-terminal of the valine residues. The resulting organogelators were found to possess gelating ability that covered a wider range of organic solvents, including alcoholic, aromatic, alicyclic hydrocarbon and polar solvents. Furthermore, attachment of longer aliphatic chains not only enhanced hydrophobic interactions between the organogelators, but also prevented crystallization of the self assembled aggregates during gelation, producing gels with a higher T_g and a lower MGC value.

EXPERIMENTAL

General. Optical rotation measurements were conducted in DMSO (unless otherwise stated) and below the MGC to avoid molecular aggregation or gelation and were measured with a Perkin Elmer 341 polarimeter. The starting materials **6** were prepared according to a literature procedure [16]. General procedures, yields and characterization data of compounds **6** (*n* = 3–20) can be found in Supporting Information File 1.

Compound 4. Trifluoroacetic acid (50 mL, 60 mmol) was added to a solution of the benzyl ester **3** [2] (6.40 g, 10.0 mol) in CH_2Cl_2 (100 mL) at 25 °C. The progress of deprotection was monitored by TLC. Upon complete deprotection (~ 12 h), the solvent was removed on a rotary evaporator. The crude product was made alkaline by the addition of aqueous $NaHCO_3$ solution to pH 8. The mixture was extracted with CH_2Cl_2 (50 mL × 3), the combined organic layers were washed with saturated NaCl solution (100 mL × 2), dried ($MgSO_4$), filtered and concentrated in vacuo to give compound **4** as a pale yellow liquid (4.3 g, 98%). $[\alpha]_D^{20}$ −177.6 (*c* 0.50, $CHCl_3$). ^1H NMR (DMSO-d_6): δ 0.85 (6H, d, *J* = 6.9, $CH(CH_3)Me$), 0.91 (6H, d, *J* = 6.9, $CH(Me)CH_3$), 1.85–1.99 (2H, m, $CH(CH_3)_2$), 3.12 (2H, d, *J* = 5.7, H_2NCHCH), 3.2–3.5 (6H, brs, NH), 5.35 (2H, s, $PhCH_2$), 7.36–7.49 (5H, m, Ar*H*), 7.98 (2H, d, *J* = 1.8, Ar*H*), 8.35 (1H, t, *J* = 1.8, Ar*H*). ^{13}C NMR (DMSO-d_6): δ 17.4, 19.5, 31.8, 60.8, 66.4, 114.4, 114.7, 128.2, 128.4, 128.6, 130.3, 136.1, 139.6, 165.4, 174.1. MS (FAB): 441 (M + H^+, 30%). HRMS (LSIMS): calcd for $C_{24}H_{32}N_4O_4$, 441.2496; found, 441.2505. Anal. found: C, 65.63; H, 7.50; N, 12.71. $C_{24}H_{32}N_4O_4$ requires C, 65.43; H, 7.32; N, 12.71.

General procedure for the preparation of bis-(*N*-alkylurea) benzyl esters 2. A mixture of *O*-succinimidyl carbamate **6** (10.0 mmol) and diisopropylethylamine (1.8 mL, 10.0 mmol) was added to a THF solution (100 mL) of the diamino benzyl ester **4** (2.2 g, 5.0 mmol). The reaction mixture was stirred at 25 °C for 4 h. The insoluble crude solid product was filtered and washed successively with boiling *n*-hexane (100 mL), acetone (100 mL) and THF (100 mL) to afford the title compound.

2 (*n* = 3). The product was obtained as a white solid (3.0 g, 94%) from *O*-succinimidyl butylcarbamate **6** (*n* = 3) (2.2 g, 10.0 mmol). $[\alpha]_D$ = +62.4 (*c* 1.01). mp 221–222 °C. ^1H NMR (DMSO-d_6): δ 0.83–0.90

(18H, m, CH(CH_3)$_2$ and $CH_2CH_2CH_3$), 1.22–1.36 (8H, m, aliphatic *H*), 1.92–1.94 (2H, m, CHC*H*(CH_3)$_2$), 2.99 (4H, q, *J* = 6, NHCH_2CH$_2$), 4.18 (2H, dd, *J* = 8.7 and 6.6, NHC*H*CH), 5.34 (2H, s, PhCH_2), 6.05 (2H, d, *J* = 4.2, urea N*H*), 6.09 (2H, t, *J* = 5.6, urea N*H*), 7.35–7.47 (5H, m, Ar*H*), 7.99 (2H, s, Ar*H*), 8.28 (1H, s, Ar*H*), 10.30 (2H, s, CON*H*Ar). ^{13}C NMR (DMSO-d_6): δ 13.7, 17.9, 19.3, 19.5, 31.2, 32.1, 38.9, 58.7, 66.4, 114.4, 114.8, 128.2, 128.4, 128.6, 130.3, 136.0, 139.6, 157.9, 165.4, 171.8. MS (FAB) 639 (M + H$^+$, 12%). HRMS (LSIMS): calcd for $C_{34}H_{50}N_6O_6$, 639.3865; found, 639.3854. Anal. found: C, 63.75; H, 7.95; N, 13.08. $C_{34}H_{50}N_6O_6$ requires C, 63.93; H, 7.89; N, 13.15.

2 (*n* = 4). The compound was obtained as a white solid (3.1 g, 94%) from *O*-succinimidyl pentylcarbamate **6** (*n* = 4) (2.3 g, 10.0 mmol). [α]$_D$ = +57.9 (*c* 1.08). mp 224–225 °C. ^{1}H NMR (DMSO-d_6): δ 0.82–0.90 (18 H, m, CH(CH_3)$_2$ and $CH_2CH_2CH_3$), 1.20–1.39 (12 H, m, aliphatic *H*), 1.91–1.95 (2H, m, CHC*H*(CH_3)$_2$), 2.97 (4H, q, *J* = 6, NHCH_2CH$_2$), 4.16 (2H, dd, *J* = 9.0 and 6.6, NHC*H*CH), 5.34 (2H, s, PhCH_2), 6.05 (2H, d, *J* = 4.5, urea N*H*), 6.08 (2H, t, *J* = 5.6, urea N*H*), 7.38–7.48 (5H, m, Ar*H*), 7.98 (2H, s, Ar*H*), 8.26 (1H, s, Ar*H*), 10.30 (2 H, s, CON*H*Ar). ^{13}C NMR (DMSO-d_6): δ 13.9, 17.9, 19.3, 21.9, 28.6, 29.6, 31.2, 39.2, 58.7, 66.4, 114.4, 114.8, 128.1, 128.2, 128.6, 130.3, 136.0, 139.6, 157.9, 165.3, 171.8. MS (FAB) 667 (M + H$^+$, 10%). HRMS (LSIMS): calcd for $C_{36}H_{54}N_6O_6$, 667.4178; found, 667.4185. Anal. found: C, 64.77; H, 8.20; N, 12.53. $C_{36}H_{54}N_6O_6$ requires C, 64.84; H, 8.16; N, 12.60.

2 (*n* = 5). The product was obtained as a white solid (3.2 g, 91%) from *O*-succinimidyl hexylcarbamate **6** (*n* = 5) (2.4 g, 10.0 mmol). [α]$_D$ = +63.6 (*c* 0.99). mp 225–226 °C. ^{1}H NMR (DMSO-d_6): δ 0.79–0.89 (18H, m, CH(CH_3)$_2$ and $CH_2CH_2CH_3$), 1.21–1.35 (16H, m, aliphatic *H*), 1.89–1.96 (2H, m, CHC*H*(CH_3)$_2$), 2.95 (4H, q, *J* = 6, NHCH_2CH$_2$), 4.17 (2H, dd, *J* = 8.4 and 6.6, NHC*H*CH), 5.31 (2H, s, PhCH_2), 6.00 (2H, d, *J* = 9, urea N*H*), 6.04 (2H, t, *J* = 5.6, urea N*H*), 7.32–7.44 (5H, m, Ar*H*), 7.96 (2H, s, Ar*H*), 8.24 (1H, s, Ar*H*), 10.20 (2H, s, CON*H*Ar). ^{13}C NMR (DMSO-d_6): δ 13.9, 17.9, 19.3, 22.1, 26.1, 29.9, 31.0, 31.2, 39, 58.7, 66.4, 114.4, 114.8, 128.1, 128.4, 128.5, 130.3, 136.0, 139.6, 157.9, 165.4, 171.8. MS (FAB) 695 (M + H$^+$, 12%). HRMS (LSIMS): calcd for $C_{38}H_{58}N_6O_6$, 695.4491; found, 695.4501. Anal. found: C, 65.26; H, 8.43; N, 11.81. $C_{38}H_{58}N_6O_6$ requires C, 65.68; H, 8.41; N, 12.09.

2 (*n* = 6). The product was obtained as a white solid (3.4 g, 94%) from *O*-succinimidyl heptylcarbamate **6** (*n* = 6) (2.6 g, 10.0 mmol).

$[\alpha]_D$ = +59.7 (c 1.05). mp 227–228 °C. ^{1}H NMR (DMSO-d_6): δ 0.81–0.91 (18H, m, CH(CH_3)$_2$ and CH$_2$CH$_2$CH_3), 1.23–1.37 (20H, m, aliphatic H), 1.90–1.97 (2H, m, CHCH(CH$_3$)$_2$), 2.98 (4H, q, J = 6, NHCH_2CH$_2$), 4.18 (2H, dd, J = 9.0 and 6.6, NHCHCH), 5.34 (2H, s, PhCH_2), 6.04 (2H, d, J = 9.3, urea NH), 6.07 (2H, t, J = 5.7, urea NH), 7.35–7.48 (5H, m, ArH), 7.99 (2H, s, ArH), 8.27 (1H, s, ArH), 10.26 (2H, s, CONHAr). ^{13}C NMR (DMSO-d_6): δ 13.9, 17.9, 19.3, 22.0, 26.3, 28.4, 30.0, 31.2, 31.3, 39, 58.7, 66.3, 114.4, 114.8, 128.1, 128.4, 128.5, 130.3, 136.0, 139.6, 157.9, 165.3, 171.8. MS (FAB) 723 (M + H$^+$, 5%). HRMS (LSIMS): calcd for $C_{40}H_{62}N_6O_6$, 723.4804; found, 723.4819. Anal. found: C, 66.42; H, 8.68; N, 11.63. $C_{40}H_{62}N_6O_6$ requires C, 66.45; H, 8.64; N, 11.62.

2 (n = 9). The product was obtained as a white solid (3.7 g, 92%) from O-succinimidyl decylcarbamate **6** (n = 9) (3.0 g, 10.0 mmol). $[\alpha]_D$ = +53.9 (c 1.00). mp 230–231 °C. ^{1}H NMR (DMSO-d_6): δ 0.82–0.91 (18H, m, CH(CH_3)$_2$ and CH$_2$CH$_2$CH_3), 1.22–1.35 (32H, m, aliphatic H), 1.91–1.97 (2H, m, CHCH(CH$_3$)$_2$), 2.98 (4 H, q, J = 6, NHCH_2CH$_2$), 4.18 (2H, dd, J = 9.0 and 6.6, NHCHCH), 5.34 (2H, s, PhCH_2), 6.03 (2H, d, J = 8.7, urea NH), 6.06 (2H, t, J = 3.8, urea NH), 7.35–7.47 (5H, m, ArH), 7.99 (2H, s, ArH), 8.26 (1H, s, ArH), 10.23 (2 H, s, CONHAr). ^{13}C NMR (DMSO-d_6): δ 13.9, 17.9, 19.3, 22.1, 26.4, 28.7, 28.8, 28.9, 29.0, 30.0, 31.3, 39, 58.7, 66.3, 114.4, 114.8, 128.1, 128.2, 128.5, 130.3, 136.0, 139.5, 157.9, 165.3, 171.8. MS (FAB) 808 (M + H$^+$, 5%). HRMS (LSIMS): calcd for $C_{46}H_{74}N_6O_6$, 807.5743; found, 807.5730. Anal. found: C, 67.97; H, 9.24; N, 10.39. $C_{46}H_{74}N_6O_6$ requires C, 68.45; H, 9.24; N, 10.41.

2 (n = 10). The product was obtained as a white solid (3.8 g, 91%) from O-succinimidyl undecylcarbamate **6** (n = 10) (3.1 g, 10.0 mmol). $[\alpha]_D$ = +52.3 (c 1.02). mp 232–233 °C. ^{1}H NMR (DMSO-d_6): δ 0.81–0.90 (18H, m, CH(CH_3)$_2$ and CH$_2$CH$_2$CH_3), 1.22–1.34 (36H, m, aliphatic H), 1.91–1.97 (2H, m, CHCH(CH$_3$)$_2$), 2.97 (4 H, q, J = 6, NHCH_2CH$_2$), 4.18 (2H, dd, J = 8.7 and 6.6, NHCHCH), 5.34 (2H, s, PhCH_2), 6.04 (2H, d, J = 5.4, urea NH), 6.07 (2H, t, J = 5.3, urea NH), 7.35–7.47 (5H, m, ArH), 7.98 (2H, s, ArH), 8.26 (1H, s, ArH), 10.27 (2H, s, CONHAr). ^{13}C NMR (DMSO-d_6): δ 14.0, 18.0, 19.4, 22.2, 26.5, 28.8, 28.9, 29.1, 29.2, 30.1, 30.8, 31.3, 31.4, 39, 58.9, 66.5, 114.6, 114.9, 128.2, 128.3, 128.7, 130.4, 136.1, 139.7, 158.1, 165.5, 171.9. MS (FAB) 836 (M + H$^+$, 5%). HRMS (LSIMS): calcd for $C_{48}H_{78}N_6O_6$, 835.6056; found, 835.6041. Anal. found: C, 68.67; H, 9.45; N, 10.05. $C_{48}H_{78}N_6O_6$ requires C, 69.03; H, 9.41; N, 10.06.

2 (n = 12). The product was obtained as a white solid (4.0 g, 90%) from *O*-succinimidyl tridecylcarbamate **6** (n = 12) (3.4 g, 10.0 mmol). $[\alpha]_D$ = +53.9 (*c* 0.99). mp 235–236 °C. ^{1}H NMR (DMSO-d_6, 100 °C): δ 0.82–0.90 (18 H, m, CH(CH_3)$_2$ and CH$_2$CH$_2$CH_3), 1.23–1.35 (44H, m, aliphatic *H*), 1.91–1.98 (2H, m, CHC*H*(CH$_3$)$_2$), 2.99 (4H, q, *J* = 7, NHCH_2CH$_2$), 4.18 (2H, dd, *J* = 9.0 and 6.6, NHC*H*CH), 5.34 (2H, s, PhCH_2), 6.02 (2H, d, *J* = 9.6, urea N*H*), 6.05 (2H, t, *J* = 6.2, urea N*H*), 7.35–7.47 (5H, m, Ar*H*), 7.98 (2H, s, Ar*H*), 8.26 (1H, s, Ar*H*), 10.20 (2H, s, CON*H*Ar). ^{13}C NMR (DMSO-d_6, 100 °C): δ 13.0, 17.3, 18.5, 21.3, 25.8, 27.9, 28.1, 28.3, 29.3, 30.4, 30.6, 39, 58.7, 65.7, 114.9, 115.0, 127.2, 127.4, 127.8, 130.0, 135.7, 138.9, 157.4, 164.9, 171.0. MS (FAB) 892 (M + H$^+$, 3%). HRMS (LSIMS): calcd for $C_{52}H_{86}N_6O_6$, 891.6682; found, 891.6691. Anal. found: C, 69.95; H, 9.79; N, 9.47. $C_{52}H_{86}N_6O_6$ requires C, 70.08; H, 9.73; N, 9.42.

2 (n = 15). The product was obtained as a white solid (4.2 g, 89%) from *O*-succinimidyl hexadecylcarbamate **6** (n= 15) (3.8 g, 10.0 mmol). $[\alpha]_D$ = +50.5 (*c* 1.05). mp 239–240 °C. ^{1}H NMR (DMSO-d_6, 100 °C): δ 0.86 (6H, t, *J* = 3.5, CH$_2$CH$_2$CH_3), 0.91 (6H, d, *J* = 6.9, CH(CH_3)Me), 0.94 (6H, d, *J* = 6.9, CHMe(CH_3)), 1.26–1.40 (56H, m, aliphatic *H*), 1.90–2.15 (2H, m, CHC*H*(CH$_3$)$_2$), 2.95–3.06 (4H, m, NHCH_2CH$_2$), 4.19 (2H, dd, *J* = 6.3 and 9, NHC*H*CH), 5.36 (2H, s, PhCH_2), 5.86 (2H, d, *J* = 9, urea N*H*), 5.93 (2H, t, *J* = 5.4, urea N*H*), 7.37–7.47 (5H, m, Ar*H*), 7.95 (2H, d, *J* = 2.1, Ar*H*), 8.26 (1H, t, *J* = 2.1, Ar*H*), 9.90 (2H, s, CON*H*Ar). ^{13}C NMR (DMSO-d_6, 100 °C): δ 13.0, 17.3, 18.6, 21.3, 25.8, 28.0, 28.2, 28.4, 29.4, 30.5, 30.6, 39, 58.6, 65.7, 114.87, 114.93, 127.3, 127.4, 127.9, 130.0, 135.7, 139.0, 157.5, 164.9, 171.0. MS (FAB) 976 (M + H$^+$, 5%). HRMS (LSIMS): calcd for $C_{58}H_{98}N_6O_6$, 975.7621; found, 975.7626. Anal. found: C, 71.46; H, 9.94; N, 8.35. $C_{58}H_{98}N_6O_6$ requires C, 71.42; H, 10.13; N, 8.61.

2 (n = 18). The product was obtained as a white solid (4.7 g, 88%) from *O*-succinimidyl nonadecylcarbamate **6** (n= 18) (4.3 g, 10.0 mmol). $[\alpha]_D$ = +52.5 (*c* 1.01). mp 243–244 °C. ^{1}H NMR (DMSO-d_6, 100 °C): δ 0.84–0.93 (18H, m, CH(CH_3)$_2$ and CH$_2$CH$_2$CH_3), 1.25–1.37 (68H, m, aliphatic *H*), 1.95–1.99 (2H, m, CHC*H*(CH$_3$)$_2$), 2.95–3.06 (4H, m, NHCH_2CH$_2$), 4.19 (2H, dd, *J* = 8.4 and 6.6, NHC*H*CH), 5.35 (2H, s, PhCH_2), 5.92 (2H, d, *J* = 8.7, urea N*H*), 5.98 (2H, t, *J* = 5.6, urea N*H*), 7.35–7.47 (5H, m, Ar*H*), 7.96 (2H, s, Ar*H*), 8.24 (1H, s, Ar*H*), 10.04 (2H, s, CON*H*Ar). ^{13}C NMR (DMSO-d_6, 100 °C): δ 13.0, 17.3, 18.6,

21.3, 25.8, 28.0, 28.2, 28.4, 29.4, 30.5, 30.6, 39, 58.6, 65.7, 114.87, 114.93, 127.3, 127.4, 127.9, 130.0, 135.7, 139.0, 157.5, 164.9, 171.0. MS (FAB) 1060 (M + H^+, 3%). HRMS (LSIMS): calcd for $C_{64}H_{110}N_6O_6$, 1069.8560; found, 1069.8550. Anal. found: C, 72.53; H, 10.57; N, 7.81. $C_{64}H_{110}N_6O_6$ requires C, 72.55; H, 10.46; N, 7.93.

2 (*n* = 20). The product was obtained as a white solid (4.8 g, 86%) from *O*-succinimidyl heneicosylcarbamate **6** (*n*= 20) (4.5 g, 10.0 mmol). $[\alpha]_D$ = +53.1 (*c* 1.00). mp 249–250 °C. ^{1}H NMR (DMSO-d_6, 100 °C): δ 0.86 (6H, t, *J* = 6.9, $CH_2CH_2CH_3$), 0.92 (6H, d, *J* = 6.9, CH(CH_3)Me), 0.95 (6H, d, *J* = 6.9, CHMe(CH_3)), 1.28–1.43 (76H, m, aliphatic *H*), 1.95–2.12 (2H, m, $CHCH(CH_3)_2$), 3.00 (4H, q, *J* = 6, $NHCH_2CH_2$), 4.19 (2H, dd, *J* = 8.1 and 6.6, NHC*H*CH), 5.36 (2H, s, $PhCH_2$), 5.82 (2H, d, *J* = 9, urea N*H*), 5.88 (2H, t, *J* = 5.6, urea N*H*), 7.37–7.47 (5H, m, Ar*H*), 7.94 (2H, d, *J* = 2.1, Ar*H*), 8.19 (1H, t, *J* = 1.8, Ar*H*), 9.80 (2H, s, CON*H*Ar). ^{13}C NMR (DMSO-d_6, 100 °C): δ 13.0, 17.3, 18.5, 20.2, 21.3, 25.8, 27.9, 28.1, 28.3, 29.3, 30.4, 30.6, 39, 58.7, 65.7, 114.9, 115.0, 127.2, 127.4, 127.8, 130.0, 135.7, 138.9, 157.5, 164.9, 171.0. MS (FAB) 1116 (M + H^+, 100%). HRMS (LSIMS): calcd for $C_{68}H_{118}N_6O_6$, 1115.9186; found, 1115.9192. Anal. found: C, 73.17; H, 10.53; N, 7.45. $C_{68}H_{118}N_6O_6$ requires C, 73.20; H, 10.66; N, 7.53.

REFERENCES

1. Terech, P.; Weiss, R. G. Chem. Rev. 1997, 97, 3133–3159. doi:10.1021/cr9700282
2. van Esch, J. H.; Feringa, B. L. Angew. Chem., Int. Ed. 2000, 39, 2263–2266. doi:10.1002/1521-3773(20000703)39:13<2263::AID-ANIE2263>3.0.CO;2-V
3. Gronwald, O.; Snip, E.; Shinkai, S. Curr. Opin. Colloid Interface Sci. 2002, 7, 148–156. doi:10.1016/S1359-0294(02)00016-X
4. Sangeetha, N. M.; Maitra, U. Chem. Soc. Rev. 2005, 34, 821–836. doi:10.1039/b417081b
5. Dastidar, P. Chem. Soc. Rev. 2008, 37, 2699–2715. doi:10.1039/b807346e
6. Vintiloiu, A.; Leroux, J.-C. J. Controlled Release 2008, 125, 179–192. doi:10.1016/j.jconrel.2007.09.014
7. Chow, H.-F.; Zhang, J.; Lo, C.-M.; Cheung, S.-Y.; Wong, K.-W. Tetrahedron 2007, 63, 363–373. doi:10.1016/j.tet.2006.10.066

8. Chow, H.-F.; Wang, G.-X. Tetrahedron 2007, 63, 7407–7418. doi:10.1016/j.tet.2007.02.037
9. Chow, H.-F.; Zhang, J. Chem.–Eur. J. 2005, 11, 5817–5831. doi:10.1002/chem.200500174
10. Chow, H.-F.; Zhang, J. Tetrahedron 2005, 61, 11279–11290. doi:10.1016/j.tet.2005.08.006
11. Steed, J. W. Chem. Soc. Rev. 2010, 39, 3686–3699. doi:10.1039/b926219a
12. van Esch, J.; Kellogg, R. M.; Feringa, B. L. Tetrahedron Lett. 1997, 38, 281–284. doi:10.1016/S0040-4039(96)02292-7
13. van Esch, J.; de Feyter, S.; Kellogg, R. M.; de Schryver, F.; Feringa, B. L. Chem.–Eur. J. 1997, 3, 1238–1243. doi:10.1002/chem.19970030811
14. van de Laan, S.; Feringa, B. L.; Kellogg, R. M.; van Esch, J. Langmuir 2002, 18, 7136–7140. doi:10.1021/la025561d
15. George, M.; Tan, G.; John, V. T.; Weiss, R. G. Chem.–Eur. J. 2005, 11, 3243–3254. doi:10.1002/chem.200401066
16. Guichard, G.; Semetey, V.; Didierjean, C.; Aubry, A.; Briand, J. P.; Rodriguez, M. J. Org. Chem. 1999, 64, 8702–8705. doi:10.1021/jo990092e

Chapter 9

SYNTHESIS AND SELF-ASSEMBLY OF 1-DEOXYGLUCOSE DERIVATIVES AS LOW MOLECULAR WEIGHT ORGANOGELATORS

Guijun Wang[1], Hao Yang[1], Sherwin Cheuk[1] and Sherman Coleman[2]

[1]Department of Chemistry, University of New Orleans, New Orleans

ABSTRACT

Low molecular weight gelators are an important class of molecules. The supramolecular gels formed by carbohydrate derived low molecular weight gelators are interesting soft materials that show great potential for many applications. Previously, we have synthesized a series of methyl 4,6-*O*-benzylidene-α-D-glucopyranoside derivatives and found that several of them are good gelators for water, aqueous mixtures of DMSO, or aqueous mixtures of ethanol. The gelation efficiency of these glycolipid derivatives is dependent upon the structures of their acyl chains. In order to understand the influence of the anomeric position of the sugar head group towards self-assembly, we synthesized a series of 1-deoxyglucose analogs, and examined their gelation properties in several solvents. Several long chain esters, including diacetylene containing esters, and aryl esters exhibited gelation in ethanol, aqueous ethanol, or aqueous

DMSO. The synthesis and characterization of these novel analogs are reported.

INTRODUCTION

In recent years, the field of low molecular weight gelators (LMWGs) has received a great deal of attention. LMWGs are an interesting class of small molecules that can form reversible supramolecular gels in organic solvents or aqueous solutions [1-9]. Non-covalent interactions such as hydrogen bonding, hydrophobic interactions, and π–π stacking are the main driving forces for the self-assembly of the gelators into 3-dimensional networks. The resulting gels may find applications as soft materials for drug delivery, enzyme immobilization, scaffolds for tissue engineering, etc. [10-14]. The structures of LMWGs span a diverse range; carbohydrates have frequently been used in the synthesis of LMWGs because they are naturally abundant and possess multiple chiral centers that can be selectively functionalized [15-37]. Sugar-based supramolecular hydrogels are being explored as biocompatible soft materials and as matrices for enzymes, DNA, and drug delivery systems [22-28]. Glucose, in particular, is a versatile building block for the preparation of various small molecule gelators [30-37], as it is relatively easy to obtain substituted products by selective functionalization of the anomeric position and the 4- and 6-hydroxy groups. We have found that further derivatization of the glucose head group to form different glycolipids can result in organogelators [34-37].

Previously, we have systematically synthesized and studied the self-assembling properties of a series of methyl 4,6-*O*-benzylidene-α-D-glucopyranoside (**1**) derivatives (Figure 1), including esters and carbamates with different functional groups. Several of these compounds proved to be effective gelators for organic solvents and aqueous solutions [34-37]. We found that the structures of the acyl chains of diester **2**, and monoesters **3** and **4**, are important for gelation. This result also indicates that the ester linked derivatives require more specific structures and many substituents are not tolerated. Typically, monoesters with alkynyl groups containing 5–7 carbons (Figure 2, **5–7**) are good gelators for water or aqueous ethanol mixtures. The monoesters presumably form an extended hydrogen bonding array between the ring oxygen and the free hydroxy groups [36].

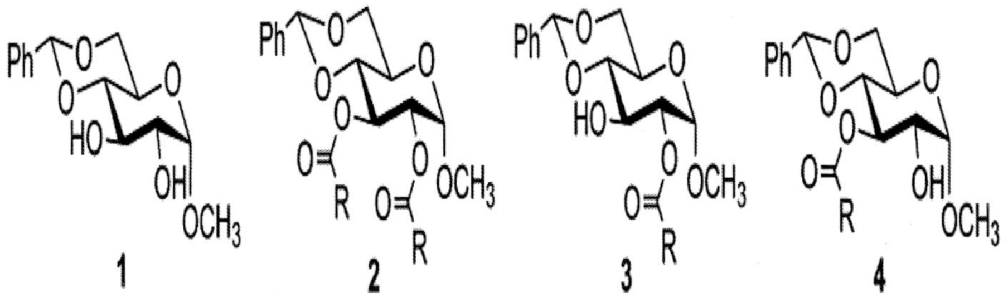

Figure 1: Structures of three ester derivatives of compound **1**.

For these compounds, the main forces that influence gelation include phenyl ring π–π interactions, hydrogen bonding, and hydrophobic interactions of the acyl chains, etc. To further understand the structural influence of the anomeric position of the sugar head group on self-assembly, we synthesized analogs using head group **8** [38](Figure 2), in which the anomeric methoxy group was replaced with a hydrogen atom and contained a similar series of acyl chains to those in compounds **2–4**. These can potentially lead to novel classes of organogelators.

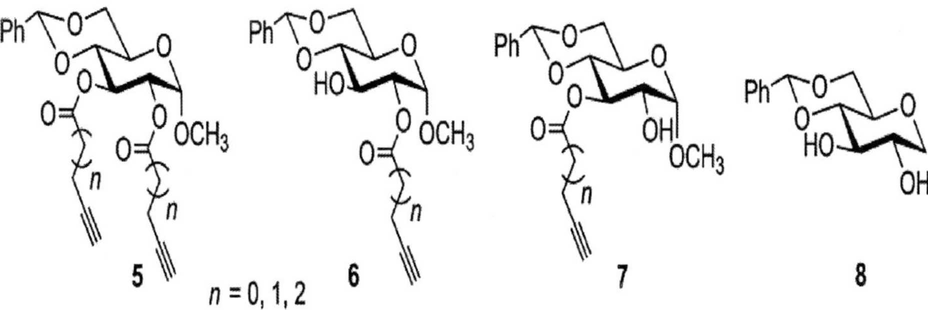

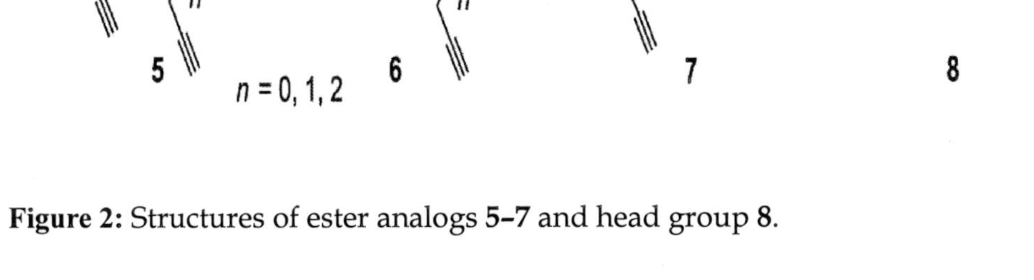

Figure 2: Structures of ester analogs **5–7** and head group **8**.

RESULTS AND DISCUSSION

In order to better understand the importance of the anomeric substituent of the sugar head group for self-assembly, we synthesized a series of esters of the head group **8** by a one pot reaction of the acid chloride with the head group (Scheme 1). In general, three products were obtained, which could be separated by flash chromatography. Our previous results had shown that esterification of head group **1** typically gave the 2-monoester as the major product, but when the head group **8** was used, the selectivity of the acylation diminished significantly. The 2- and 3-esters were obtained in similar quantities and, in some cases, the 3-ester was the major product. The difference in the acylation selectivity was possibly because of the configurations of C-1 and C-2 positions. The α-methoxy group allows intramolecular hydrogen bonding to the C-2 hydroxy to take place, which can make the 2-hydroxy group relatively more nucleophilic than the 3 position. In compound **8**, there is no α-methoxy group available for hydrogen bonding to the 2-hydroxy, so the 2- and 3-hydroxy groups were more or less equally nucleophilic [39]. The amount of diester was significantly less than the amounts of the 2- and 3-esters. Diester formation mirrors the synthesis of the diesters of compound **1**.

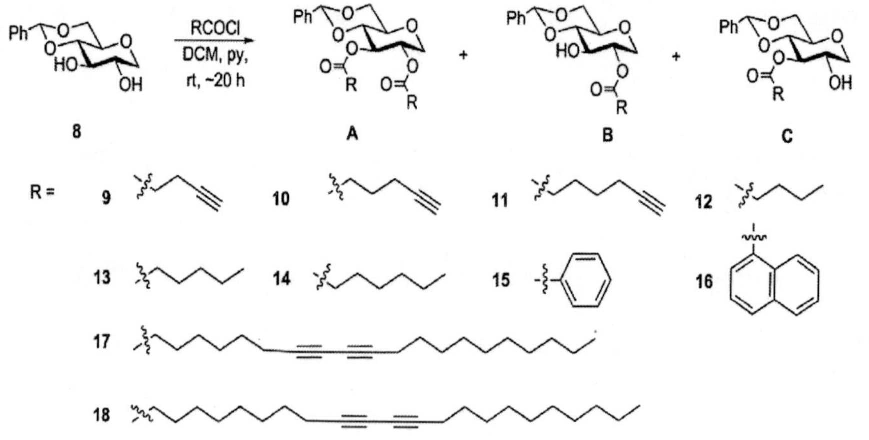

RCOCl, DCM, py, rt, ~20 h

8 → A + B + C

R = 9, 10, 11, 12, 13, 14, 15, 16, 17, 18

Scheme 1: Synthesis of a series of esters **9A–18C**.

The selection of the R groups used in this series was based on our previous results [34-36]. We synthesized the terminal acetylenes

9–11, saturated hydrocarbons **12–14**, aryl derivatives **15,16**, and two long chain di acetylene containing glycolipids **17,18**. After we obtained these compounds, we then screened them for gelation in several solvents. These results are shown in Table 1.

Table 1: The gelation test results of the compounds synthesized[a].

Compound	Hexane	H_2O	EtOH	EtOH:H_2O (1:2)	DMSO:H_2O (1:2)
9A	G 15	I	P	P	P
10A	G 20	I	S	P	P
11A	P	I	S	P	P
12A	S	P	S	P	P
13A	S	P	S	S	P
14A	S	P	S	P	P
15A	I	P	G 5	P	G 5
16A	I	P	P	P	G 20
17A	S	P	G 7	P	P
18A	P	P	G 3	P	P
9B	P	I	S	S	G 20
10B	P	P	S	P	P
11B	P	P	S	P	P
12B	P	P	S	P	P
13B	P	P	S	P	P
14B	P	P	S	P	P
15B	I	P	S	G 4	P
16B	I	P	S	P	P
17B	P	P	S	P	P
18B	P	P	S	P	P
9C	I	P	S	S	S
10C	P	I	S	P	P
11C	P	P	S	P	S
12C	P	P	S	S	P
13C	P	P	S	P	P
14C	P	P	S	P	P
15C	I	P	S	G 10	G 20
17C	P	P	S	P	P
18C	P	S	S	P	P

From the gelation test results shown in Table 1, quite a few of the diesters were good gelators for the solvents tested. In hexane, only the two short chain terminal acetylene compounds **9A** and **10A** formed gels. Several diesters were effective gelators for ethanol, including the benzoate **15A** and the long chain diacetylene compounds **17A** and **18A**. The two diaryl esters **15A** and **16A** were also able to form gels in aqueous DMSO solution. Compared to the diesters, the monoesters were not as effective as organogelators; only the 2-pentynoate **9B** was able to gelate aqueous DMSO, and the 2-benzoate **15B** was able to form a gel in ethanol. The rest of the 2-monoesters did not gelate any of the other solvents. For the 3-monoesters, compound **15C** was able to form gels in aqueous DMSO and aqueous ethanol, but none of the other esters and solvents formed gels.

The morphologies of several gels are shown in Figure 3. The hexane gel of compound **9A** formed fibrous assemblies (Figure 3A, Figure 3B). Compound **9B** showed tubular assemblies (Figure 3C) and more straight cylindrical tube or ribbons (Figure 3D) at different areas. Compound **15B** formed gels more efficiently at 4 mg/mL in the ethanol/water mixture. The morphology of the assembly showed uniform, long and narrow fibers (Figure 3E,Figure 3F).

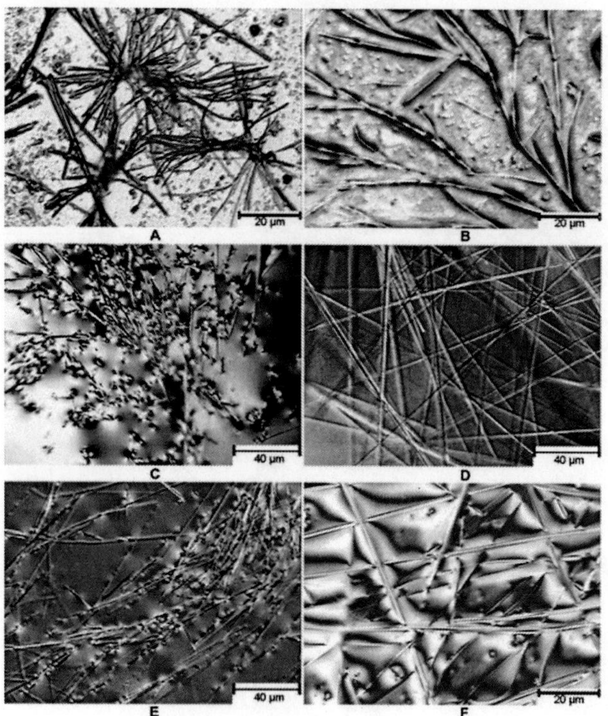

Figure 3: Optical micrographs of the gels formed by compound **9A** in hexane at 15 mg/mL (**A, B**), **9B** in DMSO/water (1:2) at 20 mg/mL (**C, D**), and **15B** in EtOH/H_2O (1:2) at 4 mg/mL (**E, F**). The images were taken with gels containing solvents (not dried gels)

It is interesting to see that the gelation trends of the diester derivatives of compound 8 are quite different compared to the corresponding compounds derived from head group 1. The gelation of the long chain diacetylene compounds were not affected adversely by changing the head group. For the monoesters, the deoxy sugar derivatives are somewhat less effective gelators in comparison to the α-methoxy sugar derivatives. The terminal alkynyl esters were among the most efficient gelators from the ester library derived from compound 1. The removal of the methoxy group resulted in a sharp reduction in gelation, but these compounds did exhibit some gelation ability. This observation indicated that the α-methoxy group is useful in the formation of a fibrillar network and may or may not be involved in the hydrogen bonding array. However, terminal alkynyl groups did show more promise than saturated hydrocarbons.

Polydiacetylenes have interesting optical and electronic properties, and diacetylene containing gels may have useful applications as advanced sensing materials [35,40,41]. The morphologies of the self-assembled structures can be retained by cross linking the diacetylene functional groups. Therefore, we have also synthesized and studied several diacetylene containing sugar lipids. Esterification of sugar head groups with diacetylene containing long chain carboxylic acids can give the desired diacetylene containing lipids. Previously we had used head group 1 to synthesize a library of diacetylene containing lipids, and found that many of them are effective gelators [35]. Using the head group 8, we also synthesized six long chain diacetylene containing glycolipids 17A–17C and 18A–18C. We found that the diesters are effective gelators for ethanol. Further characterization of the gel formed by compound 18A in ethanol proved it to be a very efficient gelator, forming gels in ethanol at concentrations lower than 1 wt %/v. In addition, the gels can also be readily polymerized with a 6 W TLC illuminating UV lamp (Figure 4) to give the typically blue colored product. The glass vial is not UV transparent, therefore the polymerization only occurred from the top of the gel (Figure 4c). When a quartz tube was used as the container, the gel turned light blue homogeneously after one min of exposure to UV light (254 nm) (Figure 4d). After three min of UV treatment, it produced a dark blue colored gel (Figure 4e). The blue gel also exhibited interesting color transition properties upon heating (Figure 4f). For a comparison of the two sugar head groups, the lipid 19 [35] was able gelate ethanol at 7 mg/mL, but it cannot be polymerized as easily with the 6 W UV lamp. This indicates that the two diacetylene chains in 19 are not aligned favorably for polymerization because of the presence of the methoxy group at the anomeric position. The ethanol gels of compound 19 formed long rods or cylindrical tubules, which may require stronger UV energy to polymerize. The optical micrographs of 18A showed fibrous assemblies composed of long intertwined thin fibers (Figure 5). The topological cross-link of the diacetylenes allowed the fibrous morphologies to be preserved.

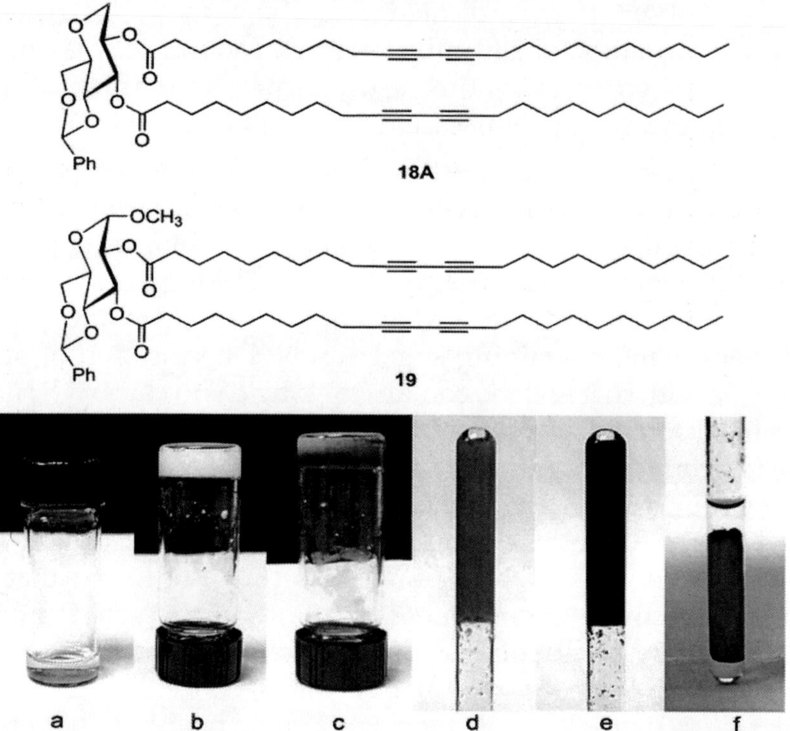

Figure 4: An ethanol gel formed by compound **18A** at <10 mg/mL. a) A clear solution when heated above 70 °C; b) a stable gel after cooling to room temperature; c) a blue gel after illuminating with UV lamp from the top of the vial in b; d) a light blue gel inside quartz tube after UV treatment for 1 min; e) a dark blue gel in quartz tube after UV treatment for 3 min; f) the blue gel in e) turned red after heating.

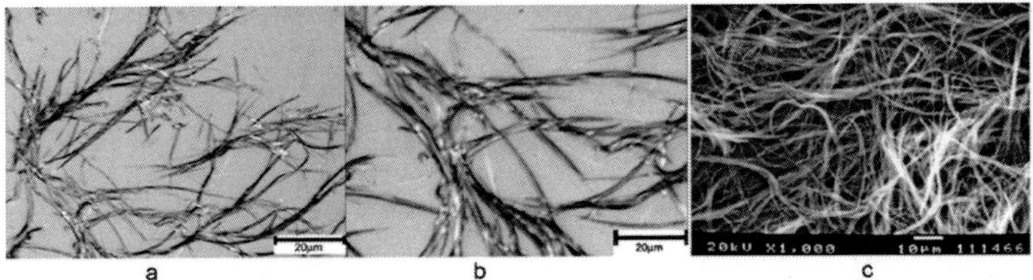

Figure 5: Optical micrographs under bright field (a, b) and scanning electron micrograph (c) of the gel formed by compound**18A** in ethanol after exposure to UV light for 3 min. Blue-purple fibers were observed in (a) and (b), and an entangled fibrous network was observed in (c).

We also carried out UV–vis studies to monitor the color transition of compound **18A** as shown in Figure 6. The gel formed by**18A** in ethanol (10 mg/mL) was treated with a 6 W UV lamp with 254 nm light for 1 min from the top of the uncovered plate. The absorption of the gel was very strong and exceeded the detection limit. The sample was split into two cells, and a small amount of ethanol was added to avoid drying. The UV–vis spectra showed two peaks at 646 nm (λ_{max}) and 590 nm, which is in agreement with the observed purple-blue color of the gel. This plate was then treated with the same UV light for another 2 min. The absorption peaks became more intense, and no new signals were observed (Figure 6A). The plate was covered with a matching glass lid and then incubated in an incubator at 35, 40, 50, and 60 °C for 10–15 min. Below 50 °C, there was a small red shift of the λ_{max}, which increased with increasing temperature. At 50 °C, the λ_{max} shifted from 650 nm to 634 nm. A short incubation at 60 °C also gave a similar spectrum but a new broad absorption at 530 nm began to appear (Figure 6B). The 530 nm absorption corresponds to the red phase of the gel. After incubating at 60 °C for 90 min, only part of the gel turned red, and the red signal at 530 nm increased significantly as shown in Figure 6C. Upon heating briefly to a higher temperature with a heat gun, the gel turned completely red, and the UV–vis spectrum (Figure 6D) showed two new strong peaks at 486 nm and 528 nm. After cooling the samples, part of the gel turned back to blue whilst the remainder stayed red.

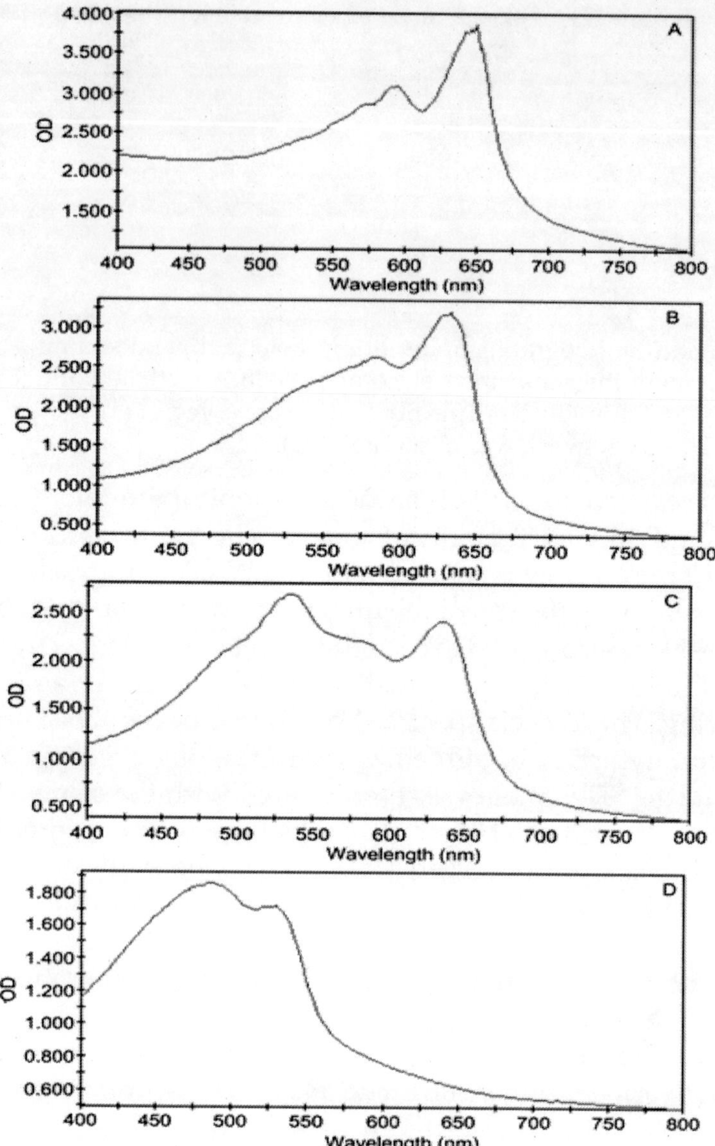

Figure 6: The UV–vis absorption spectra of the polymerized gel formed by compound **18A** in ethanol (10 mg/mL): A, at room temperature after 3 min UV irradiation (λ_{max} = 650 nm); B, at 50 °C (λ_{max} = 634 nm); C, at 60 °C incubation for 90 min (λ_{max} = 534 nm, other peaks are 582, 636 nm); D, after heating at above 60 °C till complete color change to red (λ_{max} = 486 nm together with 534 nm).

From these observations, we can see that the side chain alignment of compound **18A** is quite favorable in providing long range order for the polymerization of the diacetylene groups. Typically polydiacetylenes exhibit absorptions near 530 nm and 630 nm depending on the side chain structures, and the color transition from blue to red has also been well studied [42,43]. Our gelator **18A** polymerized readily with a TLC UV lamp and the cross linked gels only exhibited absorptions at longer wavelengths. The thermo induced color transition is probably due to an annealing effect of the blue-purple polydiacetylene gels into another stable conformation. The study indicated that the formation of diacetylene gel is an effective method to obtain polydiacetylenes with interesting optical properties.

CONCLUSION

We have synthesized a series of 1-deoxy glucose derived esters, and screened their gelation in several solvents. In comparison with the α-methoxy series with the same acyl chains, the deoxy sugar analogs, and the monoesters in particular, are less effective gelators. It appears that the presence of the α-methoxy group assists in the formation of hydrogen bonds among the compounds and with solvents. However, several diester derivatives are effective gelators, and for both the diesters and the monoesters, terminal acetylene and aromatic functional groups seemed to facilitate gelation. Therefore, even though it was found to be less effective than the α-methoxy head group **1**, the deoxy sugar head group **8** can be used for the preparation of self assembling gelators when the hydrophobic tails are chosen judicially. For instance, long chain diacetylene containing diester derivatives of **8** are effective gelators, and they can be polymerized using a TLC lamp more readily than the corresponding diesters of **1**. The polymerized diacetylene gels were dark blue with long wavelengths absorptions, these compounds may be useful as advanced stimuli responsive materials.

EXPERIMENTAL

General methods and materials

Materials and instrumentation

General chemicals and reagents were purchased from Aldrich, or VWR. The diacetylene containing fatty acids were purchased from GFS chemicals. Optical microscope images were recorded with an Olympus BX60 microscope and CCD camera. The samples were prepared as thin slices of gels placed on a cleaned glass slide, and the gels were imaged directly under the microscope. NMR spectra were recorded using a 400 MHz Varian NMR spectrometer. High resolution mass spectrometry data were measured on the Q-Tof of the Mass Spectrometry lab at the University of Illinois after the low resolution masses were confirmed. The ionization technique used was ESI (electrospray ionization). Melting points were measured using a Fisher-Jones melting point apparatus.

Optical microscopy

Optical micrographs were recorded with an Olympus BX60 microscope and CCD camera. The sample was prepared as a small piece of gel placed on a clean microscope glass slide, and the gel was imaged directly under the microscope. The program used to acquire and store the photos was Corel Photo-Paint 7.

Scanning electron microscopy

A piece of the gel was deposited on an aluminum sample holder and allowed to dry in a desiccator. The dried gel sample was coated with a thin layer of platinum (~100–150 Å) by a Denton Vacuum (model Desk II) at a reduced pressure of ~30 mTorr and a current of 45 mA for 60 sec. The sample was analyzed using a JEOL JSM 5410 scanning electron microscope with an EDAX Detecting Unit PV9757/05 ME (Model 204B+, active area = 10 mm^2).

Gelation testing

The compounds were typically tested for gelation in a 1 dram glass vial. The diacetylene containing compounds were tested in brown vials to avoid polymerization. A starting concentration of 20 mg/mL was used. The mixture was heated and sonicated until the sample was fully dissolved. The solution was then allowed to cool to room temperature for 15–20 min. If a gel is formed, then the vial is inverted; if no solvent flows while the gel is inverted, then it is called a stable gel. If the gel falls apart during inversion and by gentle shaking, then it is called an unstable gel or gel-like particulate. If a stable gel is formed, serial dilution is performed until the resulting gel is no longer stable. The concentration prior to formation of the unstable gel was recorded as the minimum gelation concentration (MGC).

UV–vis spectroscopy

The UV–vis spectra were recorded on a Bio-Tek Power Wave Micro plate Spectrophotometer using a 96-well micro plate (Corning's UV transparent micro plate #3635). A sample (about 50–100 μL) of the gel of compound **18A** in ethanol (10 mg/mL) was transferred into a cell of the plate. The plate was covered with its matching lid when not inside the plate holder. Heating was carried out using the internal incubator of the spectrophotometer (temperature range 25–50 °C) and Max 4000 incubator (temperature range 25–60 °C) from Barnstead.

General procedure for the synthesis of ester derivatives of compound 8

The three ester derivatives **A**, **B**, and **C** were synthesized in a one pot reaction by reacting the corresponding acid chloride (1.2 equiv) with head group **8** (1.0 equiv). Compound **8** was prepared according to literature procedure [38]. If the acid chloride was not commercially available, it was prepared by converting the corresponding terminal alkynyl carboxylic acid (1.2 equiv) to the acyl chloride using excess oxalyl chloride [35]. After confirming complete conversion to the acyl chloride, hexane was used to co-distill and remove the excess oxalyl chloride. The acyl chloride was added to compound **8** (1 equiv) in

DCM and 4 equiv of pyridine. The reaction mixture was left stirring under an anhydrous atmosphere for 14–20 h. The crude product was concentrated, extracted with DCM, washed with water, then brine, and the combined organic phase dried over anhydrous sodium sulfate. The crude product was isolated and purified using a gradient solvent system (hexanes and acetone), starting with 5% acetone. All yields reported are pure isolated yields, and were calculated based on compound **8**; the yields were not optimized. The following sections include the characterization data for three compounds, the rest are shown in the Supporting Information section.

Synthesis of 4-pentynoates 9A, 9B, and 9C

Compound 9A

This product was isolated as a white crystalline solid in 9.1% yield. M.p. 97–98 °C. ^{1}H NMR (400 MHz, $CDCl_3$) δ (ppm) 7.39–7.46 (m, 2H), 7.29–7.37 (m, 3H), 5.49 (s, 1H), 5.37 (t, 1H, *J* = 9.5 Hz), 5.08 (ddd~dt, 1H, *J* = 5.9, 9.5, 10.3 Hz), 4.33 (dd, 1H, *J* = 5.1, 10.3 Hz), 4.13 (dd, 1H, *J* = 5.9, 11.4 Hz), 3.71 (t, 1H, *J* = 10.3 Hz), 3.64 (t, 1H, *J* = 9.5 Hz), 3.46 (dt, 1H, *J* = 5.1, 9.5 Hz), 3.40 (t, 1H, *J* = 11.0 Hz), 2.51–2.59 (m, 4H), 2.43–2.50 (m, 4H), 1.99 (t, 1H, *J* = 2.6 Hz), 1.89 (t, 1H, *J* = 2.6 Hz). ^{13}C NMR (100 MHz, $CDCl_3$) δ (ppm) 170.8, 170.7, 136.8, 129.0, 128.1, 126.1, 101.4, 82.2, 82.0, 78.6, 72.5, 71.4, 69.8, 69.3, 68.5, 67.3, 33.2, 33.1, 14.3, 14.2. HRMS ESI calcd for $C_{23}H_{24}O_7Na$ [M + Na^+] 435.1420, found 435.1406.

Compound 9B

This product was isolated as a white crystalline solid in 14.6% yield. M.p. 100–101 °C. ^{1}H NMR (400 MHz, $CDCl_3$) δ (ppm) 7.46–7.53 (m, 2H), 7.33–7.42 (m, 3H), 5.51 (s, 1H), 4.93 (ddd~dt, 1H, *J* = 5.9, 9.5, 10.3 Hz), 4.30 (dd, 1H, *J*= 4.8, 10.6 Hz), 4.09 (dd, 1H, *J* = 5.9, 11.0 Hz), 3.85 (t, 1H, *J* = 9.2 Hz), 3.67 (t, 1H, *J* = 10.3 Hz), 3.49 (t, 1H, *J* = 9.2 Hz), 3.35 (dt, 1H, *J* = 5.1, 9.9 Hz), 3.26 (dd~t, 1H, *J* = 10.6, 11.0 Hz), 2.54–2.64 (m, 2H), 2.45–2.54 (m, 2H), 2.01 (t, 1H, *J* = 2.6 Hz). ^{13}C NMR (100 MHz, $CDCl_3$) δ (ppm) 171.1, 136.8, 129.2, 128.2, 126.2, 101.8, 82.2, 81.0, 72.4, 71.9, 70.9, 69.3, 68.5, 67.0, 33.1, 14.3. HRMS ESI calcd for $C_{18}H_{20}O_6Na$ [M + Na^+] 355.1158, found 355.1166.

Compound 9C]

This product was isolated as a white crystalline solid in 23.0% yield. M.p. 148–150 °C. ^{1}H NMR (400 MHz, $CDCl_3$) δ (ppm) 7.40–7.47 (m, 2H), 7.28–7.36 (m, 3H), 5.46 (s, 1H), 5.16 (dd~t, 1H, *J* = 9.2, 9.5 Hz), 4.30 (dd, 1H, *J* = 5.1, 10.6 Hz), 4.07 (dd, 1H, *J* = 5.9, 11.4 Hz), 3.86 (m, 1H), 3.67 (dd~t, 1H, *J* = 10.1 Hz), 3.56 (dd~t, 1H, *J* = 9.3 Hz, 1H), 3.34–3.47 (m, 2H), 2.54–2.64 (m, 2H), 2.42–2.52 (m, 2H), 1.88 (m, 1H). ^{13}C NMR (100 MHz, $CDCl_3$) δ (ppm) 172.0, 136.9, 128.9, 128.1, 126.0, 101.2, 82.3, 78.7, 76.6, 71.2, 70.6, 69.14, 69.09, 68.7, 33.3, 14.3. HRMS calcd for $C_{18}H_{20}O_7Na$ [M + Na^+] 355.1158, found 355.1147.

ACKNOWLEDGEMENTS

This research was supported by the National Science Foundation CHE#518283 and in part by the NSF REU program of AMRI at UNO for Mr. Sherman Coleman (2008 summer). We also thank the assistance of Navneet Goyal and Kristopher Williams with the preparation of the manuscript.

REFERENCES

1. Terech, P.; Weiss, R. G. *Chem. Rev.* **1997,** *97,* 3133–3160. doi:10.1021/cr9700282 Return to citation in text: [1]
2. Abdallah, D. J.; Weiss, R. G. *Adv. Mater.* **2000,** *12,* 1237–1247. doi:10.1002/1521-4095 (200009) 12:17<1237::AID-ADMA1237>3.0.CO;2-B Return to citation in text: [1]
3. Meléndez, R. E.; Carr, A. J.; Linton, B. R.; Hamilton, A. D. *Struct. Bonding* **2000,** *96,* 31–61. doi:10.1007/3-540-46591-X_2 Return to citation in text: [1]
4. Estroff, L. A.; Hamilton, A. D. *Chem. Rev.* **2004,** *104,* 1201–1218. doi:10.1021/cr0302049 Return to citation in text: [1]
5. Gronwald, O.; Snip, E.; Shinkai, S. *Curr. Opin. Colloid Interface Sci.* **2002,** *7,* 148–156. doi:10.1016/S1359-0294(02)00016-X Return to citation in text: [1]
6. George, M.; Weiss, R. G. *Acc. Chem. Res.* **2006,** *39,* 489–497. doi:10.1021/ar0500923 Return to citation in text: [1]
7. Carretti, E.; Bonini, M.; Dei, L.; Berrie, B. H.; Angelova, L. V.; Baglioni,

P.; Weiss, R. G. *Acc. Chem. Res.* **2010,** *43,* 751–760. doi:10.1021/ar900282h Return to citation in text: [1]

8. Yang, Z.; Liang, G.; Xu, B. *Acc. Chem. Res.* **2008,** *41,* 315–326. doi:10.1021/ar7001914 Return to citation in text: [1]
9. Yang, Z.; Liang, G.; Xu, B. *Soft Matter* **2007,** *3,* 515–520. doi:10.1039/b700138j Return to citation in text: [1]
10. Shah, C.; Patel, R.; Mazumder, R.; Bhattacharya, S.; Mazumder, A.; Gupta, R. N. *Pharma Rev.* **2009,** *7,* 164–166. Return to citation in text: [1]
11. Zhao, F.; Ma, M. L.; Xu, B. *Chem. Soc. Rev.* **2009,** *38,* 883–891. doi:10.1039/b806410p Return to citation in text: [1]
12. Fujita, N.; Mukhopadhyay, P.; Shinkai, S. *Annu. Rev. Nano Res.* **2006,** *1,* 385–428. doi:10.1142/9789812772374_0009 Return to citation in text: [1]
13. Vintiloiu, A.; Leroux, J.-C. *J. Controlled Release* **2008,** *125,* 179–192. doi:10.1016/j.jconrel.2007.09.014 Return to citation in text: [1]
14. Nonappa, N.; Maitra, U. *Org. Biomol. Chem.* **2008,** *6,* 657–669. doi:10.1039/B714475J Return to citation in text: [1]
15. Hafkamp, R. J. H.; Feiters, M. C.; Nolte, R. J. M. *J. Org. Chem.* **1999,** *64,* 412–426. doi:10.1021/jo981158t Return to citation in text: [1]
16. Ajayaghosh, A.; Praveen, V. K.; Vijayakumar, C. *Chem. Soc. Rev.* **2008,** *37,* 109–122. doi:10.1039/b704456a Return to citation in text: [1]
17. Vemula, P. K.; John, G. *Acc. Chem. Res,* **2008,** *41,* 769–782. doi:10.1021/ar7002682 Return to citation in text: [1]
18. John, G.; Vemula, P. K. *Soft Matter* **2006,** *2,* 909–914. doi:10.1039/b609422h Return to citation in text: [1]
19. Ghosh, R.; Chakraborty, A.; Maiti, D. K.; Puranik, V. G. *Org. Lett.* **2006,** *8,* 1061–1064. doi:10.1021/ol052963e Return to citation in text: [1]
20. Friggeri, A.; Gronwald, O.; Van Bommel, K. J. C.; Shinkai, S.; Reinhoudt, D. N. *J. Am. Chem. Soc.* **2002,** *124,* 10754–10758. doi:10.1021/ja012585i Return to citation in text: [1]
21. Chakraborty, T. K.; Jayaprakash, S.; Srinivasu, P.; Madhavendra, S. S.; Ravi Sankar, A.; Kunwar, A. C. *Tetrahedron* **2002,***58,* 2853–2859. doi:10.1016/S0040-4020(02)00158-8 Return to citation in text: [1]
22. Koshi, Y.; Nakata, E.; Yamane, H.; Hamachi, I. *J. Am. Chem. Soc.* **2006,** *128,* 10413–10422. doi:10.1021/ja0613963 Return to citation in text: [1] [2]
23. Kiyonaka, S.; Sada, K.; Yoshimura, I.; Shinkai, S.; Kato, N.; Hamachi, I. *Nat. Mater.* **2004,** *3,* 58–64. doi:10.1038/nmat1034 Return to citation in text: [1] [2]

24. Yang, Z.; Liang, G.; Ma, M.; Abbah, A. S.; Lu, W. W.; Xu, B. *Chem. Commun.* **2007,** 843–845. doi:10.1039/b616563j Return to citation in text: [1] [2]

25. Vemula, P. K.; Li, J.; John, G. *J. Am. Chem. Soc.* **2006,** *128,* 8932–8938. doi:10.1021/ja062650u Return to citation in text: [1] [2]

26. Jung, J. H.; Amaike, M.; Nakashima, K.; Shinkai, S. *J. Chem. Soc., Perkin Trans. 2* **2001,** 1938–1943.doi:10.1039/B104190H Return to citation in text: [1] [2]

27. Bhat, S.; Maitra, U. *Tetrahedron* **2007,** *63,* 7309–7320. doi:10.1016/j.tet.2007.03.118 Return to citation in text: [1] [2]

28. Suzuki, M.; Owa, S.; Shirai, H.; Hanabusa, K. *Tetrahedron* **2007,** *63,* 7302–7308. doi:10.1016/j.tet.2007.02.065 Return to citation in text: [1] [2]

29. Goyal, N.; Cheuk, S.; Wang, G. *Tetrahedron* **2010,** *66,* 5962–5971. doi:10.1016/j.tet.2010.05.071 Return to citation in text: [1]

30. Bielejewski, M.; Łapiński, A.; Luboradzki, R.; Tritt-Goc, J. *Langmuir* **2009,** *25,* 8274–8279. doi:10.1021/la900467d Return to citation in text: [1] [2]

31. Kobayashi, H.; Friggeri, A.; Koumoto, K.; Amaike, M.; Shinkai, S.; Reinhoudt, D. N. *Org. Lett.* **2002,** *4,* 1423–1426.doi:10.1021/ol025519+ Return to citation in text: [1] [2]

32. Luboradzki, R.; Gronwald, O.; Ikeda, A.; Shinkai, S. *Chem. Lett.* **2000,** *29,* 1148–1149. doi:10.1246/cl.2000.1148 Return to citation in text: [1] [2]

33. Jung, J. H.; John, G.; Masuda, M.; Yoshida, K.; Shinkai, S.; Shimizu, T. *Langmuir* **2001,** *17,* 7229–7232.doi:10.1021/la0109516 Return to citation in text: [1] [2]

34. Wang, G.; Cheuk, S.; Williams, K.; Sharma, V.; Dakessian, L.; Thorton, Z. *Carbohydr. Res.* **2006,** *341,* 705–716.doi:10.1016/j.carres.2006.01.023 Return to citation in text: [1] [2] [3] [4] [5]

35. Nie, X.; Wang, G. *J. Org. Chem.* **2006,** *71,* 4734–4741. doi:10.1021/jo052317t Return to citation in text: [1] [2] [3] [4] [5] [6] [7] [8] [9]

36. Cheuk, S.; Stevens, E. D.; Wang, G. *Carbohydr. Res.* **2009,** *344,* 417–425. doi:10.1016/j.carres.2008.12.006 Return to citation in text: [1] [2] [3] [4] [5] [6]

37. Wang, G.; Cheuk, S.; Yang, H.; Goyal, N.; Reddy, P. V. N.; Hopkinson, B. *Langmuir* **2009,** *25,* 8696–8705.doi:10.1021/la804337g Return to citation in text: [1] [2] [3] [4]

38. Nie, X.; Wang, G. *J. Org. Chem.* **2005,** *70,* 8687–8692. doi:10.1021/jo0507901 Return to citation in text: [1] [2]

39. Wang, G.; Ella-Menye, J.-R.; St. Martin, M.; Yang, H.; Williams, K.

Org. Lett. **2008,** *10,* 4203–4206.doi:10.1021/ol801316f Return to citation in text: [1]

40. Dautel, O. J.; Robitzer, M.; Lère-Porte, J.-P.; Serein-Spirau, F.; Moreau, J. J. E. *J. Am. Chem. Soc.* **2006,** *128,* 16213–16223. doi:10.1021/ja065434u Return to citation in text: [1]
41. Sada, K.; Takeuchi, M.; Fujita, N.; Numata, M.; Shinkai, S. *Chem. Soc. Rev.* **2007,** *36,* 415–435. doi:10.1039/b603555h Return to citation in text: [1]
42. Fujita, N.; Sakamoto, Y.; Shirakawa, M.; Ojima, M.; Fujii, A.; Ozaki, M.; Shinkai, S. *J. Am. Chem. Soc.* **2007,** *129,* 4134–4135. doi:10.1021/ja069307+ Return to citation in text: [1]
43. Schott, M. *J. Phys. Chem. B* **2006,** *110,* 15864–15868. doi:10.1021/jp0638437 Return to citation in text: [1]

Chapter 10

A REVIEW ON THE CHEMISTRY OF EREMANTHINE: A SESQUITERPENE LACTONE WITH RELEVANT BIOLOGICAL ACTIVITY

José C. F. Alves

Departamento de Química, Instituto de Ciências Exatas, Universidade Federal Rural do Rio de Janeiro, 23890-970 Seropédica, RJ, Brazil

ABSTRACT

The several aspects on the chemistry of eremanthine such as isolation, structural classification, biological activity, synthesis, and chemical transformations in other sesquiterpene lactones are described in this review. The main publications on this sesquiterpenolide, from its isolation of natural sources in 1972 to the current days, are included.

INTRODUCTION

Eremanthine (1) (Figure1) is a sesquiterpene lactone abundantly found in the Brazilian Compositae (Asteracea) Erymanthos elaeagnus[1] and Vanillosmopsis erythropappa [2] (Eremanthus erythropappus) [3]. This compound is a target for studies because of its biological properties, structural complexity due to the presence of a flexible seven-membered ring in its skeleton, as well as for the possibility to transform it in other potentially bioactive sesquiterpene lactones.

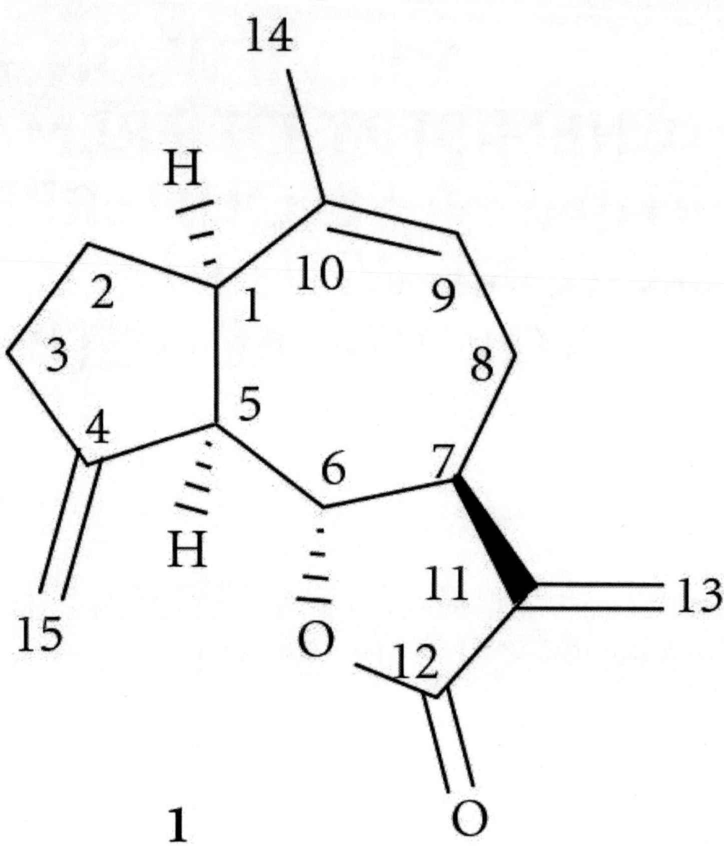

Figure 1: Structure of eremanthine (1).

This review is composed by five items which include the isolation, structural classification, biological activity, synthesis, and chemical transformations of eremanthine. The main publications on this sesquiterpenolide from its isolation of natural sources in 1972 to the current days are described in the present review. Special attention is given to the synthetic studies that were developed aiming at the preparation of potentially bioactive sesquiterpene lactones.

ISOLATION OF EREMANTHINE

The search, here in Brazil, for plant-derived inhibitors against infections by cercariae of Schist soma mansion led to the isolation and characterization of eremanthine (1) from the oil of the Compositae Eremanthus elaeagnus and Vanillosmopsis erythropappa in 1972 [1,2]. The extraction of crude oil from these vegetable species was made through the classic method of infusion of the pulverized trunk wood in organic solvent, at room temperature, followed by the evaporation stage of the extractor solvent under reduced pressure. Therefore, after evaporation of hexane that was used to extract the oil of Erymanthos elaeagnus and subsequent stage of crystallization of the crude oil with the same solvent, eremanthine (1) was obtained as colorless needles with melting point 73-74°C and $[\alpha]_D^{29}-59$ (c 1.0 in $CHCl_3$) [1]. This value of specific optical rotation was later corrected as it is going to be described soon afterwards in this review. In the case of the hexane concentrated extract obtained from the extraction of the Vanillosmopsis erythropapp oil, eremanthine (1) was obtained after elution over column chromatography of silica gel [2]. The physical data of eremanthine obtained from Vanillosmopsis erythropappa were identical to the ones of eremanthine obtained from Eremanthus elaeagnus. The only structural feature not determined in those works published in 1972 was the absolute configuration at the carbon C-1.

It was demonstrated in 1974 that vanillosmin, isolated from Vanillosmopsis erythropappa by a research group in Italy [4], also had the same structure of eremanthine (1). In this case, the oil of that vegetable species was obtained from the acetone concentrated extract of the pulverized trunk wood in infusion, followed by the subsequent stages of treatment with a mixture of MeOH-H_2O and extraction with petroleum ether. The concentrated oil obtained after these stages was eluted over column chromatography of silica gel impregnated with $AgNO_3$ to furnish vanillosmin that was crystallized in hexane as long needles with melting point 62–62.5°C, boiling point 175–180°C/0.2 mm, and −110 . There were elucidated in that work the absolute configurations at C-1 and C-5 as being both, on the basis of a series of chemical reactions.

The 1, 5-cis-fused bicycle [5.3.0]decane skeleton for was confirmed by Garcia et al. [5] through the formation of dibromoether2 when eremanthine (1) was treated with N-bromosuccinimide in dioxane containing 20% of water (Scheme1). There was demonstrated through three-dimensional analysis of 1that the formation of an ether linkage between the C-4 and C-10 positions was possible only if the five- and seven-membered rings were cis-fused. As the name eremanthine for has historical precedence over vanillosmin the latter name passed no longer to be used. It was demonstrated in that work published by Rabi et al. [5] that the value of specific optical rotation previously calculated {−59° (c 1.0, $CHCl_3$)} [1] was incorrect. A new determination was made and it was obtained (c 1.0, $CHCl_3$) being, therefore, in agreement with the value of−110° described in [4].

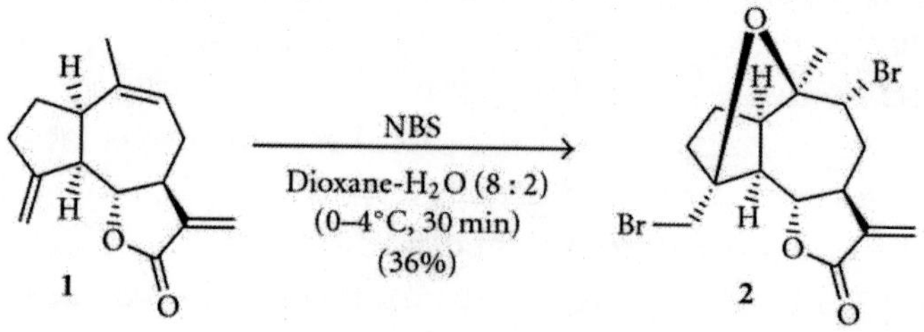

H H 1 O O

NBS

Dioxane-H_2O (8 : 2)

(0–4°C, 30 min)

(36%)

O H Br Br H O 2 O

Scheme 1:Synthesis of the dibromoether

In 1985 Lima et al. [6] published a selective method for extraction of α-methylene-γ-lactones fromVanillosmopsis erythropappa. The oil of that vegetable species was obtained by the procedure of infusion of the pulverized trunk wood left in contact with hexane and then in ethanol. After evaporation of the extractor solvents, it was obtained an oil resultant from the hexane extract and another of the ethanol extract. The ethanolic extract furnished a dark oil that was exhaustively extracted with hexane to afford an oil. The aliquots of each oil were soon afterwards submitted to the separation procedure of α-methylene-γ-lactones from the other chemical substances of the crude extracts. Such procedure used the property of the unsaturated γ-lactones undergo facile 1,4-addition with nucleophilic substances. The sequence of reactions that was used to obtain eremanthine (1) for this procedure is outlined in Scheme2. For this procedure the crude extract from Vanillosmopsis erythropappa, containing α-methylene-γ-lactones (3) in mixture with other substances, was dissolved in EtOAc and then submitted to reaction with aqueous solution of dimethylamine. The obtained mixture, containing the adducts of dimethylamine (4) together with other substances, was diluted with EtOAc and then washed with water. The separated and concentrated organic layer was submitted to reaction with acetic acid to furnish the acetates 5 in mixture with other substances. The acetates were extracted with water and the aqueous phase was washed with EtOAc, neutralized with saturated solution of sodium carbonate, and then reextracted with EtOAc. The mixture of the adducts of dimethylamine (4) was submitted, soon afterwards, to reaction with MeI to generate a mixture of iodides (6). Saturated solution of sodium carbonate was added and let to react at room temperature for 1 hour to the elimination reaction of the quaternary adducts of trimethylamine. There was obtained, after extraction of the last reaction, an exclusive mixture of α-methylene-γ-lactones (3). Filtration of this mixture of γ-lactones over column chromatography of silica gel, followed by crystallizations, furnished eremanthine (1) as majority product as well as the minority α-methylene-γ-lactones shown in Figure 2.

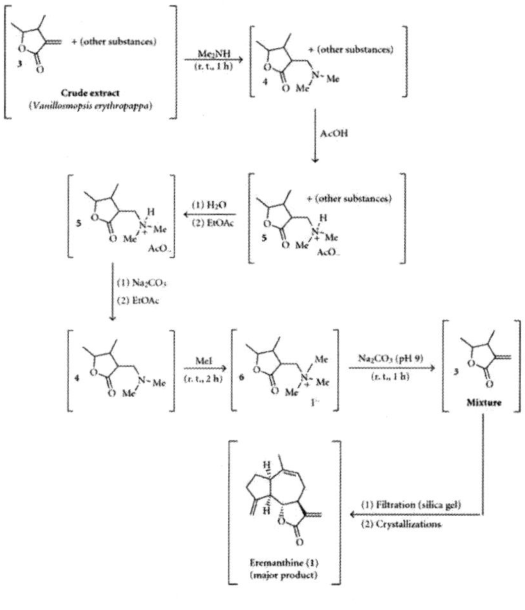

Scheme 2: Sequence of reactions used for the selective isolation of α-methylene-γ-lactones from Vanillosmopsis erythropappa

α-Cyclocostunolide (7) 4,15-α-Epoxieremanthine (8) 9,10-α-Epoxieremanthine (9) Dehydrocostus lactone (10)

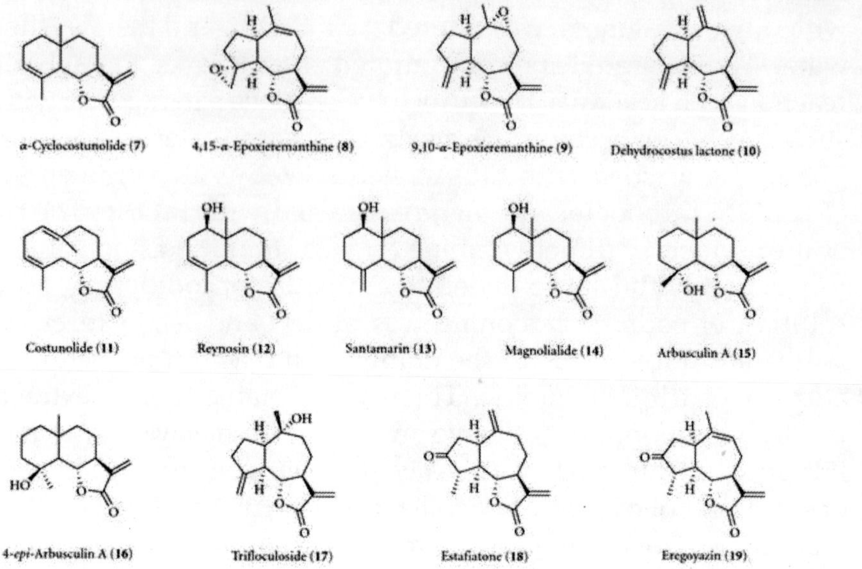

Costunolide (11) Reynosin (12) Santamarin (13) Magnolialide (14) Arbusculin A (15)

4-*epi*-Arbusculin A (16) Trifloculoside (17) Estafiatone (18) Eregoyazin (19)

Figure 2: Minority α-methylene-γ-lactones isolated from Vanillosmopsis erythropappa

Due to the importance of bioactive chemical substances present in the crude extract from Vanillosmopsis erythropappa (Eremanthus erythropappus), tree popularly known by the name of candeia, nowadays a plan to conserve this forest species is being developed [7, 8]. The wood obtained in the forests of candeiais mainly sold for the industries of essential oils [9] that extract the crude oil through the steam distillation process. For that process, the pulverized trunk wood of Eremanthus erythropappus is put inside chemical reactors of high pressure and then it is heated at high temperature with water. The steam generated in the reactors passes, soon afterwards, for industrial condensers where it is cooled and then a mixture of oil and water is collected. The densest substance (water) is removed from the mixture to furnish the concentrated oil of Eremanthus erythropappus which is mainly sold for the industry of cosmetics that uses another important chemical substance found in that oil, the α-bisabolol (20) [10] (Figure 3). The use of the concentrated oil from Eremanthus erythropappus, extracted for the modern industrial method, certainly gives good results for isolation of eremanthine

(1). Another recent method to extract the oil from Eremanthus erythropappus, using the supercritical extraction process and phase equilibrium of candeia oil with supercritical carbon dioxide, was reported by de Souza et al. [11].

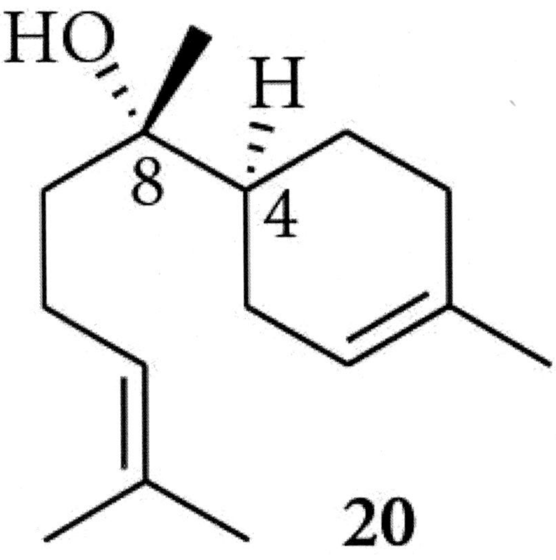

Figure 3: Structure of (-)-α-bisabolol (20).

Determination of the Absolute Stereochemistry of Eremanthine

The absolute stereochemistry of eremanthine (1) (vanillosmin) at the chiral carbons C-1, C-5, C-6, and C-7 was determined through the correlation of derivatives from the substance 1 with other lactones that possess known absolute stereochemistry in these positions. The sequences of reactions that were employed by Corbella et al. [4] to synthesize the substances used to determine the absolute stereochemistry of eremanthine are outlined in Schemes 3–8.

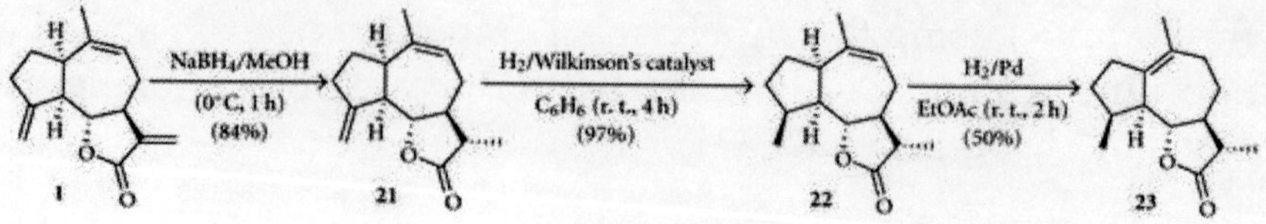

$NaBH_4/MeOH$ (0°C, 1 h) (84%)
H_2/Wilkinson's catalyst C_6H_6 (r. t., 4 h) (97%)
H_2/Pd EtOAc (r. t., 2 h) (50%)
1 21 22 23

Scheme 3: Synthesis of the lactone23 starting from eremanthine (1).

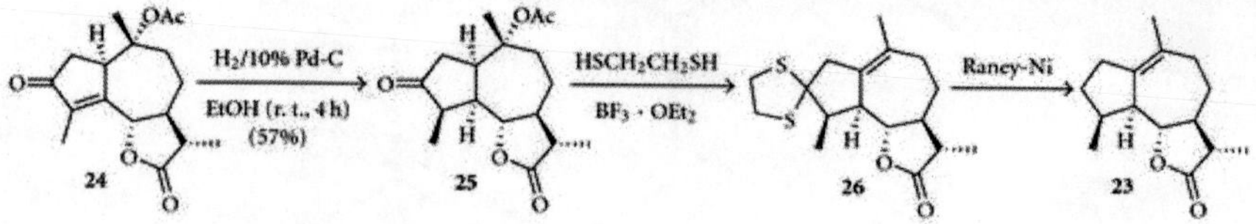

H_2/10% Pd-C EtOH (r. t., 4 h) (57%)
$HSCH_2CH_2SH$ $BF_3 \cdot OEt_2$
Raney-Ni
24 25 26 23

Scheme 4: Synthesis of the lactone23starting fromO-acetyl-isophoto-α-santonic lactone (24).

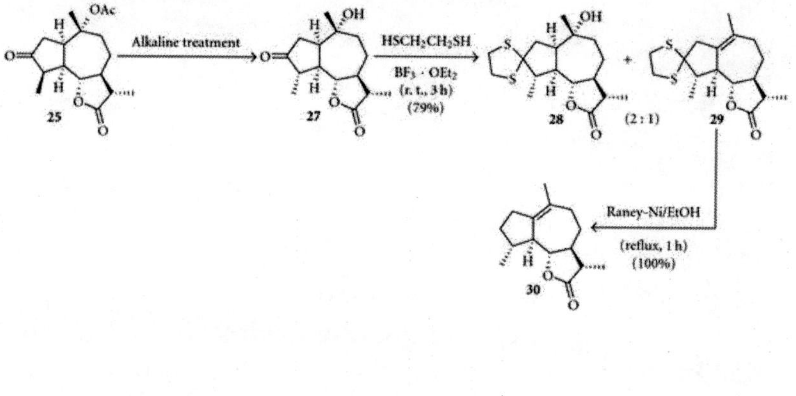

Alkaline treatment
$HSCH_2CH_2SH$ $BF_3 \cdot OEt_2$ (r. t., 3 h) (79%)
25 27 28 (2 : 1) 29
Raney-Ni/EtOH (reflux, 1 h) (100%)
30

Scheme 5:Synthesis of the lactone 30.

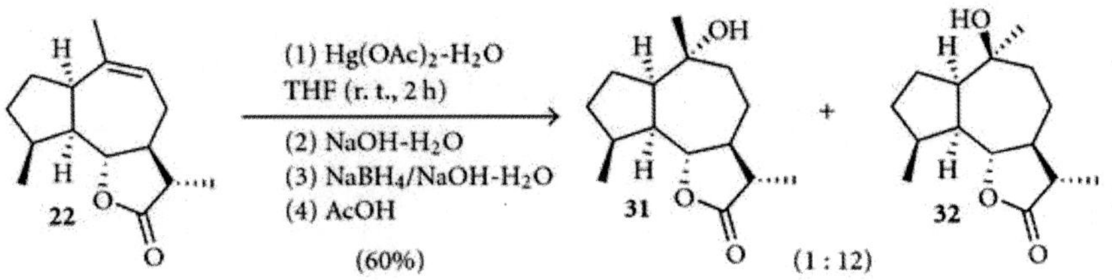

22
(1) $Hg(OAc)_2$-H_2O
THF (r. t., 2 h)
(2) NaOH-H_2O
(3) $NaBH_4$/NaOH-H_2O
(4) AcOH
(60%)
31 + 32
(1 : 12)

Scheme 6: Oxymercuration desmercuration reaction performed on tetrahydroeremanthine (22).

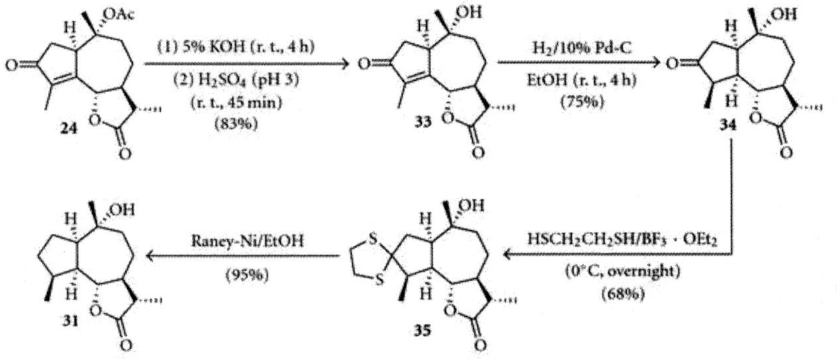

24
(1) 5% KOH (r. t., 4 h)
(2) H_2SO_4 (pH 3)
(r. t., 45 min)
(83%)
33
H_2/10% Pd-C
EtOH (r. t., 4 h)
(75%)
34
$HSCH_2CH_2SH/BF_3 \cdot OEt_2$
(0°C, overnight)
(68%)
35
Raney-Ni/EtOH
(95%)
31

Scheme 7: Synthesis of the compound 31 starting from the lactone24.

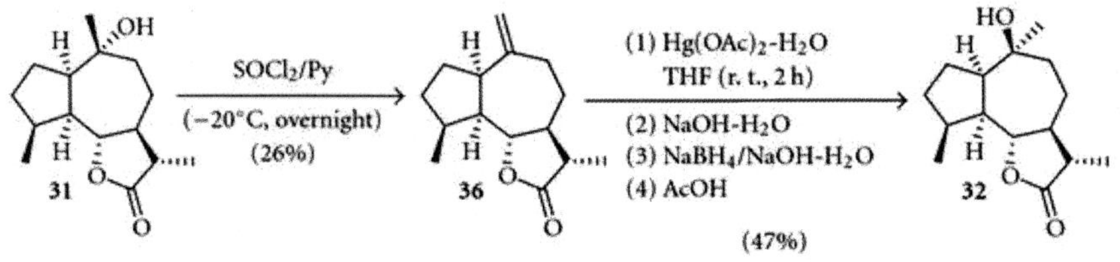

31
$SOCl_2$/Py
(−20°C, overnight)
(26%)
36
(1) $Hg(OAc)_2$-H_2O
THF (r. t., 2 h)
(2) NaOH-H_2O
(3) $NaBH_4$/NaOH-H_2O
(4) AcOH
(47%)
32

Scheme 8: Epimerization at C-10 position of the lactone 31.

As the most guaianolides possesscis fusion in the five- and seven-membered rings at the hydroazulene system, there was initially attributed the stereochemistry of eremanthine in those centers as shown in the chemical structure 1. Reaction of substance 1 with $NaBH_4$ generated a single product characterized as the lactone 21 (Scheme 3). Reduction of this substance with hydrogen and soluble Wilkinson's catalyst furnished a product identified as the compound 22. Hydrogenation of 22 in EtOAc with H_2 and Pd resulted in isomerization of the double bond C9-C10 to the tetrasubstituted position C1–C10 to afford the compound 23 as the main product of that reaction. In the following stage, the compound 23 was synthesized starting from O-acetyl-isophoto--santonic lactone (24) (Scheme 4).

Catalytic hydrogenation of lactone 24 furnished the compound 25 of known absolute stereochemistry. Reaction of substance 25 with 1, 2-ethanedithiol and $BF_3{\cdot}OEt_2$ furnished thioacetal 26, resultant from the protection of carbonyl group at C-3 position and elimination of acetic acid. Reductive removal of the thioacetal 26 with Raney-Nickel generated a compound identified as the substance 23, previously obtained from eremanthine (1).

To ensure that the acid treatment of the lactone 25 (Scheme 4) did not affect the configuration of the methyl at C-4, that is more stable in the alpha position than in the beta position, the C-4 epimer of compound 23 was prepared as described in Scheme 5. Alkaline treatment of the lactone 25 generated the compound 27 resultant from the hydrolysis of the acetate and epimerization at C-4 position. Thioacetalization of compound 27 furnished two products: one of them, the compound 28, stayed with the hydroxy group at C-10 and the other, the compound 29, resultant from elimination of water. Desulphurization of the thioacetal 29 generated the lactone 30, whose physical data were different from those of the lactone 23 obtained from eremanthine (1) (Scheme 3). This correlation confirmed the structure proposed for the compound 23 and established the absolute configuration at C-5, C-6, and C-7 positions of eremanthine as shown in the chemical structure 1.

To determine the absolute stereochemistry at C-1 position of eremanthine (1), the reactions outlined in Schemes 6–8 were performed. Starting from tetrahydroeremanthine (22), obtained from eremanthine (1) (Scheme 3), the oxymercuration desmercuration reaction shown in Scheme 6 was performed and two isomeric hydroxy-lactones 31 and 32 were obtained in the proportion of (1:12). The chemical structure of the substance 31 was confirmed through the synthesis depicted in Scheme 7.

Alkaline hydrolysis of the acetate 24 generated the lactone 33 (Scheme 7) that was submitted to catalytic hydrogenation to furnish the dihydroderivative 34. Treatment of this compound with etanedithiol and $BF_3{\cdot}OEt_2$, at low temperature, generated the thioacetal 35. This compound was reduced with Raney-Nickel to afford a product whose physical data were identical to those of the hydroxylactone 31 derived from tetrahydroeremanthine (22).

The most abundant product of the oxymercuration-desmercuration reaction on tetrahydroeremanthine (22) was the hydroxy-lactone 32 (Scheme 6), which is the C-10 epimer of substance 31. To ensure that the configuration at C-10 was the only difference between the structures 31 and 32, the product 31obtained from the conditions described in Scheme 7 was submitted to dehydration with thionyl chloride and pyridine at low temperature (Scheme 8). The anhydroderivative 36 obtained in that reaction was submitted to subsequent stage of oxymercuration-desmercuration to furnish a compound identical in all its physico-chemical properties to the hydroxy-lactone 32. These results demonstrated that the hydrogen at C-1 position of eremanthine (1) has alpha configuration and, consequently, the absolute stereochemistry of this substance is in agreement with the chemical structure shown in Figure1.

The absolute configuration of eremanthine (1) was also confirmed in an article published in 1999 by Yuuya et al. [12]. In that publication the guaianolide 1 was synthesized starting from the natural productα-santonin.

STRUCTURAL CLASSIFICATION OF EREMANTHINE

Eremanthine (1) belongs to the class of substances denominated guaianolides. These substances possess the basic skeleton of bicyclo[5.3.0]decane, characteristic of the guaiane sesquiterpenes, to which was inserted at positions C-6 and C-7 or C-7 and C-8 a γ-lactonic ring (Figure 4).

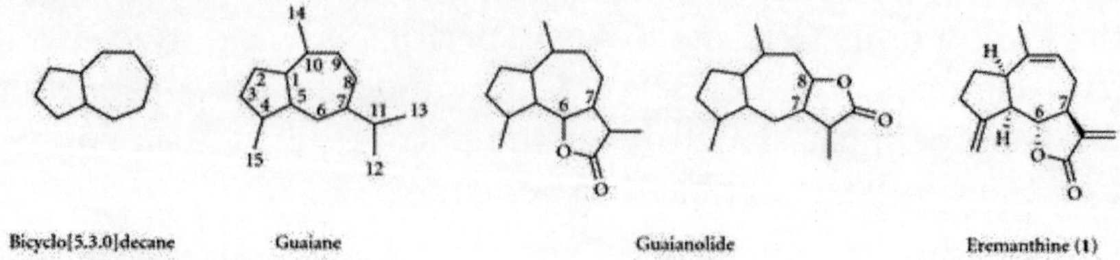

Figure 4: Bicyclo[5.3.0]decane, guaiane, guaianolide and eremanthine (1).

The Biogenesis of Sesquiterpene Lactones

Guaianolides are biogenetically derived from the farnesyl pyrophosphate (37) that passed by cyclization to generate the cyclodecadiene 38. This intermediate undergoes enzymatic oxydations to yield the germacranolide (39), a presumed precursor of the skeleton of guaianolides (Scheme 9) [13]. A detailed study on the biogenesis of sesquiterpene lactones was reported by Fischer et al. [14].

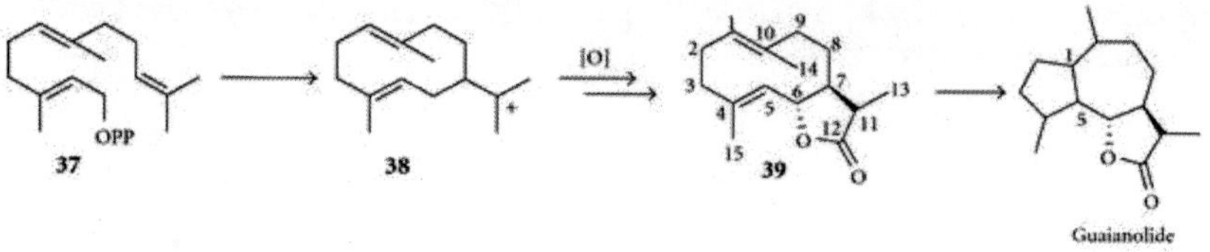

Scheme 9: Biosynthesis of guaianolides.

The skeletons of sesquiterpene lactones, with lactonic fusion at C-6 and C-7 positions, derived from germacranolides and the respective references were described in the works of Ferreira's M.S. [15] and Fantini's Ph.D. [16]. In Scheme 10 are outlined such biogenetical relationships reported in those works.

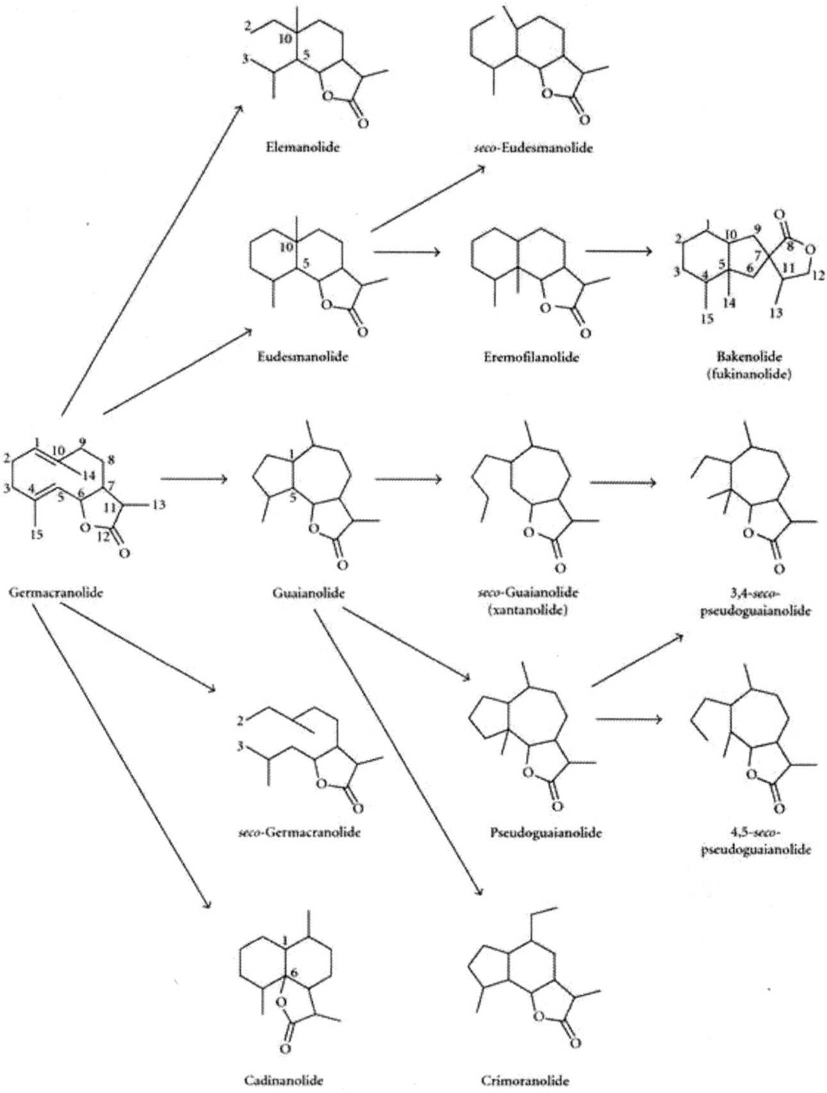

Scheme 10: Biogenesis of sesquiterpene lactones.

The biogenetical hypothesis of the guaianolides formation and their transformation into pseudoguaianolides reported by Fischer et al. [14] is outlined in Schemes 11–14. It was formulated that the skeleton of guaianolides is formed by the cyclization of a germacranolide-4,5-epoxide (40) in a chair-like transition state (Scheme 11). The intermediate cis-fused guaianolide cation (41) undergoes reaction with water to furnish the guaianolide 42 with cis fusion between the five- and seven-membered rings that is, with rare exceptions, the stereochemistry found in the most guaianolides.

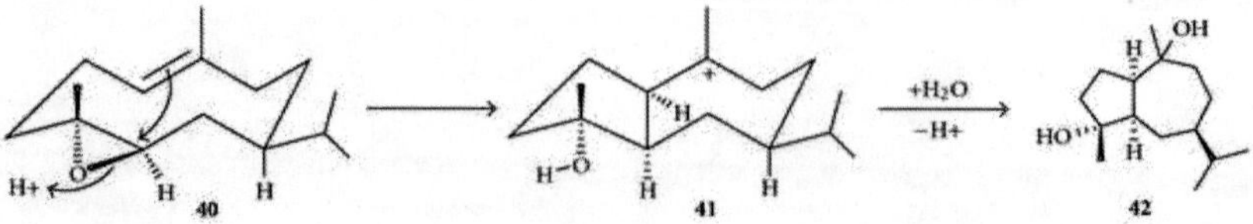

Scheme 11: Biogenesis of cis-fused guaianolides.

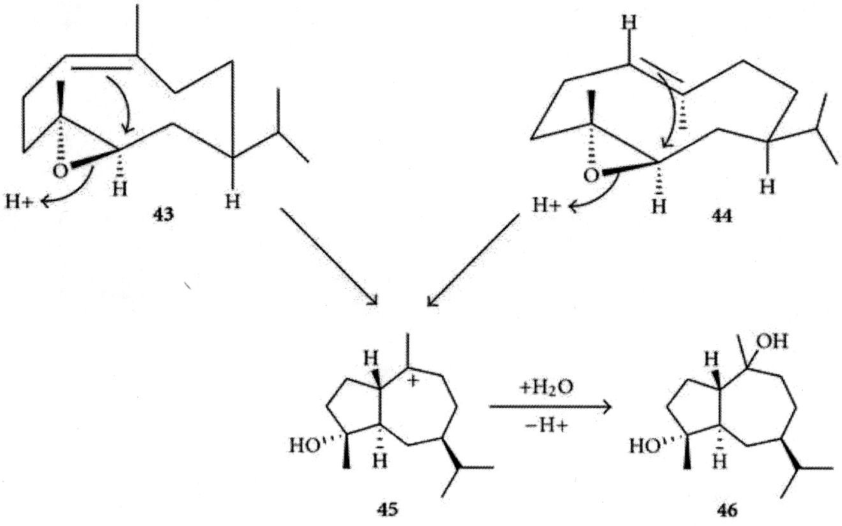

Scheme 12: Biogenesis of trans-fused guaianolides.

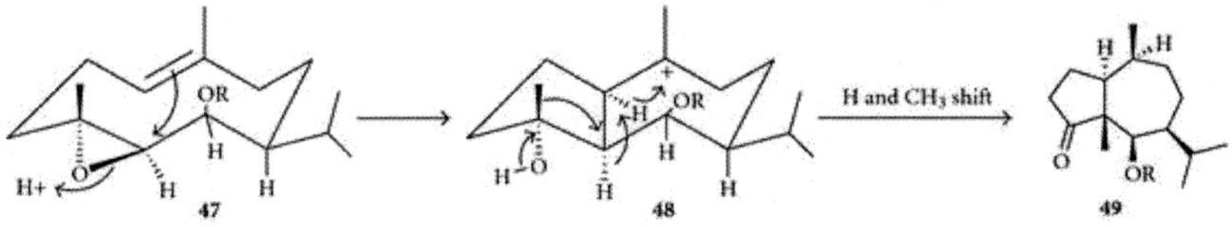

Scheme 13: Biogenesis of ambrosanolides.

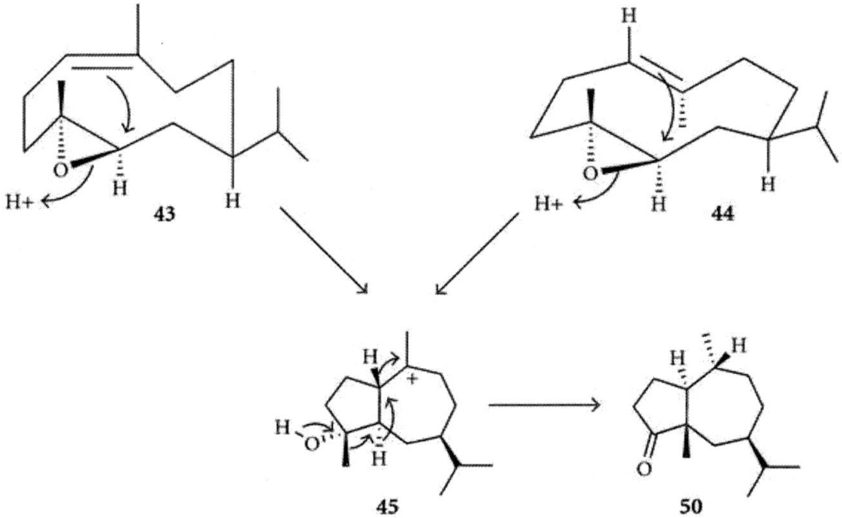

Scheme 14: Biogenesis of helenanolides.

The biogenesis for the few trans-fused guaianolides found as natural products was proposed to proceed via either the melampolide-4,5-epoxide (43) or germacranolide-4,5-epoxide (44) pathways (Scheme 12). The cyclization of these compounds should furnish the trans-fused guaianolide cation (45) which after reaction with water would result in the formation of the skeleton 46 with trans fusion between the five- and seven-membered rings.

The biogenesis for the pseudoguaianolides-denominated ambrosanolides, with C-10β methyl group and lactonic ring at C-6 and C-7 positions with the C-6 oxygen at β orientation, is outlined in Scheme 13. The cyclization of the germacranolide-4,5-epoxide (47) would give the cation guaianolide (48) which upon double hydride and methyl shift, as indicated by the arrows, gives the ambrosanolide skeleton (49). As all ambrosanolides possess cis lactonic fusion at C-6 and C-7 positions, it was proposed that a C-6-β-oxygen would be assisting the rearrangement step of the carbocation 48 to compound 49. The intramolecular frontside stabilization of the cationic center at C-10 by the C-6 oxygen at β position would allow a C-1 to C-10 hydride shift to occur before the competing step of elimination or nucleophilic attack at C-10 to form a guaianolide.

The biogenesis of the pseudoguaianolides-denominated helenanolides, with C-10α methyl group and lactonic ring at C-7 and C-8 positions with the C-8 oxygen oriented either α or β, is outlined in Scheme14. Acid-induced cyclization of the melampolide-4,5-epoxide (43) would give the cation 45 from which by the indicated shifts, the skeleton of the helenanolides (50) would be formed. An alternative route to the helenanolide intermediate (45) could involve a germacranolide 4,5-epoxide precursor (44) which would provide the same cyclised skeleton (50) which is formed via the melampolide route.

In 1981 Fischer et al. [17] published an article on the biomimetic transformation of guaianolide into pseudoguaianolide. The conditions studied in that biogenetic type in vitro conversion are outlined in Scheme 15. The reaction of epoxide 51 with $BF_3{\cdot}OEt_2$ furnished the compounds 52 and 53 and not the desired product 54 with pseudoguaianolide skeleton. It was proposed that the minor product (52) resulted from the acid-catalysed opening of the ring epoxide of 51 and the major eudesmanolide 53 from the skeleton rearrangement of that substrate catalysed by $BF_3{\cdot}OEt_2$. The nonformation of substance 54starting from the epoxide 51 reinforces the hypothesis that, to occur the rearrangements indicated by the arrows in Scheme 15, there is necessary a precursor guaianolide with a β-oxygenated function at C-6 position.

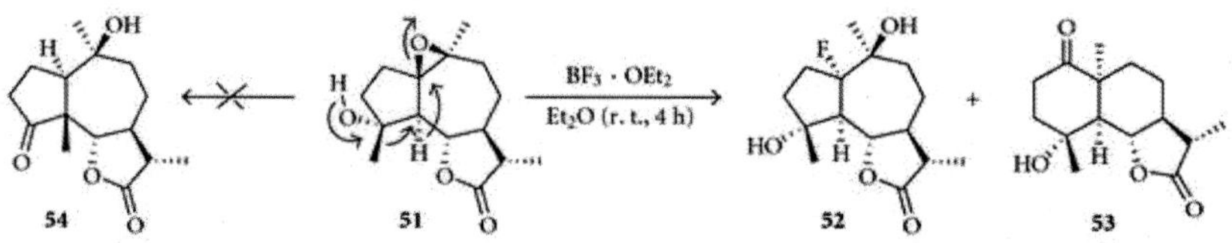

Scheme 15: Attempt of biogenetic in vitro conversion of the guaianolide 51 into pseudoguaianolide 54.

Other articles related to biomimetic transformation of guaianolide into pseudoguaianolide were reported in literature [18–20]. The study on the biomimetic transformation of eremanthine is going to be discussed in this text at the section on chemical transformations of this natural product.

Stereochemistry of Bicycle[5.3.0]Deane

The bicyclo[5.3.0]decane (hydroazulene system) is a nucleus present in a great variety of natural products such as the guaiane and pseudoguaiane sesquiterpenoids [21] and the guaiane [22–51], ingenane [52–67], daphnane [68–86], and asebotoxin [87–95] diterpenoids (Figure 5). Due to the chemical, biogenetical, and pharmacological interests that these classes of substances present, the literature reports a vast number of published works, with prominence for the reviews on sesquiterpenoids [21] and diterpenoids [96–101]. As eremanthine (1) belongs to the class of guaiane sesquiterpenoids (guaianolides), soon afterwards some considerations are going to be done on the stereochemistry of hydroazulene system present in its molecular structure.

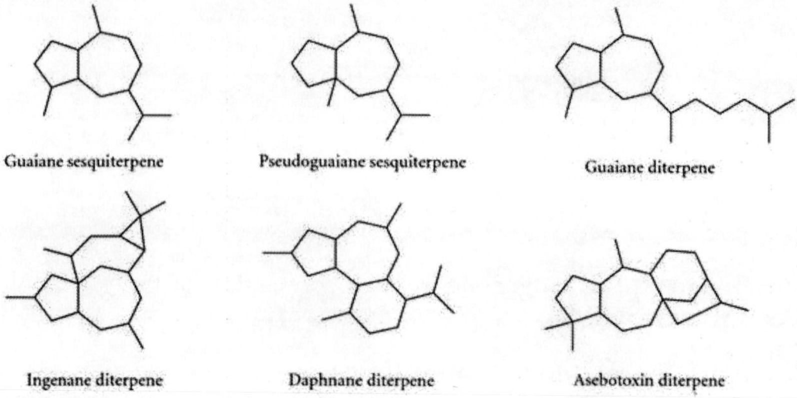

Figure 5: Hydrocarbon skeletons of guaiane and pseudoguaiane sesquiterpenes and guaiane, ingenane, daphnane and asebotoxin diterpenes.

The hydroazulene system is composed by a five-membered ring (rigid system) fused with a seven-membered ring (flexible system). Due to the flexibility of the seven-membered ring at the hydroazulene system, the study of conformational analysis of that system became necessary for a better understanding of stereoelectronic course of the reactions occurred with the classes of substances that contain this nucleus in its hydrocarbon skeleton. The conformational analysis of hydroazulenes is also important for the interpretation of spectral data from nuclear magnetic resonance (NMR) of the substances that possess this system.

The study of conformational analysis applied to cyclic systems, including the hydroazulene, was initiated in the decade of 1960 with Hendrickson's works [102–109]. Subsequent studies, in that area, were published in the decade of 1980 by Clercq [110–115]. Those works were focused on the conformational analysis of the seven-membered ring that commands the geometry of the hydroazulene system. According to the results of those published studies, the seven-membered ring can present itself in the basic conformations shown in Figures 6-7.

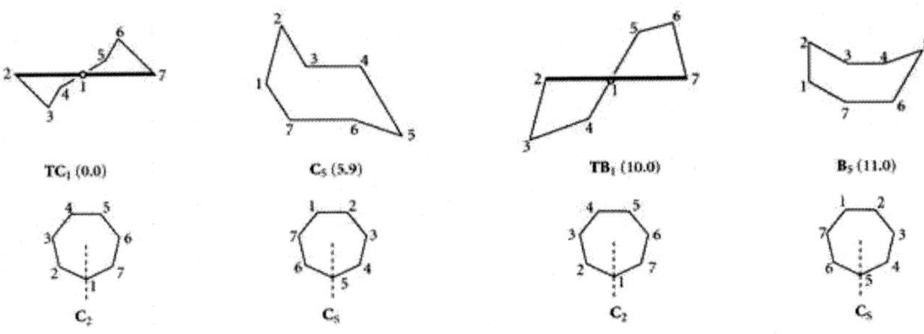

Figure 6: Conformational diagrams of the basic cycloheptane forms.

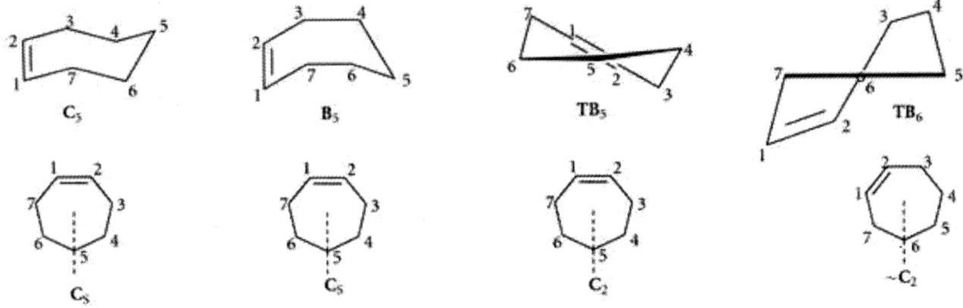

Figure 7: Conformational diagrams of the basic cycloheptene forms.

In Figure 6 the subscript used to define a particular form (twist-chair TC, chair C, twist-boat TB, and boat B) indicates the atom sectioned by the symmetry element and the strain energies are given in parenthesis in kilojoules per mol. The same annotation was used in Figure 7, with the symbol ~C_2indicating a pseudo-C_2 axis of symmetry.

According to Clercq [110] the chair and boat conformations of cycloheptane are flexible and undergo pseudorotation. The twist forms (TC and TB) with C_2-axis of symmetry are generally more stable than the chair (C) and boat (B) forms with a C_S plan of symmetry. In the case of cycloheptene, the chair form is generally the most stable conformation and the order of relative stability of conformers is the following: C > TB(C_2) > TB (~C_2) > B. The method of de Clercq

[110, 111], for systematic conformational analysis of hydroazulenes, was used in synthetic studies to explain the stereoselectivity of the reactions of sesquiterpene lactones that possess the hydroazulene skeleton in its molecular structure [115].

Nowadays, with the technological progresses, several research groups are using computation programs that were developed to aid in the visualization of the three-dimensional chemical structures of the molecules and to calculate their physical properties. It is possible to perform the conformational analyses of the studied substances, starting from the three-dimensional structures drawn by computation programs. Articles were published on the several modern techniques used to determine the conformations of the hydroazulene system from sesquiterpene lactones, including the analyses of X-ray diffraction, calculations of quantum mechanics, and molecular mechanics in combination with the NMR data [116–123].

BIOLOGICAL ACTIVITY OF EREMANTHINE

Researchers verified that animals of laboratory (mice) were protected against infections caused by cercariae of Schistosoma mansoni, when the essential oils from Eremanthus elaeagnus andVanillosmopsis erythropappa were applied on the skin of those guinea-pigs. That protective action was mainly attributed to the substance eremanthine (1), an abundant component found in those oils [1, 2]. The biological activity of eremanthine was attributed to the presence of an α-methylene-γ-lactone in its molecular structure [1]. That functional group was indicated as the main responsible for the biological activity in sesquiterpene lactones, due to its ability to react with the biological nucleophiles in a conjugate fashion [124–130]. The pharmacological results from the protective action of Eremanthus elaeagnus and Vanillosmopsis erythropappa oils were reported by Baker et al. [2].

SYNTHESIS OF EREMANTHINE

The synthesis of sesquiterpene lactones with their varied skeletons is composed by two crucial stages, which are the construction of the basic skeleton and the formation of the α-methylene-γ-lactone. In the following subitems some considerations are done about the construction of the hydroazulene skeleton characteristic of the guaiane [131] and pseudoguaiane sesquiterpenoids, the formation of the α-methylene-γ-lactone, besides the strategy and stages of the synthesis of eremanthine (1).

Basic Strategies to Construction of the Hydroazule Skeleton

The construction of the hydroazulene skeleton, reported by Heathcock et al. [132], involves four basic strategies which are shown in Scheme 16 with the respective substances that were synthesized.

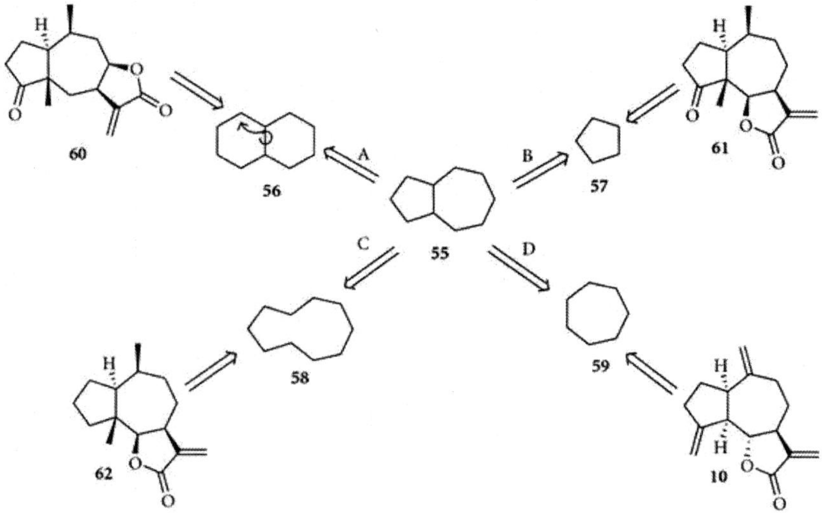

Scheme 16: Basic strategies to construction of the hydroazulene skeleton.

The synthesis of confertin (60) [133] was accomplished by the strategy A, in which the construction of the hydroazulene skeleton (55) was planned starting from a hydronaphthalene precursor (56) through a rearrangement reaction. The damsim (61) [134] could be obtained by the strategy B, whose construction of the hydroazulene skeleton was planned starting from a cyclopentane precursor (57) on which the cycloheptane ring was inserted. In the synthesis of deoxydamsim (62) [135] the construction of the hydroazulene skeleton was planned by the strategy C, using a precursor derived from the cyclodecane (58), through a transannular rearrangement reaction. Finally in strategy D, used to the synthesis of dehydrocostus lactone (10) [136], the construction of the hydroazulene skeleton was planned starting from a cycloheptane derivative (59) on which the cyclopentane ring was inserted. The most recent literature on the synthesis of sesquiterpenoids reports the preparation of a chiron derived from cycloheptenone as a building block to the synthesis of guaiane sesquiterpene natural products [137] as well as the synthetic approaches to bicyclo[5.3.0]decane sesquiterpenes [138].

Synthesis of the -Methylene- -Lactone Moiety

The α-methylene-γ-lactone is an important structural subunit found in several natural products with relevant biological activity. Due to the importance of this functional group, several researchers dedicated to the study of this structural moiety, with prominence to the various synthetic methods reported in literature [139–143].

Strategy to the Synthesis of Eremanthine

The enantioselective synthesis of eremanthine (1), developed by Yuuya et al. [12], employs the chiron approach strategy with the use of a chiral natural product as starting material on which chemical transformations are performed to reach the synthesis of the target molecule. There was used the strategyA shown in Scheme 16 to the construction of the hydroazulene skeleton of eremanthine (1), whose retrosynthetic analysis is presented in Scheme 17. In this strategy, the compound 1 was planned to be synthesized via the cationic intermediate 63, that could be generated from the mesylate 64 through solvolytic rearrangement. The mesylate 64 could be obtained from α-santonin (65).

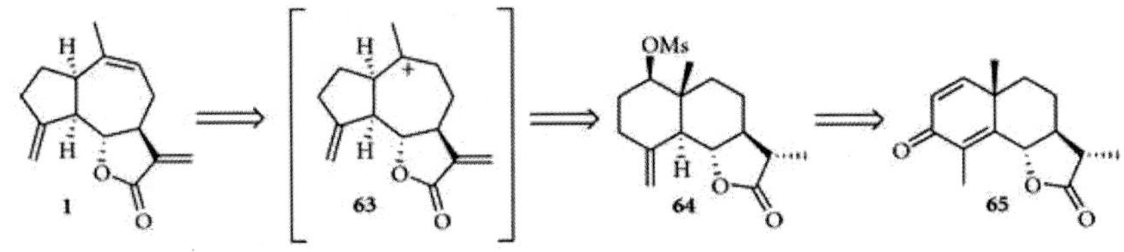

Scheme 17: Retroanalysis to the synthesis of eremanthine (1).

STAGES OF THE SYNTHESIS OF EREMANTHINE

The steps that were employed by Yuuya et al. [12] to obtain eremanthine (1) starting from α-santonin (65), by the strategy outlined in Scheme 17, are depicted in Scheme 18. The starting material used for the synthesis of key intermediate 64 was the α, β-unsaturated ketone 66, which was prepared from α-santonin (65) in six steps with an overall yield of 59% (Scheme 18). Ketalization of 66 and subsequent isomerization of double bond at C-2 by heating in ethylene glycol and TsOH gave the ketal 67 as the main product of that reaction. The acidic hydrolysis of compound 67 gave the ketone 68, which was submitted to subsequent step of dehydrogenation with DDQ and TsOH to furnish the exo-dienone 69 as the major product. Treatment of this compound with zinc amalgam in refluxing AcOH gave γ, δ-unsaturated ketone 70. Selective reduction of the C-1 carbonyl group of 70 with $LiAl(t\text{-}BuO)_3H$ furnished the desired β-alcohol 71 as the major product of that reaction. Mesylation of 71 with MsCl and pyridine gave the mesylate 64. Solvolytic rearrangement of 64 with a solution of KOAc in AcOH provided a mixture of tetra-, tri-, and disubstituted olefins 72, 21, and 73 in a respective proportion of (2:1:2.4). After separation of crude product from the reaction by preparative HPLC, a fraction with a mixture (1:1.6:4) of 72, 21, and 73 was submitted to the following step of phenylselenylation with LDA and diphenyl diselenide to afford epimeric mixtures at C-11 of phenylseleno lactones 74 and 75 along with the lactone 76. After separation by HPLC, the epimeric mixture of 75 at C-11 was submitted to oxidative elimination with H_2O_2 to furnish eremanthine (1) and the corresponding endo-unsaturated γ-lactone 77.

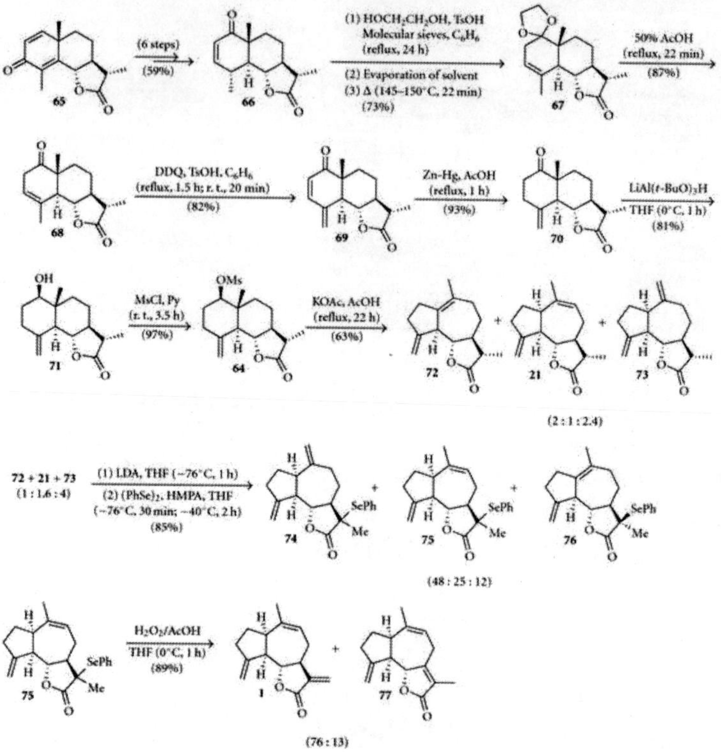

Scheme 18: Synthesis of eremanthine (1).

CHEMICAL TRANSFORMATIONS OF EREMANTHINE

As eremanthine (1) was abundantly isolated from the oil of Eremanthus elaeagnus and Vanillosmopsis erythropappa [1, 2], there was initiated a program of chemical transformations of 1 aiming at the syntheses of other biologically active derivatives as well as the preparation of less abundant naturally occurring lactones. As a result of this research program, several works of M.S. and Ph.D. degrees were accomplished using eremanthine (1) as object of study [15, 16, 144–150]. In the following subitems are described the syntheses of several eremanthine derivatives published in the period of 1972 to the current days.

The First Chemical Transformations of Eremanthine

The first chemical transformations performed with eremanthine (1) were published in 1972 by Vichnewski and Gilbert [1] and are depicted in Scheme 19.

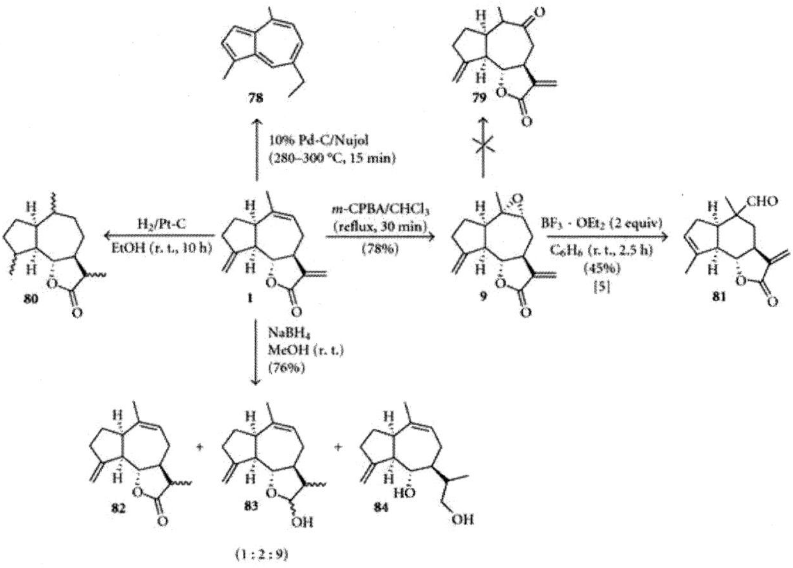

Scheme 19: The first chemical transformations of eremanthine (1).

Dehydrogenation of eremanthine (1) with palladium by heating in nujol gave chamazulene (78), and the selective epoxidation of 1 yielded the epoxide 9. Treatment of compound 9 with excess of BF_3-etherate furnished a product that was initially characterized as the ketone 79 [1]. Reinvestigation of this reaction by Garcia et al. [5] showed this structural assignment to be incorrect and the structure of the aldehyde81, resultant from a rearrangement, could be demonstrated by physical methods. Catalytic hydrogenation of eremanthine (1) resulted in the generation of a mixture of the isomers 80. Reduction of 1 with $NaBH_4$ furnished a mixture of 3 products characterized as the compounds 82, 83, and 84.

Chemical Transformations of Eremanthine for the Synthesis of Other Natural Guaianolides

In 1977 Maçaira et al. [151] published an article on the selective chemical modifications performed with eremanthine, aiming at the syntheses of other guaianolides of natural occurrence. As result of the chemical modifications performed with the substrate 1, there were obtained the synthetic derivatives depicted in Scheme 20 besides the natural product dehydrocostus lactone (10), a guaianolide initially isolated from Saussurea lappa [152].

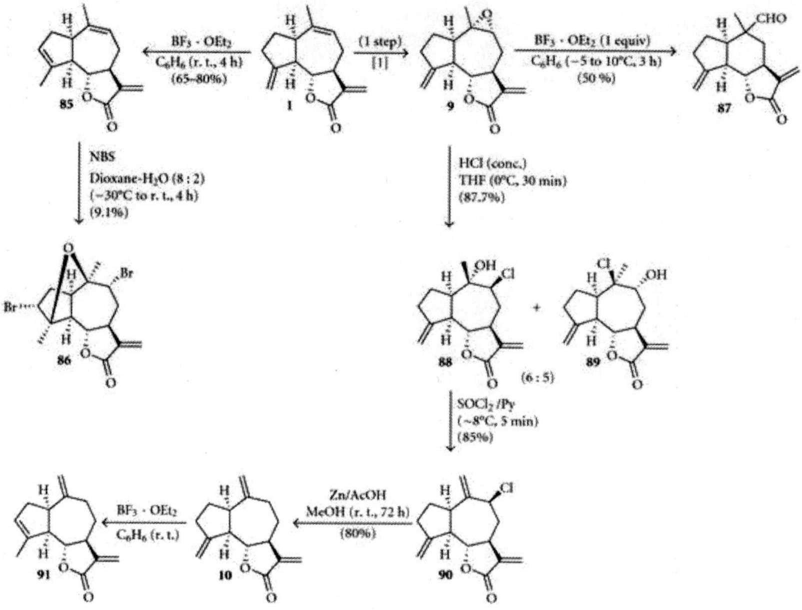

Scheme 20: Synthesis of dehydrocostus lactone (10).

The isomerization of exocyclic double bond at five-membered ring of eremanthine (1) was achieved treating this compound with excess of $BF_3 \cdot OEt_2$. In these conditions isoeremanthine (85) could be obtained in yields ranging from 65 to 80%, according to scale of the substrate, quantity of $BF_3 \cdot OEt_2$, and reaction time. To ensure that the isomerization of 1 to 85 did not result in change of the configuration at

C-1 and C-5, isoeremanthine (85) was treated with NBS in a mixture of dioxane-H_2O. The formation of the dibromoether 86 did confirm the cis fusion at the five- and seven-membered rings of isoeremanthine. The aldehyde 87 could be obtained selectively, without isomerization of exocyclic double bond at five-membered ring, by treating epoxide 9 with equimolar amount of $BF_3 \cdot OEt_2$. When epoxide 9 was treated with concentrated HCl in THF, the chlorohydrins 88 and 89 were isolated in a respective proportion of (6:5). Treatment of compound 88 with a mixture of thionyl chloride and pyridine, at low temperature, gave allylic chloride 90. Dechlorination of 90 in acid medium with zinc and MeOH furnished a compound identified as dehydrocostus lactone (10). The isomerization of exocyclic double bond at five-membered ring of 10 was achieved through the reaction of this compound with $BF_3 \cdot OEt_2$, at the same conditions used to convert eremanthine (1) into isoeremanthine (85). In those conditions isodehydrocostus lactone (91) was obtained

SYNTHESES OF EREGOYAZIN AND EREGOYAZIDIN

In order to confirm the structures of eregoyazin (19) and eregoyazidin (95), two guaianolides isolated from Eremanthus goyazensis by Vichnewski et al. [153], the syntheses of these compounds were accomplished according to the sequence of reactions depicted in Scheme 21 [153, 154]. The reaction of isoeremanthine (85) with one equivalent of bromine at −70°C gave a mixture from which the compound 92 could be isolated. Peracid oxidation of 92 from the less hindered face afforded mainly the -epoxide 93. Exposure of this epoxide to methanolic zinc resulted in debromination to furnish compound 94. Treatment of 94 with $BF_3 \cdot OEt_2$ afforded a substance identical in all aspects with eregoyazin (19). Further reduction of 19 with zinc in hot glacial acetic acid yielded eregoyazidin (95).

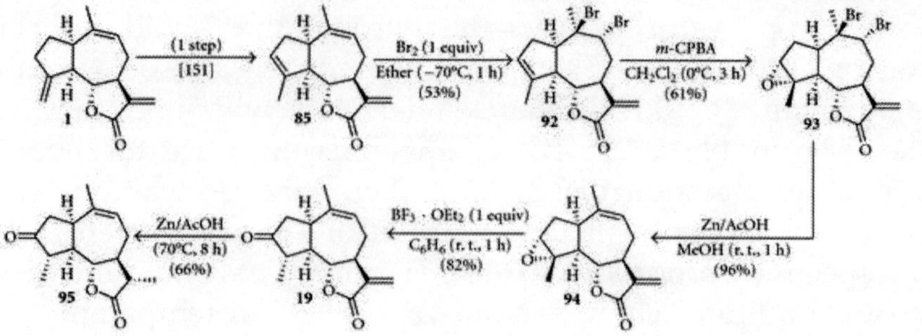

Scheme 21: Syntheses of eregoyazin (19) and eregoyazidin (95).

Synthesis of (-)-Estafiatin

The stereoselective synthesis of the natural product (-)-estafiatin (96), a sesquiterpene lactone isolated from Artemisia mexicana [155], was developed by Rabi et al. [154, 156]. The sequences of reactions used for the synthesis of this natural product are depicted in Schemes 22-23. Initially the triene 91, obtained from eremanthine (1) for the sequence of reactions outlined in Scheme 20, was submitted to epoxidation with m-chloroperbenzoic acid to furnish a mixture characterized as the epoxides 96 and 97(8 : 2,^{1}H NMR). The major isomer showed identical properties with the naturally occurring (-)-estafiatin (96) (Scheme 22) [156].

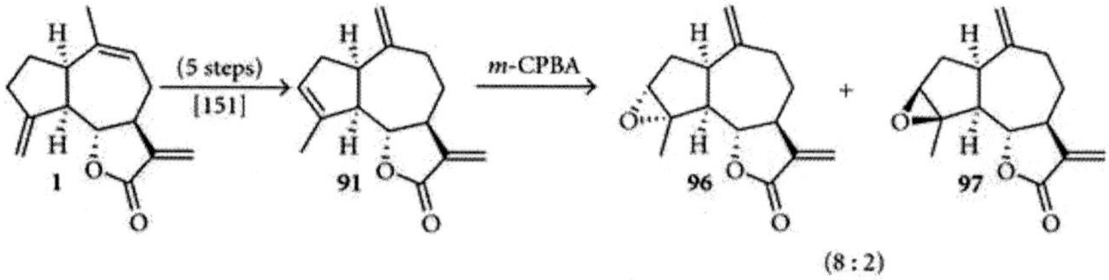

Scheme 22: Synthesis of (-)-estafiatin (96) in mixture with the isomer 97.

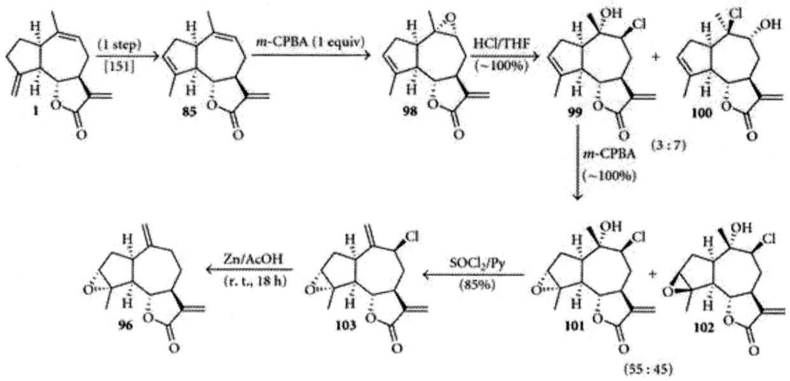

Scheme 23: Synthesis of (-)-estafiatin (96).

The compound (-)-estafiatin (96) was also obtained by the sequence of reactions depicted in Scheme 23[154, 156]. The reaction of isoeremanthine (85) with equimolar amount of m-chloroperbenzoic acid resulted in almost exclusive formation of epoxide 98 which upon reaction with HCl in THF gave a mixture of chlorohydrins 99 and 100 (3:7). Epoxidation of 99 led to a nearly equimolar mixture of epoxides 101 and 102 (55:45). Dehydration of 101 with a mixture of $SOCl_2$ and pyridine furnished 103. Dechlorination of this compound with zinc yielded a substance identified as (-)-estafiatin (96).

Reaction of Eremanthine and Isoeremanthine with Bromine

The search for a selective method to protect the most nucleophilic 9,10-double bond of eremanthine (1) led to the investigation of eletrophilic addition of bromine to this compound and also to isoeremanthine (85). The conditions that were employed by Garcia et al. [157] to study this addition reaction are described in Schemes 24–27. When eremanthine (1) was allowed to react with one equivalent of bromine at kinetic conditions, an equimolar mixture of tetrabromide 104 and 1 could be detected by ^{1}H NMR spectroscopy (Scheme 24).

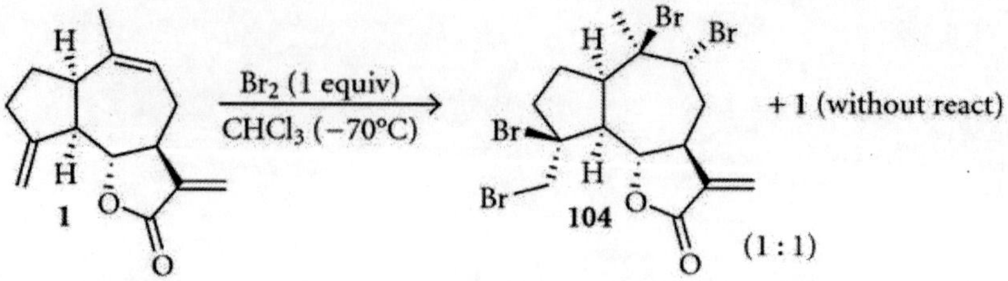

Scheme 24: Reaction of eremanthine (1) with equimolar amount of bromine.

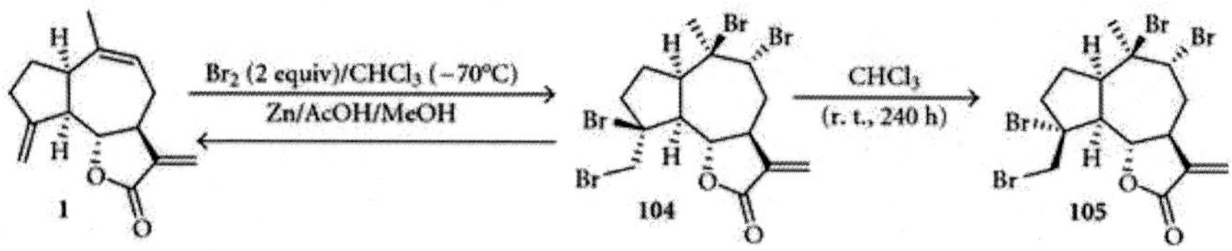

Scheme 25: Reaction of eremanthine (1) with excess of bromine.

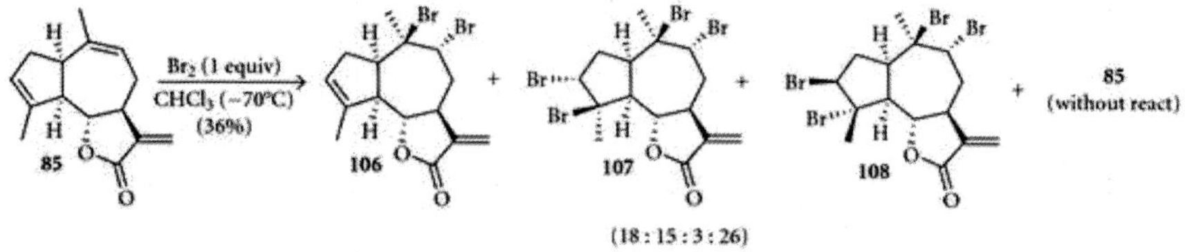

Scheme 26: Reaction of isoeremanthine (85) with bromine in chloroform.

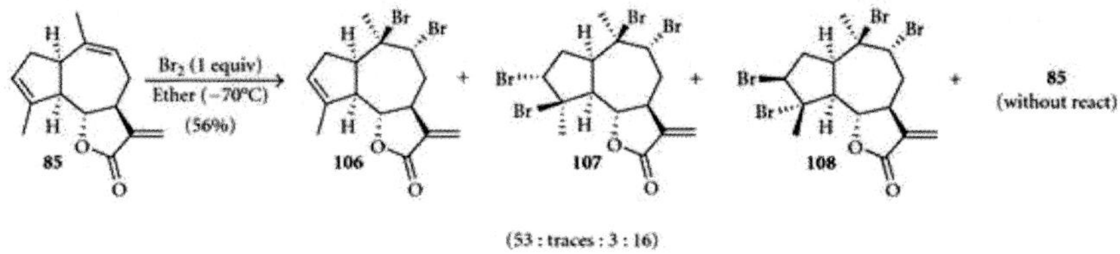

Scheme 27: Reaction of isoeremanthine (85) with bromine in ether.

Addition of two equivalents of bromine to eremanthine (1) resulted in the exclusive formation of tetrabromide 104. Reaction of this compound with zinc in methanol regenerated eremanthine (1) in quantitative yield. When a solution of compound 104 in $CHCl_3$ was left at room temperature during 240 hours, the tetrabromide 105 was obtained (Scheme 25). The inversion of configuration at C-4 in this reaction was attributed to stereoelectronic repulsion between the β-oriented bromine atoms at C-4 and C-10 at the substrate 104.

A complex mixture of products was obtained when isoeremanthine (85) was submitted to reaction with one equivalent of bromine at kinetic conditions (Schemes 26-27). A detailed discussion on the probable mechanisms for these reactions of bromine addition is presented in the article published in 1980 [157].

Biomimetic Transformations of Eremanthine

Chemical transformations of eremanthine (1) were studied by Rodrigues [148] aiming to obtain subsidies for the biogenetic hypothesis of pseudoguaianolides formation. The sequence of reactions developed in that study is depicted in Scheme 28. Epoxidation of eremanthine (1) gave the diepoxide109 which was submitted to reaction with KI and acetic acid to furnish a mixture of epoxide 110 and allylic alcohol 111. Catalytic hydrogenation of these compounds gave, respectively, the epoxide 112 and diol 113. Treatment of compound 112 with $HClO_4$ resulted in the formation of diol 113 instead of the pseudoguaianolide 114. Reaction of diol 113 with the acids BF_3OEt_2 or p-TsOH furnished a mixture of dienes 115 and 116 instead of the desired compound 117 with pseudoguaianolide skeleton. The nonformation of the substances

114 and 117 with pseudoguaianolide skeleton reinforces, once again, the hypothesis that the formation of those compounds with lactonic fusion at the C-6 and C-7 positions need a β-oxygenated function at the C-6 position of the precursor guaianolide for the occurrence of the rearrangements preconized by the biogenesis hypothesis of the ambrosanolides (Scheme 13).

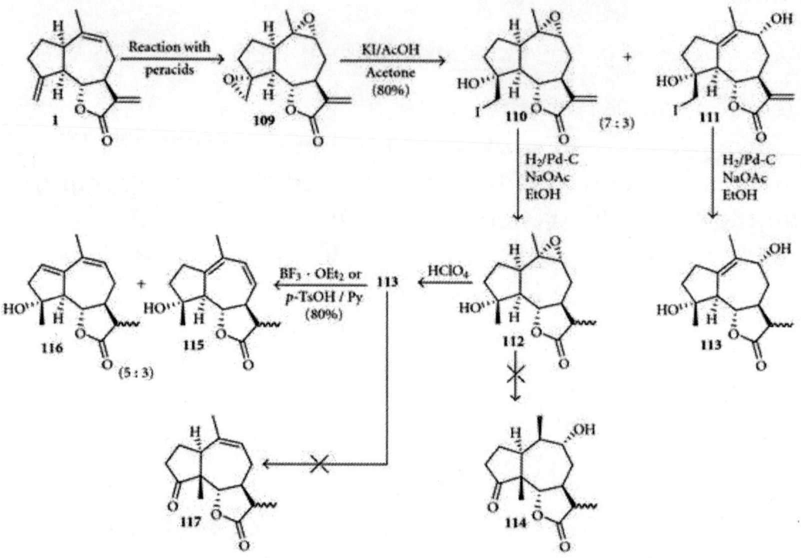

Scheme 28: Biomimetic transformations of eremanthine (1).

Study of 13C NMR Spectroscopy on Eremanthine Derivatives

In 1981 da Silva et al. [158] published an article with the chemical shifts assigned to the carbons of naturally occurring guaianolides eremanthine (1), dehydrocostus lactone (10), eregoyazin (19), eregoyazidin (95), and other semisynthetic lactones derived from eremanthine (1). A detailed discussion correlating the data of ^{13}C NMR with the probable conformations at the seven-membered ring of the investigated guaianolides is presented in the article. The sequences of reactions employed to synthesize the eremanthine derivatives used in the experiments of ^{13}C NMR spectroscopy of that article are depicted in Scheme 29. The lactones 2, 9, 10, 19, 21, 85, 90, 94, and 95 were obtained by the sequences of reactions outlined in Schemes 1 and 19–21. The others were prepared from known substances by standard simple procedures. The compounds 118, 119, 120, and 121 were prepared by reaction of eremanthine (1) or isoeremanthine (85) with MeOH/Na_2CO_3 or Me_2NH/EtOAc. The reaction of eremanthine (1) with m-chloroperbenzoic acid in $CHCl_3$ at room temperature yielded a mixture of epoxides 9, 8, and 109. The compounds 98 and 122 were obtained by reaction of isoeremanthine (85) with m-chloroperbenzoic acid in $CHCl_3$ at low temperature in a respective proportion of (62:21). The bromohydrin 123 was prepared by reaction of dibromoether 2 with zinc in refluxing MeOH.

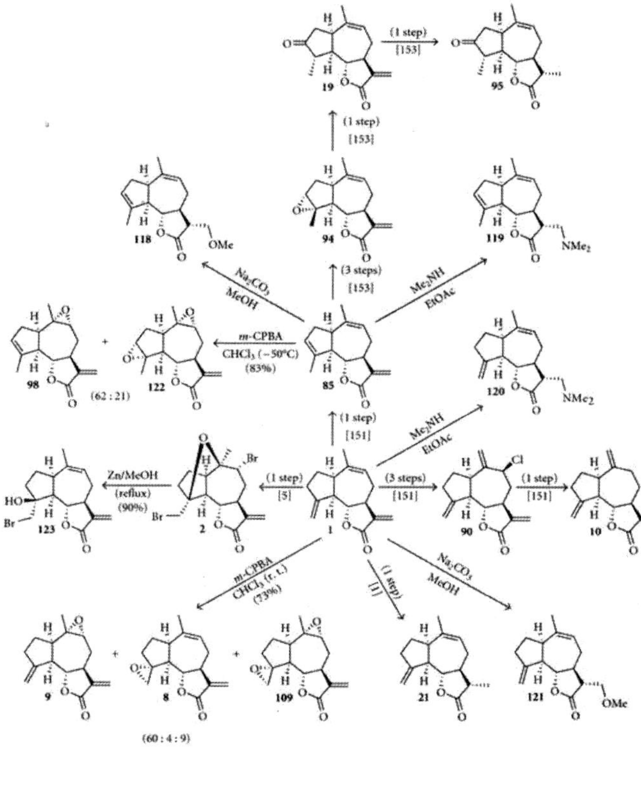

Scheme 29: Syntheses of eremanthine derivatives to the study of ^{13}C NMR spectroscopy.

Studies on the Chemical Reactivity of Epoxides Derived from Eremanthine

The syntheses of guaianolides precursors of the series $\Delta^{1,10}$ and $\Delta^{10,14}$ starting from epoxides derived from eremanthine (1) were studied by Ferreira [15] (Schemes 30-31). The epoxidation of eremanthine (1) with peracetic acid during 5 hours gave a mixture of epoxides 9, 8, and 109 in a respective proportion of (35:2:5) (Scheme 30). When this same reaction was performed during 96 h, the diepoxide 109 was obtained as a single product. Treatment of a diluted solution of diepoxide 109 in acetone with KI and AcOH furnished a mixture of iodohydrin epoxide 110 and iodohydrin allylic alcohol 111 in a respective proportion of (13:12). When diepoxide 109 was submitted to this same reaction in a concentrated solution of acetone, the iodohydrin 111 could be isolated as a single product. The iodohydrin 124 was obtained by treatment of monoepoxide 8 with KI and AcOH in acetone. The transformation of iodohydrin 111 into epoxide 125 was achieved by an intramolecular reaction of nucleophilic substitution by using a solution of Na_2CO_3 in ethanol. Treatment of 111 in THF with an aqueous solution of $HClO_4$ yielded the diene 126 and the acetylation of 111 with Ac_2O and pyridine gave allylic acetate 127. When the iodohydrin 111 was submitted to elimination reaction with zinc in refluxing MeOH, the desired triene 128 was obtained along with another product identified as the oxacycloguaiane methyl ester 129. Acetylation of allylic alcohol 128 with acetic anhydride and DMAP yielded the allylic acetate 130. The opening of epoxide 9 was studied using different conditions (Scheme31). The reaction of 9 in THF with an aqueous solution of $HClO_4$ yielded allylic alcohol 131 and the opening of epoxide 9 with aluminum oxide (neutral, acid or basic) in refluxing hexane furnished the desired product 131 accompanied by another compound identified as the aldehyde 87. Acetylation of allylic alcohol 131 with Ac_2O and pyridine gave the acetate 132. The protection of α-methylene-γ-lactone of 131 was achieved by reaction of this compound with methanol and Na_2CO_3 to afford the adduct 133. Attempts to deoxygenate the C-9 position of compounds 131 and 133 were performed with limited success. The reaction of allylic alcohol 131 with thionyl chloride and DMF furnished a product identified by attempt as allylic chloride 134. On the other hand, the reaction of 133 with triphenylphosphine and

CCl_4 gave a mixture of desired product 135 accompanied by another compound identified as the conjugate diene 136. Dechlorination of 135 with zinc and AcOH in refluxing MeOH furnished a mixture of two products with the same R_f of authentic samples of the methanol adducts of eremanthine (121) and dehydrocostus lactone (137).

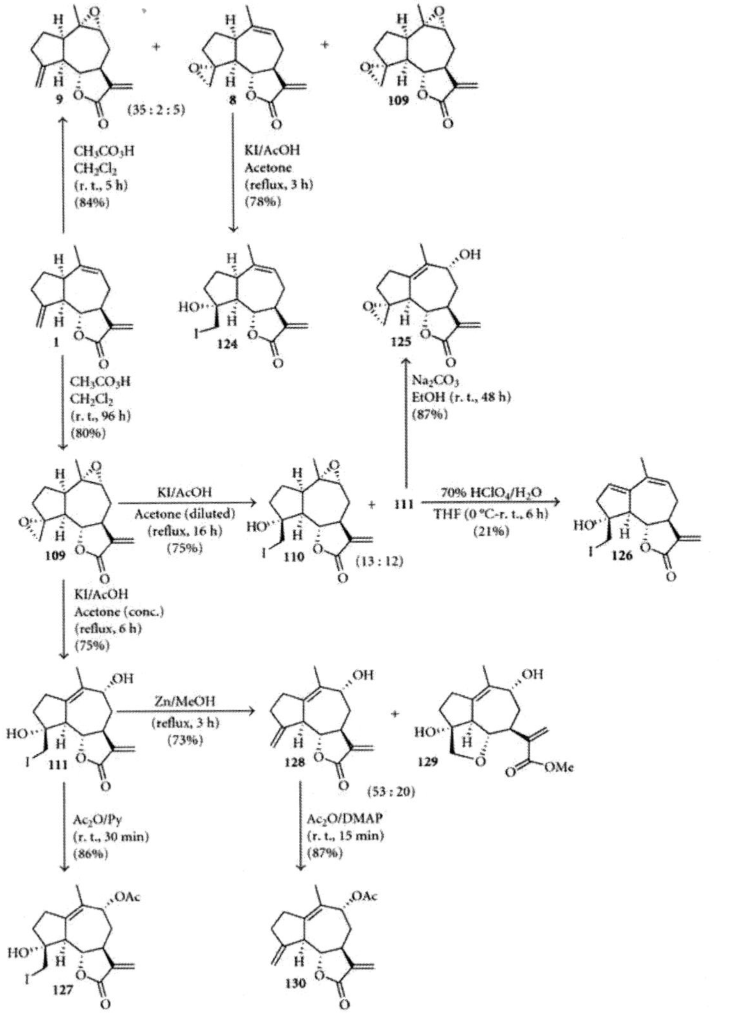

Scheme 30: Study on the chemical reactivity of epoxides derived from eremanthine (1).

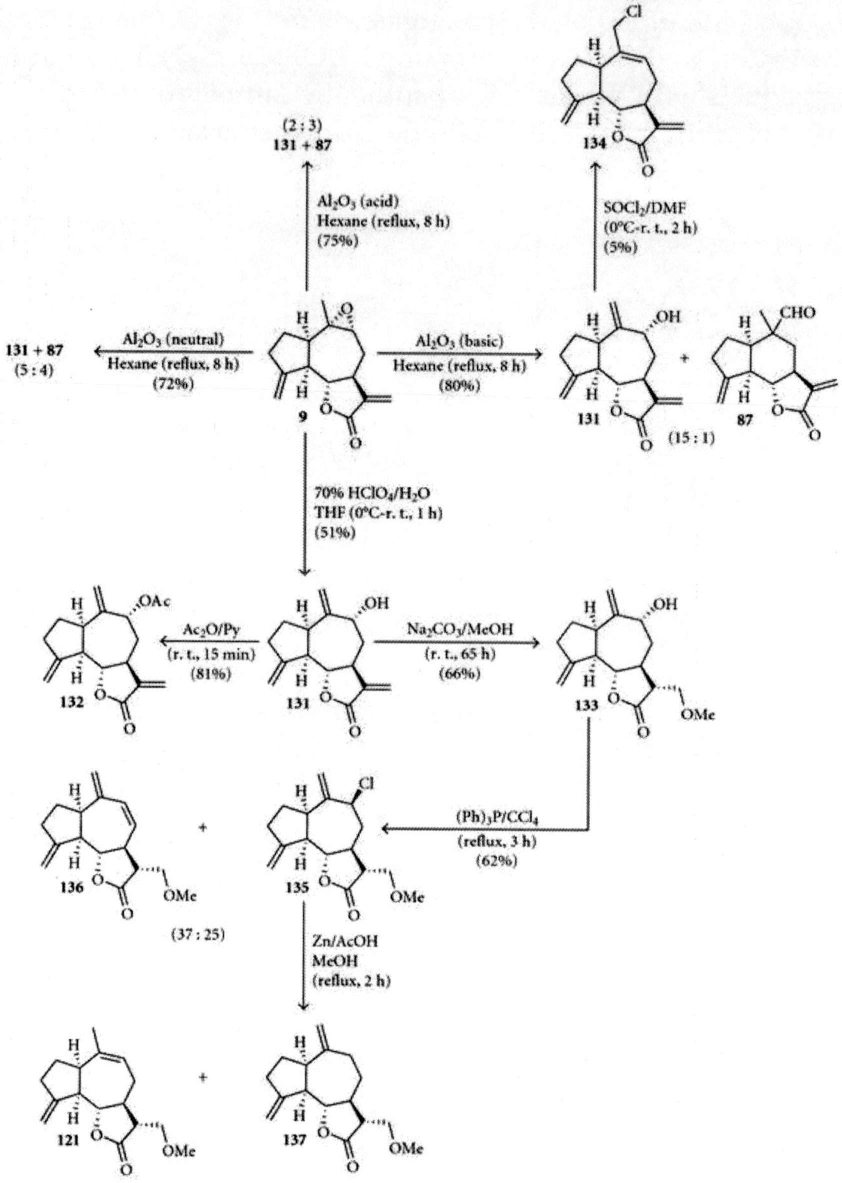

(2 : 3)
131 + 87

Al_2O_3 (acid)
Hexane (reflux, 8 h)
(75%)

Cl
H
H
134

$SOCl_2$/DMF
(0°C-r. t., 2 h)
(5%)

131 + 87
(5 : 4)

Al_2O_3 (neutral)
Hexane (reflux, 8 h)
(72%)

9

Al_2O_3 (basic)
Hexane (reflux, 8 h)
(80%)

OH
131
+
CHO
87
(15 : 1)

70% $HClO_4/H_2O$
THF (0°C-r. t., 1 h)
(51%)

OAc
132

Ac_2O/Py
(r. t., 15 min)
(81%)

OH
131

Na_2CO_3/MeOH
(r. t., 65 h)
(66%)

OH
133
OMe

136
OMe
+
Cl
135
OMe
(37 : 25)

$(Ph)_3P/CCl_4$
(reflux, 3 h)
(62%)

Zn/AcOH
MeOH
(reflux, 2 h)

121
OMe
+
137
OMe

Scheme 31: Reaction conditions studied to the opening of epoxide 9.

Study on the Chemical Reactivity of Eremanthine: The -Methylene- -Lactone Moiety

With the objective of extending the study on the chemical transformations of eremanthine (1) to the lactonic ring, Fantini [16] developed her Ph.D. thesis focusing the α-methylene-γ-lactone of that substance and of its synthetic derivatives. The syntheses of compounds used in that study are outlined in Schemes 32–35.

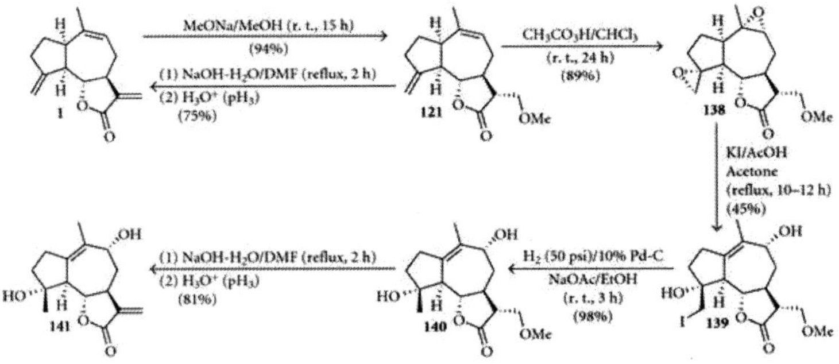

Scheme 32: Synthesis of the diol 141.

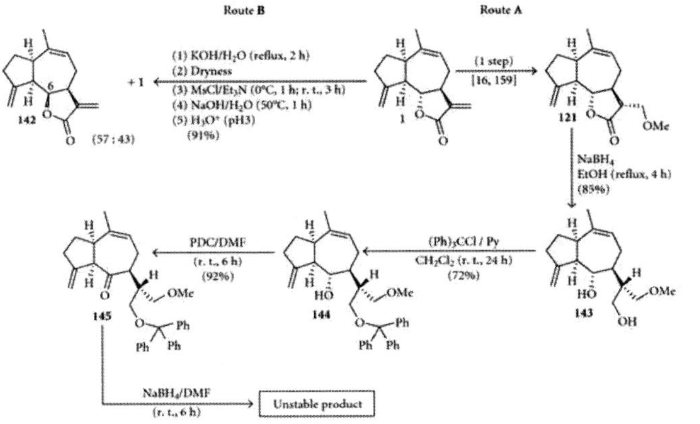

Scheme 33: Synthesis of 6-epi-eremanthine (142).

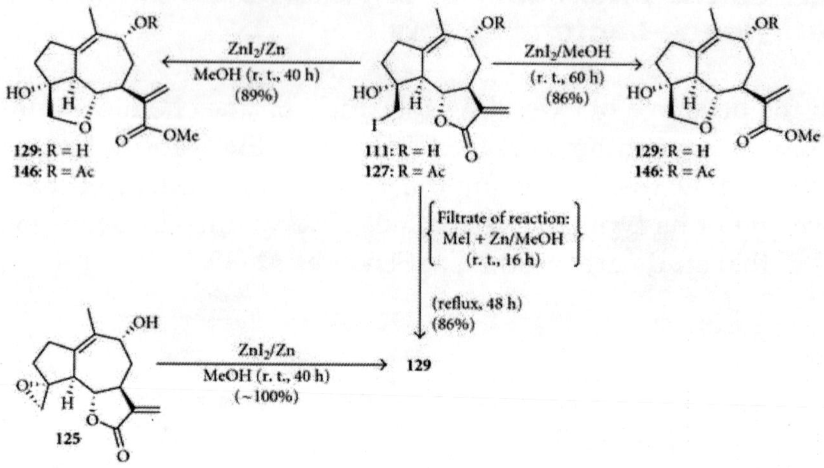

Scheme 34: Methanolysis of the iodohydrins 111 and 127 and epoxide 125.

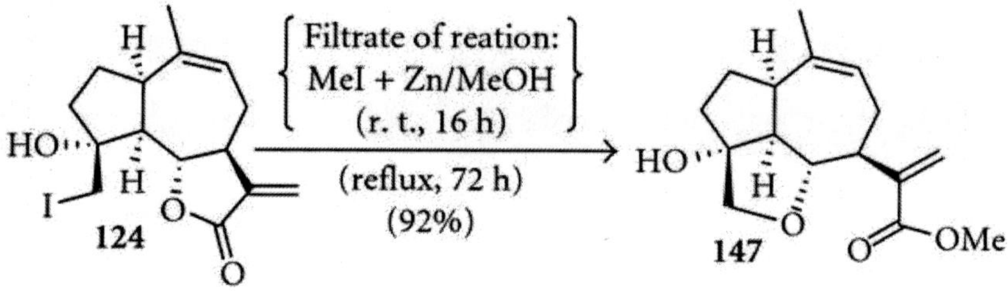

Scheme 35: Methanolysis of the iodohydrin 124.

The search for a protective group of α-methylene-γ-lactones resistant in certain reaction conditions, for example, catalytic hydrogenation, led to the synthesis of methanol adduct of eremanthine (121). It was verified that in basic conditions (aqueous NaOH, DMF, reflux), the -methylene--lactone could be regenerated in high yields [159]. The use of methoxyl as protective group for the α-methylene-γ-lactone of eremanthine (1) was accomplished with success during the synthesis of diol 141 (Scheme 32) [16]. The reaction of eremanthine (1) with a solution of MeONa in methanol gave the adduct 121. Treatment of this compound with a solution of peracetic acid in chloroform yielded diepoxide 138. The cleavage of oxiranic rings of 138 with KI and acetic acid in refluxing acetone furnished iodohydrin 139. Catalytic hydrogenation of 139 with hydrogen, Pd-C, and sodium acetate in EtOH yielded diol 140. This substance was submitted to subsequent step of treatment with an aqueous solution of NaOH in refluxing DMF to give, after aqueous acid work up, the -methylene--lactone 141 as the main product of this reaction.

The inversion of configuration at C-6 position of eremanthine (1), aiming at the synthesis of 6-epi-eremanthine (142), was studied by the two sequences of reactions depicted in Scheme 33 [16, 160]. In the route A, the inversion of configuration at C-6 position of 1 was planned by the method of oxidation-reduction of secondary hydroxy group in this position. The reduction of carboxy group of the lactonic ring at methanol adduct 121 with $NaBH_4$ in ethanol gave diol 143. In the next step, this compound was submitted to protection of primary hydroxyl with $(Ph)_3CCl$ and pyridine in CH_2Cl_2 to furnish compound 144. The oxidation of secondary hydroxyl at 144 was performed with pyridinium dichromate in DMF to afford the ketone 145. Treatment of 145 with sodium borohydride in DMF furnished an unstable product of difficult purification. With this unsatisfactory result, the synthesis of142 was studied by the route B. The inversion of configuration at C-6 of eremanthine (1) was achieved by displacement of intermediate mesylate generated in this position to afford a mixture of 1 and the unstable 6-epi-eremanthine (142) (43:57, ^1H NMR). The instability of 142 was attributed to steric effects at the hydroazule system.

After the synthesis of triene 128, obtained in mixture with oxacycloguaiane methyl ester 129 (Scheme30), there was initiated a study aiming at to optimize the formation of compound 129 [15, 16, 161, 162]. As a result of this study, the furanic derivatives 129, 146, and 147 were obtained (Schemes 34-35). There is a discussion evaluating the structures of reactive eremanthine derivatives (111, 124, 125, and 127) in comparison with other inert compounds obtained from eremanthine (1) as well as the reactive species responsible for the methanolysis of the lactonic ring in that reaction as reported in an article published in 1986 by Fantini et al. [163].

The reaction of iodohydrin 111 with zinc in refluxing methanol (Scheme 30) showed that furanic derivative 129 was formed after the triene 128. It was also verified that triene 128 did not generate the compound 129. Starting from these observations, it was presumed that the reactive species responsible for the formation of compound 129 was a by-product from the reaction of iodohydrin 111 with zinc (IZnOH) [16]. As an attempt to simulate the reactional system of iodohydrin 111 with zinc in methanol, there was developed a reagent that in fact converted directly the substrate 111 into oxacycloguaiane 129. This reagent was the filtrate of a mixture containing MeI, zinc, and methanol left in contact for 16 h at room temperature [15]. When iodohydrin 111 was submitted to reaction with this reagent under reflux during 48 h, the oxacycloguaiane 129 was obtained as a single product (Scheme 34). When iodohydrin124 was submitted to similar conditions, the compound 147 was obtained after 72 h of reaction (Scheme35) [15]. In the following stage this reaction was performed with ZnI_2. When iodohydrins 111, and 127were allowed to react with ZnI_2 in methanol at room temperature, the respective furanic derivatives 129and 146 were obtained (Scheme 34). It was also verified that the addition of zinc to reactional mixture accelerated the formation of these compounds. When the iodohydrins 111, 127 and epoxide 125 were allowed to react with a mixture of ZnI_2, zinc, and methanol at room temperature for 40 h, there was verified the total conversion of these substrates to respective products 129 and 146 (Scheme 34).

Synthesis of the Trienes 128 and 130

Besides the reactions of her Ph.D. thesis [16], described in the previous subitem (6.9), Fantini developed the synthesis of the trienes 128 and 130 starting from the respective iodohydrins 111 and 127 as a Researcher Professor of the Department of Chemistry at the Rural Federal University of Rio de Janeiro (UFRRJ). The optimized conditions of those reactions were reported in a congress abstract [164] and are depicted in Scheme 36.

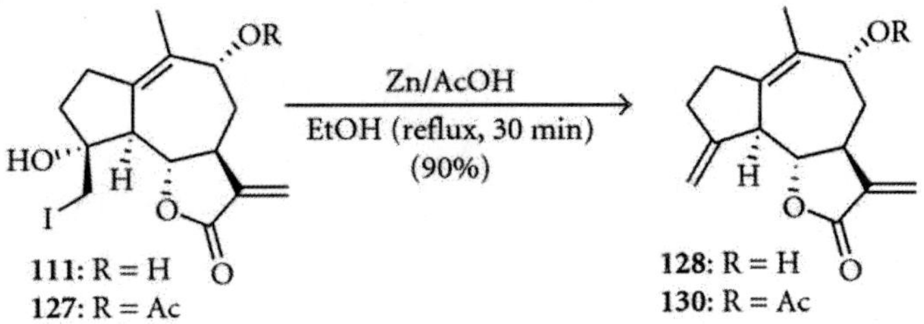

Zn/AcOH
EtOH (reflux, 30 min)
(90%)

111: R = H
127: R = Ac

128: R = H
130: R = Ac

Scheme 36: Synthesis of the trienes 128 and 130.

Study of the Inversion Reaction of the Lactonic Fusion on Eremanthine Derivatives and Synthesis of Micheliolide

The last studies with eremanthine (1) were performed by Alves [150]. These studies were developed aiming at the syntheses of substrates derived from the lactone 1 with different structural features to be used in the next stage of inversion of the lactonic fusion employing the conditions depicted in Scheme33 (route B), aiming to get epimers more stable than 6-epi-eremanthine (142) previously synthesized [16, 160]. That study of inversion of the lactonic fusion was also planned to obtain a substance with the necessary structural requirements for an eventual investigation of the biomimetic transformation of guaianolide into pseudoguaianolide, discussed in the text of this review at the subitems 3.1 (Scheme 13) and 6.6 (Scheme 28). There was also studied the synthesis of micheliolide, a naturally occurring substance initially isolated from Michelia compressa with relevant biological activity [165]. In the following subitems are described the sequences of reactions that were developed aiming to attain these objectives [150, 166–170] as well as the results of the study from the reactions of catalytic hydrogenation and methanol addition to α-methylene-γ-lactone of eremanthine derivatives [171].

Study on the Synthesis of Micheliolide

The synthesis of micheliolide (155) was studied by the sequences of reactions depicted in Schemes 37 and 38. Initial attempts to deoxygenate the C-9 position of acetate 148 and iodohydrin 139 with concomitant hydrogenolysis of the bond C15-I in these compounds were unsuccessful [150]. When the allylic acetate 148 was submitted to photolysis conditions in a solution of HMPA and water, using the procedure reported by Deshayes et al. [172], an untreatable mixture of products was obtained instead of the desired compound 149 (Scheme 37). On the other hand, treatment of iodohydrin 139 with MsCl and Et_3N in dichloromethane, by the procedure of Crossland and Servis [173], generated a mixture of conjugate dienes 151 and 152 instead of the intermediate mesylate at C-9 position (150), which would be used in the next step of hydrogenolysis with $NaBH_3CN$ in HMPA by the procedure of Hutchins et al. [174], aiming to obtain the target molecule 149. The author is disregarding in this review the preliminary results reported before about this reaction [166]. On that occasion this reaction was described in the following way: treatment of 139 with MsCl/Et_3N/CH_2Cl_2 (r. t.) followed by $NaBH_3CN$/HMPA (r. t.) furnished dienes ($\Delta^{1,2}$, $\Delta^{9,10}$) (^{1}H NMR 200 MHz, $CDCl_3$, 5.85, m, 1H, C2-H) [166]. On that time, in which that abstract was written with the results from scientific initiation of the author, his supervisor described the ^{1}H NMR data of only one elimination product (compound 151). For the occasion of the M.S., the author obtained the ^{1}H NMR spectrum of the crude product from the mesylation reaction described in Scheme 37 and he detected the presence of the conjugate dienes 151 and152 (1:1) [150], concluding that the mesylate 150 is formed and it is instantly converted to the final products of elimination (151 and 152). Therefore, the subsequent and unnecessary step of reduction with $NaBH_3CN$ was not performed on that occasion of his academic formation (M.S.). It is important to emphasize in this point that the strategy aiming to deoxygenate the C-9 position of allylic acetate 148and allylic mesylate 150 was unsuccessful due to the following reasons. In the case of the photolysis reaction, the method described by Deshayes et al. [172] is applied to nonactivated carboxylic esters. Evidently the allylic acetate at the substrate 148 is an activated moiety but the chemistry is an experimental science and the author, in that time, should test this reaction with the substance 148 to confirm experimentally the data of literature and know how to execute the reaction in a photochemical reactor. In the case of the attempt to isolate the allylic mesylate 150, this was impossible because electron-withdrawing groups such as OSO_2R increase the acidity of the hydrogen that is lost in any elimination mechanism (E1, E2, and E1cB) when those groups are conjugated with double bond [175, page 893]. Once again in that time of scientific initiation in which this reaction was performed for the first time, the author had little knowledge of advanced organic chemistry and he should make that reaction to confirm the data described in literature on conjugate eliminations [175, page 900], in which allylic mesylate undergoes elimination reaction to furnish conjugate dienes. That result was also important to confirm the elimination reactions previously occurred with the allylic derivatives of eremanthine 113(Scheme 28) and 111 (Scheme 30) yielding the respective conjugate dienes 115-116 [148] and 126 [15]. With those unsatisfactory results, the synthesis of compound 149 was attempted by catalytic hydrogenation of iodohydrin 139 with NaOAc in EtOH, using a longer reaction time and a higher hydrogen pressure than those commonly used to get diol 140 (Scheme 32). In those reaction conditions (60 psi of hydrogen, r. t., 48 h), the desired compound 149 was obtained as the minor product in mixture with diol 140 and compound 153 (Scheme 37) [168, 169].

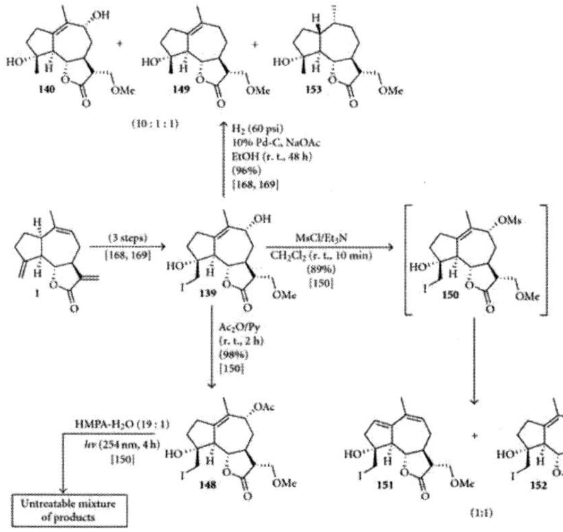

Scheme 37: Initial attempts to the synthesis of compound 149.

In the next stage, the hydrogenolysis reaction aiming at the synthesis of compound 149 was performed with the substrate 140 (Scheme 38), using the hydrogenolysis conditions of allylic alcohols reported by House [176, page 23]. When diol 140 in EtOH was submitted to hydrogenation (55 psi of hydrogen) with Pd-C, a compound identified as 153 was obtained as a result of hydrogenolysis of the bond C9-OH and hydrogenation of tetrasubstituted double bond C1-C10. Attempt to perform only hydrogenolysis of the bond C9-OH on diol 140 without reduction of double bond C1-C10 was carried out using a less reactive catalyst (PdS-C) than Pd-C in EtOH under low hydrogen pressure (5 psi).

153 + 149 (1 : 8)
NaOH-H2O
DMF (reflux, 3 h)
(80%)
[168, 169]
155
H_2 (5 psi)/10% Pd-C
EtOH (r. t., 15 min)
(94%)
[168, 169]
139
(1 step)
[168, 169]
140
H_2 (55 psi)/10% Pd-C
EtOH (r. t., 30 min)
(98%)
[168, 169]
153
NaOH-H_2O
DMF (reflux, 2.5 h)
(85%)
[168, 169]
154
H_2 (5 psi)/5% PdS-C
EtOH (r. t., 1.5 h)
(100%)
[150]
140 + 149 + 153
(4 : 3 : 3)

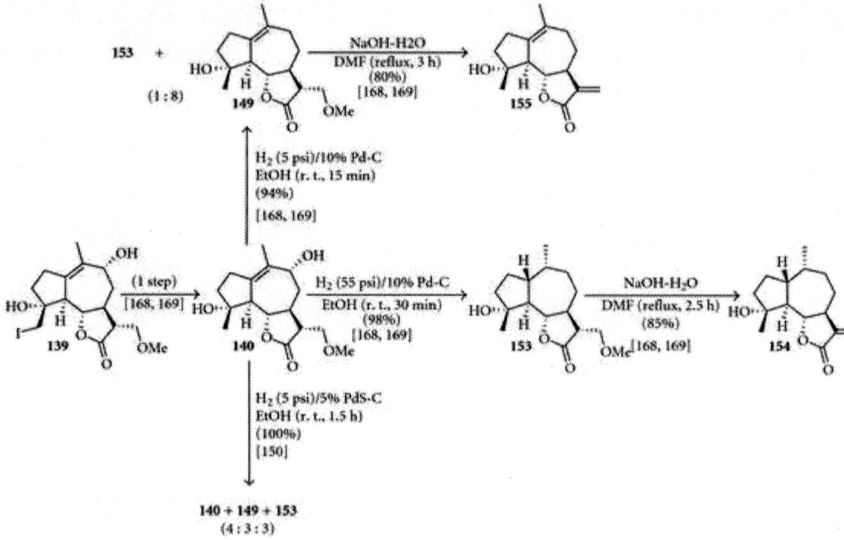

Scheme 38: Syntheses of micheliolide (155) and 1R, 10R-dihydromicheliolide (154).

After the reaction time (1.5 h, r. t.), the substrate 140 was recovered in mixture with 149 and 153 in a respective proportion of (4:3:3) [150]. Hydrogenation of allylic alcohol 140 using a low hydrogen pressure (5 psi) during a short reaction time (15 min) generated the target compound 149 as the main product of this reaction in mixture with the lactone 153 in a respective proportion of (8:1), according to ^1H NMR spectrum of crude product from that reaction. The probable causes from the low reactivity of allylic alcohol 139 in catalytic hydrogenation reaction with NaOAc, in opposition to the high reactivity of the similar allylic alcohol 140 without the use of NaOAc, were reported in a recently published work [171]. Elimination of methanol on the compounds 149 and 153 generated the respective α-methylene-γ-lactones micheliolide (155) and 1R,10R-dihydromicheliolide (154) (Scheme 38).

Study of the Inversion Reaction of the Lactonic Fusion on Eremanthine Derivatives

The inversion of configuration at C-6 position on eremanthine derivatives was studied with the substrates 85, 130, 154, and 156–158 shown in Scheme 39. These substances were prepared from eremanthine (1) by simple standard procedures and were described in recent articles [168–170]. After exposure of these substrates to reaction conditions of inversion of the lactonic fusion, epimeric mixtures at C-6 were obtained in different proportions (Scheme 39). The variations in the proportions of products with cis lactonic fusion obtained in that work were attributed to steric effects at the hydroazulene system. A detailed discussion on the probable causes of variation in the proportions of products with cis lactonic fusion displayed in Scheme 39 was reported in a recent article and in its supplementary information [170]. The author would like to make a correction on the major product of the inversion reaction of lactonic fusion of the allylic acetate 156 (Scheme 39) previsously described in a congress abstract [167]. The major product obtained in that reaction is the allylic acetate 163 and not the allylic alcohol 162 as it was written in an equivocal way in that abstract [167].

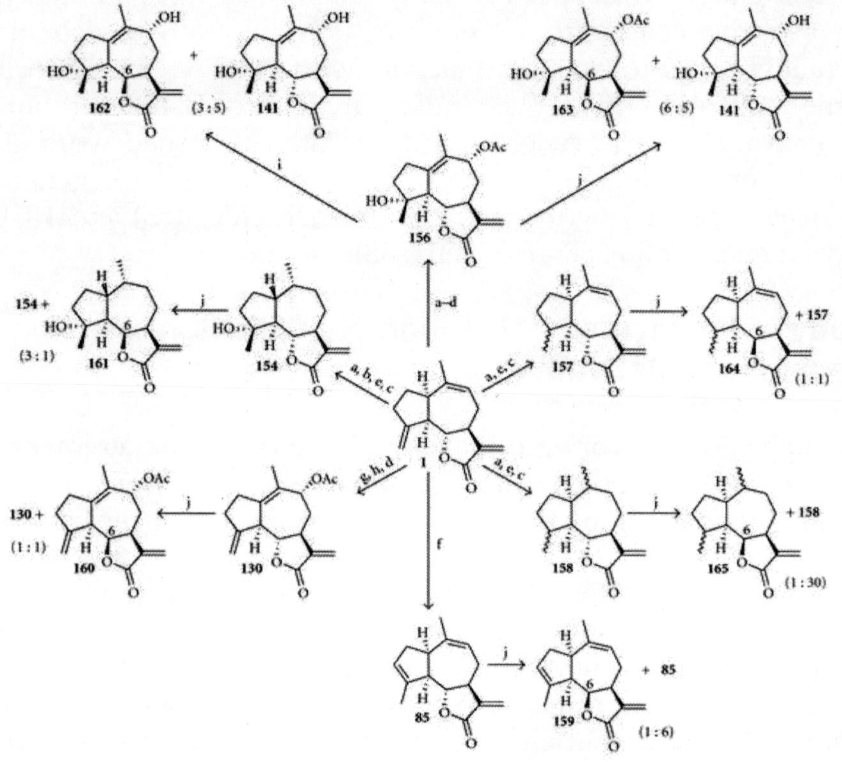

Scheme 39: Syntheses of the substrates 85, 130, 154, and 156–158 and their epimers at C-6 positions (159–165). (a) MeONa, MeOH; (b) (i) AcO_2H, CH_2Cl_2; (ii) KI, AcOH, acetone; (iii) H_2, Pd-C, NaOAc, EtOH; (c) NaOH-H_2O, DMF; (d) Ac_2O, pyridine; (e) H_2, Pd-C, EtOH; (f) $BF_3.OEt_2$, benzene; (g) (i) AcO_2H, $CHCl_3$; (ii) KI, AcOH, acetone; (h) Zn, AcOH, EtOH; (i) (i) KOH-H_2O (ii) Dryness; (iii) MsCl, Et_3N, DMSO; (iv) NaOH-H_2O; (v) HCl-H_2O; (j) (i) KOH-H_2O (ii) Dryness; (iii) MsCl, Et_3N, THF; (iv) NaOH-H_2O; (v) HCl-H_2O.

An intriguing result was obtained in the reaction of diol 140 with aqueous NaOH in refluxing DMF (Scheme 40). The product of this reaction was identified by ^{1}H NMR as a mixture of epimers 141 and162 (3:1). After acetylation of this mixture with Ac_2O and pyridine, the major product 156 was separated and then used in reaction of inversion of the lactonic fusion (Scheme 39) [170]. The author would like to explain that the allylic acetate 156 was obtained for the first time in his works of scientific initiation [166] and M.S. [150] and not as it was previously described in an equivocal way [167]. It was reported in the congress abstract [167] that the substance 156 had been obtained in a work of Ph.D. [16], but the compound obtained in that thesis [16] was the diol 141 (Scheme 32). Reinvestigation of the methanol elimination on diol 140 using MeCN as the solvent of reaction resulted in generation of the epimeric mixture 141 and 162. The major product of that reaction was the epimer 162 with cis lactonic fusion obtained in a proportion of (6:5) (^{1}H NMR) in relation to compound 141 (Scheme 40) [170]. The allylic alcohol 162 and its correspondent allylic acetate 163 possess the necessary structural requirements to unchain the rearrangements preconized by the hypothesis of the biotransformation of guaianolide into pseudoguaianolide (ambrosanolide) reported by Fischer et al. [14].

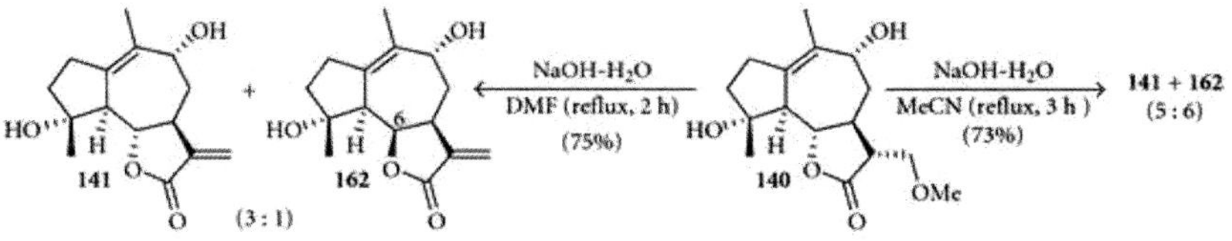

Scheme 40: Inversion of configuration at C-6 position of the diol 141 occurred at the step of methanol elimination on adduct 140.

It was verified at the step of inversion of the lactonic fusion on isoeremanthine (85) that the use of concentrated solutions of that substrate in THF generated 6-epi-isoeremanthine (159) in mixture with a minor product identified by ^{1}H NMR as the compound 166 (5:1) (Scheme 41) [170]. The allylic alcohol133, was obtained when iodohydrin 139 was treated with zinc and acetic acid in refluxing EtOH (Scheme 42). The speculative mechanism of this reaction was published in a recent article [170].

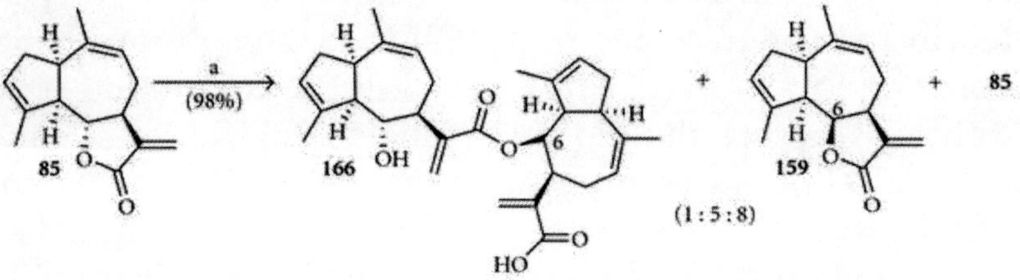

Scheme 41: Synthesis of the compound 166 in mixture with the lactone 159: (a) (i) KOH-H_2O; (ii) Dryness; (iii) MsCl, Et_3N, THF (conc.); (iv) NaOH-H_2O; (v) HCl-H_2O.

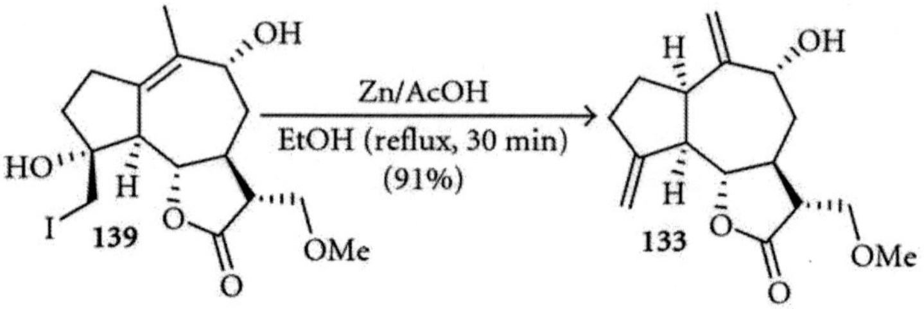

Scheme 42: Synthesis of the allylic alcohol 133.

Study of Catalytic Hydrogenation and Methanol Addition to *α-Methylene-γ-Lactone* of Eremanthine Derivatives

Besides the sequences of reactions described in the previous subitem (6.11), Alves [171] also studied the reactivity of allylic derivatives 127, 128, 133, and 140 in catalytic hydrogenation reactions as well as the methanol addition to α-methylene-γ-lactone of iodohydrin 111 (Scheme 43). Catalytic hydrogenation of iodohydrin acetate 127 in EtOH with hydrogen and Pd-C yielded a single product identified by ^{1}H NMR as the allylic acetate 167. On the other hand, catalytic hydrogenation of allylic alcohol 128 in EtOH with hydrogen and Pd-C furnished a complex mixture of substances. After a meticulous analysis of the ^{1}H NMR and ^{13}C NMR spectra, in combination with the calculations of molecular modeling, it was verified that the mixture obtained in the reaction was composed by intermediates that did not totally react and two products were resultant from hydrogenation of the reactive functions of the substrate 128, characterized as the isomers 168 (major) and 169 (minor). Hydrogenation of allylic alcohol 133 afforded a mixture of products characterized by ^1H NMR as the compounds 170 and 171 (5:1) in mixture with traces of the lactones 172 and 173. The stereochemistry of methyl groups C-14 and C-15 at the major product 170 was determined by NOE experiment. The catalytic hydrogenation reaction of allylic alcohol140 using the catalyst Pt-C and low hydrogen pressure generated the compound 153 in quantitative yield. The treatment of iodohydrin 111 with a solution of NaOMe in methanol at room temperature furnished a single product characterized by ^1H NMR as the dimethoxylated compound 174, as result of methanol addition to α-methylene-γ-lactone and nucleophilic substitution at C-15.

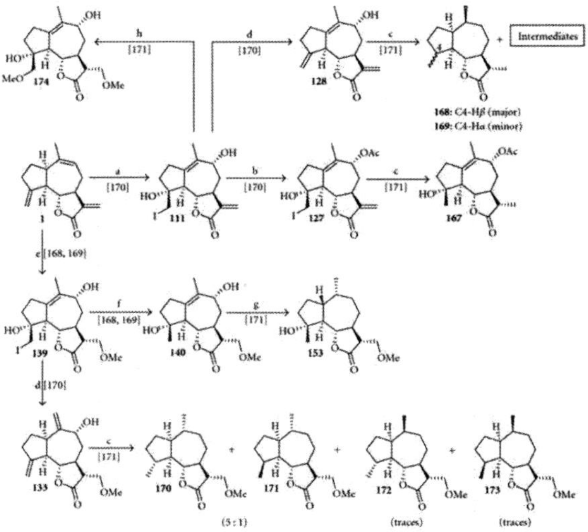

Scheme 43: Preparation of the substrates 111, 127, 128, 133, and 140 and subsequent reactions of catalytic hydrogenation as well as methanol addition to α-methylene-γ-lactone of the iodohydrin 111. (a) (i) AcO_2H, $CHCl_3$; (ii) KI, AcOH, acetone; (b) Ac_2O, pyridine; (c) H_2, Pd-C, EtOH; (d) Zn, AcOH, EtOH; (e) (i) MeONa, MeOH; (ii) AcO_2H, CH_2Cl_2; (iii) KI, AcOH, acetone; (f) H_2, Pd-C, NaOAc, EtOH; (g) H_2, Pt-C, EtOH; (h) MeONa, MeOH.

The catalytic hydrogenation of allylic alcohol 140 in mixture with its epimer at C-11 position (175), followed by the step of methanol elimination, generated a mixture of compounds 154 and 176 (Scheme44). This result suggests that the addition of hydrogen to double bond C1-C10 on this mixture of allylic alcohols is induced by the group CH_2OMe at C-11 position. A detailed discussion on the reactivity of allylic derivatives from eremanthine shown in Schemes 43-44, in catalytic hydrogenation reaction, was presented in a recently published work including analysis of molecular modeling with the use of molecular mechanic tools (MM2 calculation) [171].

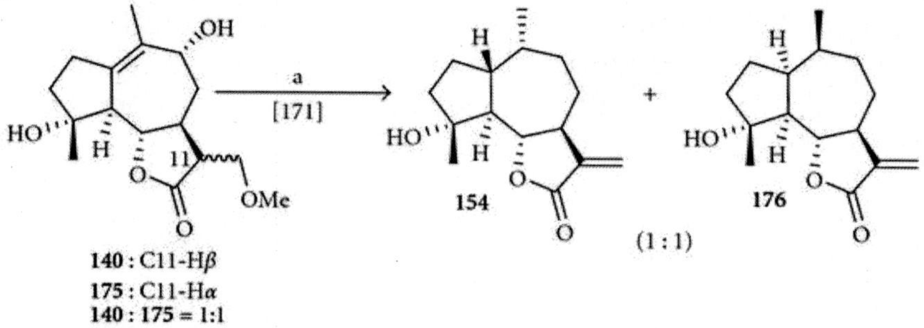

140 : C11-Hβ
175 : C11-Hα
140 : 175 = 1:1

154 + 176 (1 : 1)

Scheme 44: Synthesis of the compound 176 in mixture with the lactone 154: (a) (i) H_2, Pd-C, EtOH; (ii) NaOH-H_2O, DMF.

CONCLUSIONS

This review about the chemistry of eremanthine (1) has demonstrated the usefulness of the chiron approach to the syntheses of other sesquiterpene lactones derived from 1, by the use of a building block with well-established stereocenters as starting material. From the described synthetic studies it was possible to know the reactivity of several functional groups at the compound 1 as well as in other derivatives obtained from 1. Although the chemistry of eremanthine and its derivatives has been quite explored, there is still opportunity for new discoveries and syntheses of new substances by using this natural sesquiterpenoid as starting material

ACKNOWLEDGMENTS

J. C. F. Alves thanks FAPERJ and CNPq for the fellowships to develop the project "Chemical transformations of natural substances. I-Studies with eremanthine" [177], Professor Dr. Edna C. Fantini (in memoriam) for the supervision of the research project, and the Rural Federal University of Rio de Janeiro (UFRRJ) for the reception during the period in which the project was developed

REFERENCES

1. W. Vichnewski and B. Gilbert, "Schistosomicidal sesquiterpene lactone from Eremanthus elaeagnus," Phytochemistry, vol. 11, no. 8, pp. 2563–2566, 1972. View at Scopus
2. P. M. Baker, C. C. Fortes, E. G. Fortes et al., "Chemoprophylactic agents in schistosomiasis: eremanthine, costunolide, α-cyclocostunolide and bisabolol," Journal of Pharmacy and Pharmacology, vol. 24, no. 11, pp. 853–857, 1972. View at Scopus
3. M. S. Silvério, O. V. Sousa, G. Del-Vechio-Vieira, M. A. Miranda, F. C. Matheus, and M. A. C. Kaplan, "Pharmacological properties of the ethanol extract from Eremanthus erythropappus (DC.) McLeisch (Asteraceae)," Brazilian Journal of Pharmacognosy, vol. 18, no. 3, pp. 430–435, 2008.View at Publisher · View at Google Scholar
4. A. Corbella, P. Gariboldi, G. Jommi, F. Orsini, and G. Ferrari, "Structure and absolute stereochemistry of vanillosmin, a guaianolide from Vanillosmopsis erythropappa,"Phytochemistry, vol. 13, no. 2, pp. 459–465, 1974. View at Scopus
5. M. Garcia, A. J. R. da Silva, P. M. Baker, B. Gilbert, and J. A. Rabi, "Absolute stereochemistry of eremanthine, a schistosomicidal sesquiterpene lactone from Eremanthus elaeagnus,"Phytochemistry, vol. 15, no. 2, pp. 331–332, 1976. View at Scopus
6. P. D. D. B. Lima, M. Garcia, and J. A. Rabi, "Selective extraction of α-methylene-γ-lactones. Reinvestigation of Vanillosmopsis erythropappa," Journal of Natural Products, vol. 48, no. 6, pp. 986–988, 1985. View at Scopus
7. J. F. M. Pérez, J. R. S. Scolforo, A. D. de Oliveira, J. M. de Mello, L. F. R. Borges, and J. F. Camolesi, "Management system for native candeia forest (Eremanthus erythropappus (DC.) MacLeish)—the option for selective cutting," Cerne, vol. 10, no. 2, pp. 257–273, 2004.
8. A. D. de Oliveira, I. S. A. Ribeiro, J. R. S. Scolforo, J. M. de Mello, F.

W. Acerbi Jr., and J. F. Camolesi, "Market chain analysis of candeia timber (Eremanthus erythropappus)," Cerne, vol. 15, no. 3, pp. 257–264, 2009. View at Scopus

9. One of the industries of essential oils that extract the oil from Eremanthus erythropappus is Citróleo Indústria e Comércio de Óleos Essenciais Ltda, http://www.citroleo.com.br/.

10. G. P. P. Kamatou and A. M. Viljoen, "A review of the application and pharmacological properties of α-bisabolol and α-bisabolol-rich oils," Journal of the American Oil Chemists› Society, vol. 87, no. 1, pp. 1–7, 2010. View at Publisher · View at Google Scholar · View at Scopus

11. A. T. de Souza, T. L. Benazzi, M. B. Grings et al., "Supercritical extraction process and phase equilibrium of Candeia (Eremanthus erythropappus) oil using supercritical carbon dioxide," The Journal of Supercritical Fluids, vol. 47, no. 2, pp. 182–187, 2008. View at Publisher · View at Google Scholar · View at Scopus

12. S. Yuuya, H. Hagiwara, T. Suzuki et al., "Guaianolides as immunomodulators. Synthesis and biological activities of dehydrocostus lactone, mokko lactone, eremanthin, and their derivatives,"Journal of Natural Products, vol. 62, no. 1, pp. 22–30, 1999. View at Publisher · View at Google Scholar · View at PubMed · View at Scopus

13. J. W. de Kraker, M. C. R. Franssen, M. Joerink, A. de Groot, and H. J. Bouwmeester, "Biosynthesis of costunolide, dihydrocostunolide, and leucodin. Demonstration of cytochrome P450-catalyzed formation of the lactone ring present in sesquiterpene lactones of chicory," Plant Physiology, vol. 129, no. 1, pp. 257–268, 2002. View at Publisher · View at Google Scholar · View at PubMed · View at Scopus

14. N. H. Fischer, E. J. Olivier, and H. D. Fischer, "The biogenesis and chemistry of sesquiterpene lactones," in Progess in the Chemistry of Organic Natural Products, W. Herz, H. Grisebach, and G. W. Kirby, Eds., vol. 38, pp. 47–390, Springer, New York, NY, USA, 1979.

15. J. L. P. Ferreira, Studies on the chemical reactivity of epoxides derived from eremanthine, M.Sc. dissertation, NPPN, Universidade Federal do Rio de Janeiro, Rio de Janeiro, Brazil, 1985.

16. E. C. Fantini, Study on the chemical reactivity of eremanthine. The α-methylene-γ-lactone moiety, Ph.D. thesis, Universidade Federal do Rio de Janeiro, Rio de Janeiro, Brazil, 1985.

17. N. H. Fischer, Y. F. Wu-Shih, G. Chiari, F. R. Fronczek, and S. F. Watkins, "Molecular structure of a cis-decalin-type eudesmanolide and its formation from a guaianolide-1(10)-epoxide," Journal of Natural Products, vol. 44, no. 1, pp. 104–110, 1981. View at Scopus

18. J. E. Barquera-Lozada and G. Cuevas, "Biogenesis of sesquiterpene lactones pseudoguaianolides from germacranolides: theoretical study on the reaction mechanism of terminal biogenesis of 8-epiconfertin," The Journal of Organic Chemistry, vol. 74, no. 2, pp. 874–883, 2009. View at Publisher · View at Google Scholar · View at Scopus
19. A. Ortega and E. Maldonado, "A one step transformation of 4α,5β-epoxygermacranolide into pseudoguaianolide," Heterocycles, vol. 29, no. 4, pp. 635–638, 1989. View at Scopus
20. M. J. Bordoloi, R. P. Sharma, and J. C. Sarma, "Biomimetic transformation of a guaianolide to a pseudoguaianolide," Tetrahedron Letters, vol. 27, no. 38, pp. 4633–4634, 1986. View at Scopus
21. B. M. Fraga, "Natural sesquiterpenoids," Natural Product Reports, vol. 27, no. 11, pp. 1681–1708, 2010. View at Publisher · View at Google Scholar
22. D. N. Cavalcanti, M. A. V. Gomes, A. C. Pinto, C. M. de Rezende, R. C. Pereira, and V. L. Teixeira, "Effects of storage and solvent type in a lipophylic chemical profile of the seaweedDictyota menstrualis," Brazilian Journal of Oceanography, vol. 56, no. 1, pp. 51–57, 2008. View at Scopus
23. J. C. De-Paula, L. B. Bueno, D. N. Cavalcanti, Y. Yoneshigue-Valentin, and V. L. Teixeira, "Diterpenes from the brown alga Dictyota crenulata," Molecules, vol. 13, no. 6, pp. 1253–1262, 2008. View at Publisher · View at Google Scholar · View at Scopus
24. M. A. Vallim, V. L. Teixeira, and R. C. Pereira, "Feeding-deterrent properties of diterpenes ofDictyota mertensii (phaeophyceae, dictyotales)," Brazilian Journal of Oceanography, vol. 55, no. 3, pp. 223–229, 2007.
25. X. Zhang, H. Wang, J. Sheng, and X. Luo, "A new guaiane diterpenoid from Euphorbia wallichii,"Natural Product Research, vol. 20, no. 1, pp. 89–92, 2006. View at Publisher · View at Google Scholar · View at PubMed · View at Scopus
26. S. A. Kolesnikova, A. I. Kalinovsky, S. N. Fedorov, L. K. Shubina, and V. A. Stonik, "Diterpenes from the far-eastern brown alga Dictyota dichotoma," Phytochemistry, vol. 67, no. 19, pp. 2115–2119, 2006. View at Publisher · View at Google Scholar · View at PubMed · View at Scopus
27. J. W. Blunt, B. R. Copp, M. H. G. Munro, P. T. Northcote, and M. R. Prinsep, "Marine natural products," Natural Product Reports, vol. 23, no. 1, pp. 26–78, 2006. View at Publisher · View at Google Scholar · View at PubMed · View at Scopus
28. J. W. Blunt, B. R. Copp, M. H. G. Munro, P. T. Northcote, and M. R.

Prinsep, "Marine natural products," Natural Product Reports, vol. 22, no. 1, pp. 15–61, 2005. View at Publisher · View at Google Scholar · View at PubMed · View at Scopus

29. M. A. Vallim, J. C. de Paula, R. C. Pereira, and V. L. Teixeira, "The diterpenes from Dictyotacean marine brown algae in the tropical Atlantic American region," Biochemical Systematics and Ecology, vol. 33, no. 1, pp. 1–16, 2005. View at Publisher · View at Google Scholar · View at Scopus

30. P. Siamopoulou, A. Bimplakis, D. Iliopoulou et al., "Diterpenes from the brown algae Dictyota dichotoma and Dictyota linearis," Phytochemistry, vol. 65, no. 14, pp. 2025–2030, 2004. View at Publisher · View at Google Scholar · View at PubMed · View at Scopus

31. S. R. Gedara, O. B. Abdel-Halim, S. H. El-Sharkawy, O. M. Salama, T. W. Shier, and A. F. Halim, "Cytotoxic hydroazulene diterpenes from the brown alga Dictyota dichotoma," Zeitschrift für Naturforschung Section C, vol. 58, no. 1-2, pp. 17–22, 2003. View at Scopus

32. S. E. N. Ayyad, O. B. Abdel-Halim, W. T. Shier, and T. R. Hoye, "Cytotoxic hydroazulene diterpenes from the brown alga Cystoseira myrica," Zeitschrift für Naturforschung Section C, vol. 58, no. 1-2, pp. 33–38, 2003. View at Scopus

33. V. L. Teixeira, D. N. Cavalcanti, and R. C. Pereira, "Chemotaxonomic study of the diterpenes from the brown alga Dictyota menstrualis," Biochemical Systematics and Ecology, vol. 29, no. 3, pp. 313–316, 2001. View at Publisher · View at Google Scholar · View at Scopus

34. R. C. Pereira, D. N. Cavalcanti, and V. L. Teixeira, "Effects of secondary metabolites from the tropical Brazilian brown alga Dictyota menstrualis on the amphipod Parhyale hawaiensis,"Marine Ecology Progress Series, vol. 205, pp. 95–100, 2000. View at Scopus

35. M. Gavagnin and A. Fontana, "Diterpenes from marine opisthobranch molluscs," Current Organic Chemistry, vol. 4, no. 12, pp. 1201–1248, 2000. View at Publisher · View at Google Scholar · View at Scopus

36. D. J. Faulkner, "Marine natural products," Natural Product Reports, vol. 15, no. 2, pp. 113–158, 1998. View at Scopus

37. R. Durán, E. Zubía, M. J. Ortega, and J. Salvá, "New diterpenoids from the alga Dictyota dichotoma," Tetrahedron, vol. 53, no. 25, pp. 8675–8688, 1997. View at Publisher · View at Google Scholar · View at Scopus

38. G. M. König, A. D. Wright, O. Sticher, and H. Ruegger, "Four new hydroazulenoid diterpenes from the tropical marine brown alga Dictyota volubilis," Planta Medica, vol. 59, no. 2, pp. 174–178, 1993. View at Scopus

39. V. L. Teixeira, S. A. D. S. Almeida, and A. Kelecom, "Chemosystematic and biogeographic studies of the diterpenes from the marine brown alga Dictyota dichotoma," Biochemical Systematics and Ecology, vol. 18, no. 2-3, pp. 87–92, 1990. View at Scopus

40. V. L. Teixeira and A. Kelecom, "A chemotaxonomic study of diterpenes from marine brown algae of the genus Dictyota," Science of the Total Environment, vol. 75, no. 2-3, pp. 271–283, 1988. View at Scopus

41. J. T. Vázquez, M. Chang, K. Nakanishi, E. Manta, C. Pérez, and J. D. Martín, "Structure of hydroazulenoid diterpenes from a marine alga and their absolute configuration based on circular dichroism," The Journal of Organic Chemistry, vol. 53, no. 20, pp. 4797–4800, 1988. View at Scopus

42. S. de Rosa, S. de Stefano, and N. Zavodnik, "Hydroazulenoid diterpenes from the brown algaDictyota dichotoma var. Implexa," Phytochemistry, vol. 25, no. 9, pp. 2179–2181, 1986. View at Scopus

43. A. Kelecom and V. L. Teixeira, "Diterpenes of marine brown algae of the family dictyotaceae: their possible role as defense compounds and their use in chemotaxonomy," The Science of the Total Environment, vol. 58, no. 1-2, pp. 109–115, 1986. View at Scopus

44. H. H. Sun, F. J. McEnroe, and W. Fenical, "Acetoxycrenulide, a new bicyclic cyclopropane-containing diterpenoid from the brown seaweed Dictyota crenulata," The Journal of Organic Chemistry, vol. 48, no. 11, pp. 1903–1906, 1983. View at Scopus

45. J. Finer, J. Clardy, W. Fenical et al., "Structures of dictyodial and dictyolactone, unusual marine diterpenoids," The Journal of Organic Chemistry, vol. 44, no. 12, pp. 2044–2047, 1979. View at Scopus

46. A. E. Greene, "Synthesis of (+)-pachydictyol-A," Tetrahedron Letters, vol. 19, no. 9, pp. 851–854, 1978. View at Scopus

47. B. Danise, L. Minale, R. Riccio, et al., "Further perhydroazulene diterpenes from marine organisms," Experientia, vol. 33, no. 4, pp. 413–415, 1977.

48. D. J. Faulkner, B. N. Ravi, J. Finer, and J. Clardy, "Diterpenes from Dictyota dichotoma,"Phytochemistry, vol. 16, no. 7, pp. 991–993, 1977. View at Scopus

49. L. Minale and R. Riccio, "Constituents of the digestive gland of the molluscs of the genus Aplysia. I. Novel diterpenes from Aplysia depilans," Tetrahedron Letters, vol. 17, no. 31, pp. 2711–2714, 1976. View at Scopus

50. E. Fattorusso, S. Magno, L. Mayol et al., "Dictyol A and B, two novel diterpene alcohols from the brown alga Dictyota dichotoma," Journal

of the Chemical Society, Chemical Communications, no. 14, pp. 575–576, 1976. View at Publisher · View at Google Scholar · View at Scopus

51. D. R. Hirschfeld, W. Fenical, G. H. Y. Lin, R. M. Wing, P. Radlick, and J. J. Sims, "Marine natural products. VIII. Pachydictyol A, an exceptional diterpene alcohol from the brown alga,Pachydictyon coriaceum," Journal of the American Chemical Society, vol. 95, no. 12, pp. 4049–4050, 1973. View at Scopus
52. Y. Ye, X. Q. Li, and C. P. Tang, "Natural products chemistry research 2006›s progress in China,"Chinese Journal of Natural Medicines, vol. 6, no. 1, pp. 70–78, 2008. View at Publisher · View at Google Scholar · View at Scopus
53. Q. L. Dang, Y. H. Choi, G. J. Choi et al., "Pesticidal activity of ingenane diterpenes isolated fromEuphorbia kansui against Nilaparvata lugens and Tetranychus urticae," Journal of Asia-Pacific Entomology, vol. 13, no. 1, pp. 51–54, 2010. View at Publisher · View at Google Scholar · View at Scopus
54. Q. C. Wu, Y. P. Tang, A. W. Ding, F. Q. You, L. Zhang, and J. A. Duan, "^{13}C-NMR data of three important diterpenes isolated from Euphorbia species," Molecules, vol. 14, no. 11, pp. 4454–4475, 2009. View at Publisher · View at Google Scholar · View at Scopus
55. E. Sulyok, A. Vasas, D. Rédei, G. Dombi, and J. Hohmann, "Isolation and structure determination of new 4,12-dideoxyphorbol esters from Euphorbia pannonica Host," Tetrahedron, vol. 65, no. 20, pp. 4013–4016, 2009. View at Publisher · View at Google Scholar · View at Scopus
56. Z. Q. Lu, M. Yang, J. Q. Zhang et al., "Ingenane diterpenoids from Euphorbia esula,"Phytochemistry, vol. 69, no. 3, pp. 812–819, 2008. View at Publisher · View at Google Scholar ·View at PubMed · View at Scopus
57. A. R. Jassbi, "Chemistry and biological activity of secondary metabolites in Euphorbia from Iran,"Phytochemistry, vol. 67, no. 18, pp. 1977–1984, 2006. View at Publisher · View at Google Scholar ·View at PubMed · View at Scopus
58. J. K. Cha and O. L. Epstein, "Synthetic approaches to ingenol," Tetrahedron, vol. 62, no. 7, pp. 1329–1343, 2006. View at Publisher · View at Google Scholar · View at Scopus
59. M. E. Krafft, Y. Y. Cheung, S. A. Kerrigan, and K. A. Abboud, "Synthesis of 'inside-outside' medium-sized rings via ring-closing metathesis," Tetrahedron Letters, vol. 44, no. 4, pp. 839–843, 2003. View at Publisher · View at Google Scholar · View at Scopus
60. J. H. Rigby and M. Fleming, "Construction of the ingenane core

using an Fe(III) or Ti(IV) Lewis acid-catalyzed intramolecular [6+4] cycloaddition," Tetrahedron Letters, vol. 43, no. 48, pp. 8643–8646, 2002. View at Publisher · View at Google Scholar · View at Scopus

61. M. Blanco-Molina, G. C. Tron, A. Macho et al., "Ingenol esters induce apoptosis in Jurkat cells through an AP-1 and NF-κB independent pathway," Chemistry and Biology, vol. 8, no. 8, pp. 767–778, 2001. View at Publisher · View at Google Scholar · View at Scopus
62. J. H. Rigby, H. Jingdan, and M. J. Heeg, "Synthetic studies on the ingenane diterpenes. Construction of an ABC tricycle exhibiting trans-intrabridgehead stereochemistry," Tetrahedron Letters, vol. 39, no. 16, pp. 2265–2268, 1998. View at Publisher · View at Google Scholar · View at Scopus
63. J. A. Marco, J. F. Sanz-Cervera, F. J. Ropero, J. Checa, and B. M. Fraga, "Ingenane and lathyrane diterpenes from the latex of Euphorbia acrurensis," Phytochemistry, vol. 49, no. 4, pp. 1095–1099, 1998. View at Publisher · View at Google Scholar · View at Scopus
64. J. A. Marco, J. F. Sanz-Cervera, and A. Yuste, "Ingenane and lathyrane diterpenes from the latex ofEuphorbia canariensis," Phytochemistry, vol. 45, no. 3, pp. 563–570, 1997. View at Publisher ·View at Google Scholar · View at Scopus
65. J. D. Winkler, B. C. Hong, A. Bahador, M. G. Kazanietz, and P. M. Blumberg, "Synthesis of ingenol analogs with affinity for protein kinase C," Bioorganic & Medicinal Chemistry Letters, vol. 3, no. 4, pp. 577–580, 1993. View at Publisher · View at Google Scholar · View at Scopus
66. G. Brooks, A. T. Evans, D. P. Markby, M. E. Harrison, M. A. Baldwin, and F. J. Evans, "An ingenane diterpene from belizian Mabea excelsa," Phytochemistry, vol. 29, no. 5, pp. 1615–1617, 1990. View at Scopus
67. P. A. Wender, C. L. Hillemann, and M. J. Szymonifka, "An approach to the tiglianes, daphnanes, and ingenanes via the divinylcyclopropane rearrangement," Tetrahedron Letters, vol. 21, no. 23, pp. 2205–2208, 1980. View at Scopus
68. J. Y. Hong, J. W. Nam, E. K. Seo, and S. K. Lee, "Daphnane diterpene esters with anti-proliferative activities against human lung cancer cells from Daphne genkwa," Chemical & Pharmaceutical Bulletin, vol. 58, no. 2, pp. 234–237, 2010. View at Publisher · View at Google Scholar · View at Scopus
69. L. Pan, X. F. Zhang, Y. Deng, Y. Zhou, H. Wang, and L. S. Ding, "Chemical constituents investigation of Daphne tangutica," Fitoterapia, vol. 81, no. 1, pp. 38–41, 2010. View at Publisher· View at Google Scholar · View at Scopus

70. S. A. Ayatollahi, A. Shojaii, F. Kobarfard, M. Nori, M. Fathi, and M. I. Choudhari, "Terpens from aerial parts of Euphorbia splendida," Journal of Medicinal Plant Research, vol. 3, no. 9, pp. 660–665, 2009. View at Scopus

71. P. Y. Hayes, S. Chow, M. J. Somerville, J. J. de Voss, and M. T. Fletcher, "Pimelotides A and B, diterpenoid ketal-lactone orthoesters with an unprecedented skeleton from Pimelea elongata,"Journal of Natural Products, vol. 72, no. 12, pp. 2081–2083, 2009. View at Publisher · View at Google Scholar · View at Scopus

72. R. Yazdanparast and A. Meshkini, "3-hydrogenkwadaphnine, a novel diterpene ester fromDendrostellera lessertii, its role in differentiation and apoptosis of KG1 cells," Phytomedicine, vol. 16, no. 2-3, pp. 206–214, 2009. View at Publisher · View at Google Scholar · View at PubMed ·View at Scopus

73. C. V. Diogo, L. Félix, S. Vilela et al., "Mitochondrial toxicity of the phyotochemicals daphnetoxin and daphnoretin—relevance for possible anti-cancer application," Toxicology in Vitro, vol. 23, no. 5, pp. 772–779, 2009. View at Publisher · View at Google Scholar · View at Scopus

74. B. Y. Park, B. S. Min, K. S. Ahn et al., "Daphnane diterpene esters isolated from flower buds ofDaphne genkwa induce apoptosis in human myelocytic HL-60 cells and suppress tumor growth in Lewis lung carcinoma (LLC)-inoculated mouse model," Journal of Ethnopharmacology, vol. 111, no. 3, pp. 496–503, 2007. View at Publisher · View at Google Scholar · View at PubMed · View at Scopus

75. L. Pan, X. F. Zhang, H. F. Wu, and L. S. Ding, "A new daphnane diterpene from Daphne tangutica," Chinese Chemical Letters, vol. 17, no. 1, pp. 38–40, 2006. View at Scopus

76. R. Yazdanparast and M. A. Moosavi, "Daphnane-type diterpene esters as powerful agents for the treatment of leukemia," Medical Hypotheses, vol. 67, no. 6, pp. 1472–1473, 2006. View at Publisher · View at Google Scholar · View at PubMed · View at Scopus

77. A. Tempeam, N. Thasana, C. Pavaro, W. Chuakul, P. Siripong, and S. Ruchirawat, "A new cytotoxic daphnane diterpenoid, rediocide G, from Trigonostemon reidioides," Chemical & Pharmaceutical Bulletin, vol. 53, no. 10, pp. 1321–1323, 2005. View at Publisher · View at Google Scholar · View at Scopus

78. Z. J. Zhan, C. Q. Fan, J. Ding, and J. M. Yue, "Novel diterpenoids with potent inhibitory activity against endothelium cell HMEC and cytotoxic activities from a well-known TCM plant Daphne genkwa," Bioorganic and Medicinal Chemistry, vol. 13, no. 3, pp. 645–655, 2005.

View at Publisher · View at Google Scholar · View at PubMed · View at Scopus

79. A. T. Tchinda, A. Tsopmo, M. Tene et al., "Diterpenoids from Neoboutonia glabrescens(Euphorbiaceae)," Phytochemistry, vol. 64, no. 2, pp. 575–581, 2003. View at Publisher · View at Google Scholar · View at Scopus
80. W. He, M. Cik, L. van Puyvelde et al., "Neurotrophic and antileukemic daphnane diterpenoids from Synaptolepis kirkii," Bioorganic and Medicinal Chemistry, vol. 10, no. 10, pp. 3245–3255, 2002. View at Publisher · View at Google Scholar · View at Scopus
81. J. R. Carney, J. M. Krenisky, R. T. Williamson et al., "Maprouneacin, a new daphnane diterpenoid with potent antihyperglycemic activity from Maprounea africana," Journal of Natural Products, vol. 62, no. 2, pp. 345–347, 1999. View at Publisher · View at Google Scholar · View at Scopus
82. F. Abe, Y. Iwase, T. Yamauchi et al., "Minor daphnane-type diterpenoids from Wikstroemia retusa," Phytochemistry, vol. 47, no. 5, pp. 833–837, 1998. View at Publisher · View at Google Scholar · View at Scopus
83. F. Abe, Y. Iwase, T. Yamauchi, K. Kinjo, and S. Yaga, "Daphnane diterpenoids from the bark ofWikstroemia retusa," Phytochemistry, vol. 44, no. 4, pp. 643–647, 1997. View at Publisher · View at Google Scholar · View at Scopus
84. P. C. B. Page, D. C. Jennens, and H. McFarland, "An IMDA approach to tigliane and daphnane diterpenoids: generation of the tetracyclic ring system of the tiglianes," Tetrahedron Letters, vol. 38, no. 39, pp. 6913–6916, 1997. View at Publisher · View at Google Scholar · View at Scopus
85. P. C. B. Page, D. C. Jennens, and H. McFarland, "An IMDA approach to tigliane and daphnane diterpenoids: introduction of the C-12, C-13 C ring oxygenation of phorbol," Tetrahedron Letters, vol. 38, no. 30, pp. 5395–5398, 1997. View at Publisher · View at Google Scholar · View at Scopus
86. Y. Shiryo, K. Kazuhiko, H. Hiroya, M. Naochika, A. Fumiko, and Y. Tatsuo, "Diterpenoids with the daphnane skeleton from Wikstroemia retusa," Phytochemistry, vol. 32, no. 1, pp. 141–143, 1992. View at Scopus
87. T. Terai, K. Osakabe, M. Katai et al., "Preparation of 9-hydroxy grayanotoxin derivatives and their acute toxicity in mice," Chemical & Pharmaceutical Bulletin, vol. 51, no. 3, pp. 351–353, 2003.View at Scopus

88. M. Shimizu, Y. Nakagawa, Y. Sato, et al., "Studies on endophytic actinomycetes (I) Streptomycessp. isolated from rhododendron and its antifungal activity," Journal of General Plant Pathology, vol. 66, no. 4, pp. 360–366, 2000.

89. L. Q. Wang, B. Y. Ding, G. W. Qin, G. Lin, and K. F. Cheng, "Grayanoids from Pieris formosa,"Phytochemistry, vol. 49, no. 7, pp. 2045–2048, 1998. View at Publisher · View at Google Scholar ·View at Scopus

90. M. Sato, Y. Katsube, M. Katai, J. Katakawa, and T. Tetsumi, "Crystal and molecular structure of asebotoxin IV," Bulletin of the Chemical Society of Japan, vol. 67, no. 3, pp. 866–868, 1994.

91. N. Harada, "Pharmacological studies on the mechanisms of asebotoxin III-induced centrogenic pulmonary hemorrhagic edema in guinea pigs," Nippon Yakurigaku Zasshi, vol. 81, no. 2, pp. 105–113, 1983.

92. K. Takeya, Y. Hotta, N. Harada, G. Itoh, and J. Sakakibara, "Asebotoxin-induced centrogenic pulmonary hemorrhage in guinea pigs," The Japanese Journal of Pharmacology, vol. 31, no. 1, pp. 137–140, 1981. View at Scopus

93. H. Hikino, M. Ogura, S. Fushiya, C. Konno, and T. Takemoto, "Stereostructure of asebotoxin VI, VIII, and IX, toxins of Pieris japonica," Chemical & Pharmaceutical Bulletin, vol. 25, no. 3, pp. 523–524, 1971. View at Scopus

94. H. Hikino, T. Ohta, M. Ogura, Y. Ohizumi, C. Konno, and T. Takemoto, "Structure activity relationship of ericaceous toxins on acute toxicity in mice," Toxicology and Applied Pharmacology, vol. 35, no. 2, pp. 303–310, 1976.

95. H. Hikino, M. Ogura, and T. Takemoto, "Stereostructure of asebotoxin VII, toxin of Pieris japonica," Chemical & Pharmaceutical Bulletin, vol. 19, no. 9, pp. 1980–1981, 1971.

96. R. J. Peters, "Two rings in them all: the labdane-related diterpenoids," Natural Product Reports, vol. 27, no. 11, pp. 1521–1530, 2010. View at Publisher · View at Google Scholar

97. F. Berrue and R. G. Kerr, "Diterpenes from gorgonian corals," Natural Product Reports, vol. 26, no. 5, pp. 681–710, 2009. View at Publisher · View at Google Scholar · View at Scopus

98. T. Busch and A. Kirschning, "Recent advances in the total synthesis of pharmaceutically relevant diterpenes," Natural Product Reports, vol. 25, no. 2, pp. 318–341, 2008. View at Publisher · View at Google Scholar · View at PubMed · View at Scopus

99. J. R. Hanson, "Diterpenoids," Natural Product Reports, vol. 24, no. 6, pp. 1332–1341, 2007. View at Publisher · View at Google Scholar ·

View at PubMed · View at Scopus

100. R. A. Keyzers, P. T. Northcote, and M. T. Davies-Coleman, "Spongian diterpenoids from marine sponges," Natural Product Reports, vol. 23, no. 2, pp. 321–334, 2006. View at Publisher · View at Google Scholar · View at PubMed · View at Scopus
101. J. R. Hanson, "Diterpenoids," Natural Product Reports, vol. 22, no. 5, pp. 594–602, 2005. View at Publisher · View at Google Scholar · View at PubMed · View at Scopus
102. J. B. Hendrickson, "Molecular geometry. I. Machine computation of the common rings," Journal of the American Chemical Society, vol. 83, no. 22, pp. 4537–4547, 1961. View at Scopus
103. J. B. Hendrickson, "Molecular geometry. II. Methyl-cyclohexanes and cycloheptanes," Journal of the American Chemical Society, vol. 84, no. 17, pp. 3355–3359, 1962. View at Scopus
104. J. B. Hendrickson, "Sesquiterpenes-IV. Conformational analysis in the perhydroazulenic sesquiterpenes," Tetrahedron, vol. 19, no. 9, pp. 1387–1396, 1963. View at Scopus
105. J. B. Hendrickson, "Molecular geometry. IV. The medium rings," Journal of the American Chemical Society, vol. 86, no. 22, pp. 4854–4866, 1964. View at Scopus
106. J. B. Hendrickson, "Molecular geometry. V. Evaluation of functions and conformations of medium rings," Journal of the American Chemical Society, vol. 89, no. 26, pp. 7036–7043, 1967.View at Scopus
107. J. B. Hendrickson, "Molecular geometry. VI. Methyl-substituted cycloalkanes," Journal of the American Chemical Society, vol. 89, no. 26, pp. 7043–7046, 1967. View at Scopus
108. J. B. Hendrickson, "Molecular geometry. VII. Modes of interconversion in the medium rings,"Journal of the American Chemical Society, vol. 89, no. 26, pp. 7047–7061, 1967. View at Scopus
109. J. B. Hendrickson, R. K. Boeckman, J. D. Glickson, and E. Grunwald, "Molecular geometry. VIII. Proton magnetic resonance studies of cycloheptane conformations," Journal of the American Chemical Society, vol. 95, no. 2, pp. 494–505, 1973.
110. P. J. de Clercq, "Systematic conformational analysis. General method for rapid conformational evaluation. Its application to the hydroazulene system," The Journal of Organic Chemistry, vol. 46, no. 4, pp. 667–675, 1981. View at Scopus
111. P. J. de Clercq, "Systematic conformational analysis. Torsion constraint evaluation in cyclic systems," Tetrahedron, vol. 37, no. 24, pp. 4277–4286, 1981. View at Scopus

112. P. J. de Clercq, "Systematic conformational analysis. A microcomputer method for the semiquantitative evaluation of polycyclic systems containing five-, six- and seven-membered rings. 1. Program characteristics," Tetrahedron, vol. 40, no. 19, pp. 3717–3727, 1984. View at Scopus

113. P. J. de Clercq, "Systematic conformational analysis. A microcomputer method for the semiquantitative evaluation of polycyclic systems containing five-, six- and seven-membered rings. 2. Scope and limitations," Tetrahedron, vol. 40, no. 19, pp. 3729–3738, 1984. View at Scopus

114. J. Hoflack and P. J. de Clercq, "The sca program: an easy way for the conformational evaluation of polycyclic molecules," Tetrahedron, vol. 44, no. 21, pp. 6667–6676, 1988. View at Scopus

115. P. J. de Clercq, "Systematic conformational analysis. General method for rapid conformational evaluation. Its application to the hydroazulene system," The Journal of Organic Chemistry, vol. 46, no. 4, pp. 667–675, 1981, Supplementary material and references cited therein.

116. Z. Hassan, H. Hussain, V. U. Ahmad et al., "Absolute configuration of 1β,10β-epoxydesacetoxymatricarin isolated from Carthamus oxycantha by means of TDDFT CD calculations," Tetrahedron Asymmetry, vol. 18, no. 24, pp. 2905–2909, 2007. View at Publisher ·View at Google Scholar · View at Scopus

117. S. Bercion, T. Buffeteau, L. Lespade, and M. A. C. D. Martin, "IR, VCD, ^{1}H and ^{13}C NMR experimental and theoretical studies of a natural guaianolide: unambiguous determination of its absolute configuration," Journal of Molecular Structure, vol. 791, no. 1–3, pp. 186–192, 2006.View at Publisher · View at Google Scholar · View at Scopus

118. F. A. Macías, V. M. I. Viñolo, F. R. Fronczek, G. M. Massanet, and J. M. G. Molinillo, "11,16 Oxetane lactones. Spectroscopic evidences and conformational analysis," Tetrahedron, vol. 62, no. 33, pp. 7747–7755, 2006. View at Publisher · View at Google Scholar · View at Scopus

119. S. Milosavljevic, I. Juranic, V. Bulatovic et al., "Conformational analysis of guaianolide-type sesquiterpene lactones by low-temperature NMR spectroscopy and semiempirical calculations,"Structural Chemistry, vol. 15, no. 3, pp. 237–245, 2004. View at Publisher · View at Google Scholar

120. K. Schorr, A. J. García-Piñeres, B. Siedle, I. Merfort, and F. B. da Costa, "Guaianolides fromViguiera gardneri inhibit the transcription factor NF-κB," Phytochemistry, vol. 60, no. 7, pp. 733–740, 2002. View at

Publisher · View at Google Scholar · View at Scopus

121. T. J. Schmidt, "Helenanolide type sesquiterpene lactones. Part 1. Conformations and molecular dynamics of helenalin, its esters and 11,13-dihydro derivatives," Journal of Molecular Structure, vol. 385, no. 2, pp. 99–112, 1996. View at Publisher · View at Google Scholar · View at Scopus
122. T. J. Schmidt, F. R. Fronczek, and Y.-H. Liu, "Helenanolide-type sesquiterpene lactones. Part 2. The molecular conformations of arnifolin and some related helenanolides as determined by X-ray crystallographic and NMR spectroscopic analyses," Journal of Molecular Structure, vol. 385, no. 2, pp. 113–121, 1996. View at Publisher · View at Google Scholar
123. M. de Bernardi, L. Garlaschelli, L. Toma, G. Vidari, and P. Vita-Finzi, "The chemical basis of hot-tasting and yellowing of the mushrooms Lactarius chrysorrheus and L. scrobiculatus,"Tetrahedron, vol. 49, no. 7, pp. 1489–1504, 1993. View at Publisher · View at Google Scholar ·View at Scopus
124. M. T. Scotti, M. B. Fernandes, M. J. P. Ferreira, and V. P. Emerenciano, "Quantitative structure-activity relationship of sesquiterpene lactones with cytotoxic activity," Bioorganic & Medicinal Chemistry, vol. 15, no. 8, pp. 2927–2934, 2007. View at Publisher · View at Google Scholar · View at PubMed · View at Scopus
125. H. Matsuda, T. Kagerura, I. Toguchida, H. Ueda, T. Morikawa, and M. Yoshikawa, "Inhibitory effects of sesquiterpenes from bay leaf on nitric oxide production in lipopolysaccharide-activated macrophages: structure requirement and, role of heat shock protein induction," Life Sciences, vol. 66, no. 22, pp. 2151–2157, 2000. View at Scopus
126. T. J. Schmidt, "Toxic activities of sesquiterpene lactones: structural and biochemical aspects,"Current Organic Chemistry, vol. 3, no. 6, pp. 577–608, 1999. View at Scopus
127. E. Rodriguez, G. H. N. Towers, and J. C. Mitchell, "Biological activities of sesquiterpene lactones," Phytochemistry, vol. 15, no. 11, pp. 1573–1580, 1976. View at Scopus
128. S. M. Kupchan, M. A. Eakin, and A. M. Thomas, "Tumor inhibitors. 69. Structure-cytotoxicity relationships among the sesquiterpene lactones," Journal of Medicinal Chemistry, vol. 14, no. 12, pp. 1147–1152, 1971. View at Scopus
129. R. L. Hanson, H. A. Lardy, and S. M. Kupchan, "Inhibition of phosphofructokinase by quinone methide and α-methylene lactone tumor inhibitors," Science, vol. 168, no. 3929, pp. 378–380, 1970. View at Scopus

130. S. M. Kupchan, D. C. Fessler, M. A. Eakin, and T. J. Giacobbe, "Reactions of alpha methylene lactone tumor inhibitors with model biological nucleophiles," Science, vol. 168, no. 3929, pp. 376–378, 1970. View at Scopus

131. B. T. Zhuzbaev, S. M. Adekenov, and V. V. Veselovsky, "Approaches to the total synthesis of sesquiterpenoids of the guaiane series," Russian Chemical Reviews, vol. 64, no. 2, pp. 187–200, 1995.

132. C. H. Heathcock, C. M. Tice, and T. C. Germroth, "Synthesis of sesquiterpene antitumor lactones. 10. Total synthesis of (±)-parthenin," Journal of the American Chemical Society, vol. 104, no. 22, pp. 6081–6091, 1982. View at Scopus

133. C. H. Heathcock, E. G. DelMar, and S. L. Graham, "Synthesis of sesquiterpene antitumor lactones. 9. The hydronaphthalene route to pseudoguaianes. Total synthesis of (±)-confertin," Journal of the American Chemical Society, vol. 104, no. 7, pp. 1907–1917, 1982. View at Scopus

134. P. de Clercq and M. Vandewalle, "Total synthesis of (±)-damsin," The Journal of Organic Chemistry, vol. 42, no. 21, pp. 3447–3450, 1977. View at Scopus

135. J. A. Marshall and W. R. Snyder, "Total synthesis of (±)-4-deoxydamsin. Structure correlation of pseudoguaianolide sesquiterpenes," The Journal of Organic Chemistry, vol. 40, no. 11, pp. 1656–1659, 1975. View at Scopus

136. J. H. Rigby and J. Z. Wilson, "Total synthesis of guaianolides: (±)-dehydrocostus lactone and (±)-estafiatin," Journal of the American Chemical Society, vol. 106, no. 26, pp. 8217–8224, 1984. View at Scopus

137. T. J. Brocksom, U. Brocksom, and F. P. Barbosa, "The enantioselective synthesis of (R)-(+)-6-isopropenyl-3-methyl-2-cycloheptenone," Journal of the Brazilian Chemical Society, vol. 17, no. 4, pp. 792–796, 2006. View at Scopus

138. D. A. Foley and A. R. Maguire, "Synthetic approaches to bicyclo[5.3.0] decane sesquiterpenes,"Tetrahedron, vol. 66, no. 6, pp. 1131–1175, 2010. View at Publisher · View at Google Scholar ·View at Scopus

139. J. Méndez-Andino and L. A. Paquette, "Tandem development of aqueous indium chemistry and ring-closing metathesis as a general route to fused-ring α-methylene-γ-butyrolactones," Advanced Synthesis and Catalysis, vol. 344, no. 3-4, pp. 303–311, 2002. View at Publisher · View at Google Scholar · View at Scopus

140. N. Petragnani, H. M. C. Ferraz, and G. V. J. Silva, "Advances in the synthesis of α-methylenelactones," Synthesis, no. 3, pp. 157–183, 1986.

141. H. M. R. Hoffmann and J. Rabe, "Synthesis and biological activity of α-methylene-γ-butyrolactones," Angewandte Chemie International Edition in English, vol. 24, no. 2, pp. 94–110, 1985. View at Scopus

142. P. A. Grieco, "Methods for the synthesis of α-methylene lactones," Synthesis, no. 2, pp. 67–82, 1975. View at Scopus

143. R. B. Gammill, C. A. Wilson, and T. A. Bryson, "Synthesis of α-methylene-γ-butyrolactones,"Synthetic Communications, vol. 5, no. 4, pp. 245–268, 1975.

144. M. Garcia, Study on the chemical reactivity of eremanthine, M.Sc. dissertation, NPPN, Universidade Federal do Rio de Janeiro, Rio de Janeiro, Brazil, 1975.

145. A. J. R. da Silva, Techniques of nuclear magnetic resonance applied to the study of some sesquiterpenes, M.Sc. dissertation, NPPN, Universidade Federal do Rio de Janeiro, Rio de Janeiro, Brazil, 1976.

146. F. W. L. Machado, Structural modifications of isoeremanthine. Synthesis of eregoyazin and eregoyazidin, M.Sc. dissertation, NPPN, Universidade Federal do Rio de Janeiro, Rio de Janeiro, Brazil, 1977.

147. L. A. Maçaira, Structural modifications of eremanthine. Synthesis of dehydrocostus lactone and estafiatin, M.Sc. dissertation, NPPN, Universidade Federal do Rio de Janeiro, Rio de Janeiro, Brazil, 1978.

148. A. A. S. Rodrigues, Biomimetic transformations of costunolide and eremanthine, M.Sc. dissertation, NPPN, Universidade Federal do Rio de Janeiro, Rio de Janeiro, Brazil, 1979.

149. P. D. D. B. Lima, Development of a new specific method for the isolation of α-methylene lactones. Reinvestigation of Vanillosmopsis erythropappa Sch. Bip., M.Sc. dissertation, NPPN, Universidade Federal do Rio de Janeiro, Rio de Janeiro, Brazil, 1983.

150. J. C. F. Alves, Inversion of the lactonic fusion on eremanthine derivatives and synthesis of micheliolide, M.Sc. dissertation, Universidade Federal Rural do Rio de Janeiro, Rio de Janeiro, Brazil, 1993.

151. L. A. Maçaira, M. Garcia, and J. A. Rabi, "Chemical transformations of abundant natural products. 3. Modifications of eremanthin leading to other naturally occurring guaianolides," The Journal of Organic Chemistry, vol. 42, no. 26, pp. 4207–4209, 1977. View at Scopus

152. S. B. Mathur, S. V. Hiremath, G. H. Kulkarni et al., "Terpenoids-LXX. Structure of dehydrocostus lactone," Tetrahedron, vol. 21, no. 12, pp. 3575–3590, 1965. View at Scopus

153. W. Vichnewski, F. W. L. Machado, J. A. Rabi, R. Murari, and W. Herz, "Eregoyazin and eregoyazidin, two new guaianolides from

Eremanthus goyazensis," The Journal of Organic Chemistry, vol. 42, no. 24, pp. 3910–3913, 1977. View at Scopus

154. J. A. Rabi, M. Garcia, L. A. Maçaira, and F. W. L. Machado, "Chemical transformations of eremanthine: a way for the synthesis of cercareacide agents," Anais da Academia Brasileira de Ciências, vol. 49, no. 4, pp. 563–565, 1977.
155. F. Sánchez-Viesca and J. Romo, "Estafiatin, a new sesquiterpene lactone isolated from Artemisia mexicana (Willd)," Tetrahedron, vol. 19, no. 8, pp. 1285–1291, 1963. View at Scopus
156. L. A. Maçaira, F. W. L. Machado, M. Garcia, and J. A. Rabi, "Unambiguous transformation of eremanthin into (-)-estafiatin," Tetrahedron Letters, vol. 21, no. 9, pp. 773–776, 1980. View at Scopus
157. M. Garcia, F. W. L. Machado, L. A. Maçaira, and J. A. Rabi, "The reaction of eremanthin and isoeremanthin with bromine. Unprecedented simultaneous addition of Br_2 to two isolated but topographically related double bonds," Tetrahedron Letters, vol. 21, no. 9, pp. 777–780, 1980.View at Scopus
158. A. J. R. da Silva, M. Garcia, P. M. Baker, and J. A. Rabi, "^{13}C NMR spectra of natural products. 1-guaianolides," Organic Magnetic Resonance, vol. 16, no. 3, pp. 230–233, 1981.
159. E. C. Fantini and J. A. Rabi, "Methoxyl group as protector of α-methylene-γ-lactones," Ciência e Cultura, vol. 35, no. 7, suplemento, p. 403 (49-D.2.3), 1983.
160. E. C. Fantini and J. A. Rabi, "Synthesis of 6-epi-eremanthine," Ciência e Cultura, vol. 37, no. 7, suplemento, p. 417 (27-D.2.3), 1985.
161. J. L. P. Ferreira, E. C. Fantini, and J. A. Rabi, "Participation of neighboring-group on the methanolysis of α-methylene-γ-lactones derived from eremanthine," Ciência e Cultura, vol. 35, no. 7, suplemento, p. 403 (48-D.2.3), 1983.
162. E. C. Fantini, J. L. P. Ferreira, and J. A. Rabi, "Participation of neighboring-groups on the methanolysis of α-methylene-γ-lactones derived from eremanthine. II-Reaction conditions and structural requirements," Ciência e Cultura, vol. 36, no. 7, suplemento, p. 487 (73-D.2.3), 1984.
163. E. C. Fantini, J. L. P. Ferreira, and J. A. Rabi, "Metal ion promoted methanolysis of sesquiterpene lactones leading to O6,15-cycloguaiane methyl esters," Journal of Chemical Research (Synopses), no. 8, pp. 298–299, 1986.
164. E. C. Fantini and J. A. Rabi, "Elimination versus formation of 15,6-α-oxidos on 4-α-hydroxy-15-iodine derived from eremanthine,"

Ciência e Cultura, vol. 38, no. 7, suplemento, p. 521 (55-D.2.3), 1986.

165. M. Ogura, G. A. Cordell, and N. R. Farnsworth, "Anticancer sesquiterpene lactones of Michelia compressa (magnoliaceae)," Phytochemistry, vol. 17, no. 5, pp. 957–961, 1978. View at Scopus

166. J. C. F. Alves, J. A. Rabi, and E. C. Fantini, "Trans-cis inversion of the lactonic fusion on eremanthine derivatives. I—preliminary studies," in Abstracts of the 14a Reunião Anual da Sociedade Brasileira de Química, (IC-32), Caxambu, Brazil, 1991.

167. J. C. F. Alves and E. C. Fantini, "Trans-cis inversion of the lactonic fusion on eremanthine derivatives. II—study with substrates bearing double bonds at the positions 1,10; 3,4 and 9,10; 1,10 and 4,15," in Abstracts of the 16a Reunião Anual da Sociedade Brasileira de Química, (QO-24), Caxambu, Brazil, 1993.

168. J. C. F. Alves and E. C. Fantini, "Chemical transformations of eremanthine. Synthesis of micheliolide and 1(R),10(R)-dihydromicheliolide," Journal of the Brazilian Chemical Society, vol. 16, no. 4, pp. 749–755, 2005.

169. J. C. F. Alves and E. C. Fantini, "Erratum: Chemical transformations of eremanthine: Synthesis of micheliolide and 1(R),10(R)-dihydromicheliolide," Journal of the Brazilian Chemical Society, vol. 21, no. 5, p. 946, 2010.

170. J. C. F. Alves and E. C. Fantini, "Study of the inversion reaction of the lactonic fusion on eremanthine derivatives," Journal of the Brazilian Chemical Society, vol. 18, no. 3, pp. 643–664, 2007. View at Scopus

171. J. C. F. Alves, "Study of catalytic hydrogenation and methanol addition to α-methylene-γ-lactone of eremanthine derivatives," Organic Chemistry International, vol. 2010, Article ID 603436, 11 pages, 2010. View at Publisher · View at Google Scholar

172. H. Deshayes, J. P. Pete, C. Portella, and D. Scholler, "Photolysis of carboxylic esters: conversion of alcohols into alkanes," Journal of the Chemical Society, Chemical Communications, no. 11, pp. 439–440, 1975. View at Publisher · View at Google Scholar · View at Scopus

173. R. K. Crossland and K. L. Servis, "A facile synthesis of methanesulfonate esters," The Journal of Organic Chemistry, vol. 35, no. 9, pp. 3195–3196, 1970. View at Scopus

174. R. O. Hutchins, D. Kandasamy, C. A. Maryanoff, D. Masilamani, and B. E. Maryanoff, "Selective reductive displacement of alkyl halides and sulfonate esters with cyanoborohydride reagents in hexamethylphosphoramide," The Journal of Organic Chemistry, vol. 42, no. 1, pp. 82–91, 1977.View at Scopus

175. J. March, Advanced Organic Chemistry, John Wiley & Sons, New York, NY, USA, 3rd edition, 1985.

176. H. O. House, Modern Synthetic Reactions, W. A. Benjamin, Menlo Park, California, USA, 2nd edition, 1972.

177. J. C. F. Alves would like to explain that the incorrectness previously committed in the writing of works about chemical transformations of eremanthine [150, 166–167] were, whenever possible, corrected and revised in works published on this decade [168–170].

Chapter 11

REGIONAL-SCALE OZONE DEPOSITION TO NORTH-EAST ATLANTIC WATERS

L. Coleman, S. Varghese, O. P. Tripathi, S. G. Jennings, and C. D. O'Dowd

School of Physics and Centre for Climate and Air Pollution Studies, National University of Ireland, Galway, Ireland

ABSTRACT

A regional climate model is used to evaluate dry deposition of ozone over the North East Atlantic. Results are presented for a deposition scheme accounting for turbulent and chemical enhancement of oceanic ozone deposition and a second non-chemical, parameterised gaseous dry deposition scheme. The first deposition scheme was constrained to account for sea-surface ozone-iodide reactions and the sensitivity of modelled ozone concentrations to oceanic iodide concentration was investigated. Simulations were also performed using nominal reaction rate derived from in-situ ozone deposition measurements and using a preliminary representation of organic chemistry. Results show insensitivity of ambient ozone concentrations modelled by the chemical-enhanced scheme to oceanic iodide concentrations, and iodide reactions alone cannot account for observed deposition velocities. Consequently, we suggest a missing chemical sink due to reactions of ozone with organic matter at the air-sea interface. Ozone loss rates are estimated to be in the range of 0.5–6 ppb per day. A potentially significant ozone-driven flux of iodine to the atmosphere is estimated to be in the range of 2.5–500 M molec , leading to a mixing-layer enhancement of organo-iodine concentrations of 0.1–22.0 ppt, with an average increase in the N.E. Atlantic of around 4 ppt per day.

INTRODUCTION

Ozone plays a key role in atmospheric chemistry, absorbing harmful UV rays in the stratosphere whilst simultaneously acting as a greenhouse gas (radiative forcing of tropospheric ozone is around 25% that of CO_2 [1]), and acting as a harmful pollutant in the troposphere [2–4]. Influencing the oxidising capacity of the atmosphere as a powerful oxidising agent, it is the dominant precursor to the ubiquitous hydroxyl radical which acts as an atmospheric cleansing agent by determining the lifetime of important atmospheric trace gases. It is vital, therefore, that tropospheric ozone concentrations are realistically simulated in modelling both air pollution and chemistry-climate interactions.

600–1000 Tg O_3 year^{-1} is removed from the troposphere via dry deposition [5]. The deposition of ozone to water surfaces is small compared to deposition to land [6]: typically, the rate of dry deposition of ozone to the continent is nearly six times faster than the dry deposition rate of ozone to the ocean [7]. However, considering 70% of the globe has ocean coverage, the loss of marine boundary layer ozone via oceanic dry deposition still represents a significant sink for the global ozone budget. Obtaining an accurate prediction of ozone flux to the sea is imperative not only in predicting ambient ozone concentrations, but also because of the biogeochemical consequences of ozone reactions in the sea surface. For example, recent work [8, 9] has shown that ozonation of iodide in the sea surface can result in the formation of reactive organoiodine products that result in iodocarbon emissions from the sea surface. These iodocarbons photodissociate rapidly form iodine atoms which are known to catalytically destroy ozone [10], resulting in further reduction of marine boundary layer ozone levels. Also, ozonation of iodine atoms results in formation of iodine oxide (IO) radicals which have the potential to lead to new marine aerosol formation [11]. This newly discovered mechanism could have significant biogeochemical consequences, in terms of the feedback mechanism in halogen-mediated ozone destruction, the halogen source in coastal areas which could account for the hitherto unexplained elevated levels of IO observed at Cape Verde [12], and the subsequent formation of new particles which could influence the solar radiation budget.

There is still major uncertainty regarding the amount of ozone lost to the ocean by dry deposition [13]. The dry deposition of ozone to the ocean involves a number of complex processes: physical, chemical, and biological. Transport of ozone through the atmosphere depends on surface roughness, wind speed, and atmospheric turbulence; transport across the quasilaminar boundary layer is limited by the diffusivity of the gas in air, while the ocean-surface uptake of ozone depends on the water-side turbulence conditions, ozone solubility, and availability of reacting chemicals in the surface layer [14, 15]. Up to now, these processes have been poorly understood and have not generally been considered in deposition schemes commonly applied in atmospheric chemistry models. The difficulty in the definition, and therefore the parameterisation of these processes is exacerbated by the scarcity of in situ measurements of ozone dry deposition velocities. Wesely and Hicks [16] found deposition velocities of ozone to the sea surface to vary between 0.01 and 0.05 cm s^{-1}. However, the eddy correlation studies of Gallagher et al. [17] carried out in the North Sea indicate deposition velocities as high as 0.1 cm s^{-1}.

The rate of gaseous dry deposition to a surface is parameterised by deposition velocity, (cm s^{-1}). is used to calculate the downward flux of ozone to the ocean, = where C denotes surface level gas concentration. is computed (in analogy with electrical transport) as the reciprocal of the sum of the resistances encountered by the gas on its journey to the surface sink. Using the standard resistance model to compute deposition velocity, resistances to gaseous deposition are atmospheric resistance, quasilaminar resistance and surface resistance [14].

The gaseous dry deposition scheme of Wesely [18] is widely applied in air quality, atmospheric chemistry, and chemistry climate models to represent gaseous dry depositional sinks. In its evaluation of deposition velocity, this scheme calculates explicitly only aerodynamic resistance and resistance to transfer across the quasilaminar surface layer. Surface resistance is calculated as a series of smaller resistances which are generally provided by look-up tables, with values differing according to land use type, species in question and season. In this system, derivation of surface resistance to deposition at water surfaces is neglected and surface resistance is set to a constant value of 2000 m^{-1} s. The same surface resistance is applied to all water surfaces, irrespective of water body classification, climate, or meteorological conditions, thereby grossly simplifying the complex mechanisms involved in surface transfer. Surface resistance is the most significant parameter in the case of ozone deposition to the ocean [19] and in reality, surface resistance is greatly diminished by turbulence in the ocean surface and by the presence of ocean-depleting chemicals in the ocean surface—namely iodide [20, 21], and organic matter [22], for example, chlorophyll [23] and DMS [5, 20]. Due to the highly reactive nature of ozone, these ocean surface reactions can have a very significant enhancing effect on oceanic ozone deposition and the chemical reactions between ozone and organic matter and chlorophyll have yet to be parameterised and constrained in dry deposition models.

Due to it's parameterisation of surface resistance, the dry deposition scheme of Wesely [18] underestimates deposition rate of ozone to the ocean at both high- and low-wind speeds compared to observations [20].

The advanced Fairall et al. [13] scheme described in Section 2.1 is an explicit parameterisation of ozone deposition to the ocean, allowing for enhanced ozone deposition due to ocean surface turbulence and ocean surface layer reactions.

The Fairall et al. scheme has recently been incorporated into a global model, with analysis, discussion on role of controlling parameters, and validation the parameterisation conducted by Ganzeveld et al. [5]. In this study, the parameterisation is scaled to include reactions of ozone with iodide, dimethyl sulfide (DMS), alkenes, and organic chemistry. The study finds the role of biogeochemistry to dominate in computation of deposition velocities in the tropical and subtropical regions whereas the turbulent forcing of ozone deposition dominated over biogeochemical factors in the mid to high latitudinal regions. Simulation results indicate a small sensitivity of marine boundary layer ozone concentrations to varying biogeochemical and turbulent conditions, despite the wide range of deposition rates simulated using the Fairall et al. [13] scheme which is scaled to include organic reactions. This occurs partly due to interaction of different factors affecting deposition velocity. For example, in tropical regions, the reducing effect of high water temperature values on ozone solubility in water counteracted higher chemical ozone transfer due to elevated iodide concentrations. As a result, applying enhanced iodide concentrations in coastal regions did not explain discrepancies between observed and simulated . Furthermore, the lack of sensitivity of simulated boundary layer ozone concentration to dry deposition velocity illustrates the role of compensating effects in climate models due to atmospheric transport and chemistry which maintain ozone concentrations within the marine boundary layer, despite the temporal and spatial variability in oceanic ozone uptake. The global model study also found the use of the advanced Fairall et al. [13] scheme in the model only slightly reduced (6%) the total dry deposition flux of ozone to the ocean compared to simulations ran using the Wesely [18] scheme with constant surface-resistance. This shows that the Wesely [18] scheme is appropriate for modelling ozone deposition on a global scale for an annual period, but it is thought that due to the North East. Atlantic region being both organically active and prone to turbulent conditions, use of the chemically enhanced Fairall et al. [13] scheme on ozone concentrations in this area may be more significant due to regional-scale, short-term effects than indicated by findings of the global model study of Ganzeveld et al. [5]. Also, as the effect of temperature on molecular diffusivity is accounted for in this study, it is thought that the limiting effect of ozone water solubility on computed deposition velocities will be less pronounced than found in the global model study (see Section 2.2.3). Furthermore, in contrast with the global model simulations, in this study the effect of salinity on ozone solubility, diffusivity, and ozone-iodide reaction rates are included. Our analysis shows that effect of salinity on ozone-iodide reactions can enhance reaction rates by as much as 23% (Section

2.3.1). Including the effect of salinity on key parameters accounts for further variations between results the global model study and results of regional model simulations presented here.

For this study, the Fairall et al. [13] parameterisation has been adapted to account for enhanced deposition due to iodide-ozone reactions in the ocean-surface layer. Furthermore, a mechanistic scaling to account for marine organic chemistry using oceanic chlorophyll concentrations as a proxy for organic activity has been included in the scheme. The parameterisation is then incorporated into the regional climate model with tracer extensions (REMOTE) [24, 25] and the sensitivity of simulated ozone flux and ambient ozone concentrations to varying oceanic iodide concentrations and inclusion of organic chemistry is investigated.

METHODS

Details of the Fairall et al. [13] Parameterisation of Oceanic Ozone Deposition

The ozone deposition parameterisation of Fairall et al. [13] calculates oceanic ozone dry deposition velocity by integration of the turbulent-molecular transport equation in the ocean, while accounting for increased deposition of ozone to the ocean due to sea surface turbulence and due to surface-level chemical reactions. The sea-surface turbulence is a function of wind speed, water density, and the buoyancy flux of water [26].

Chemical reactions are integrated into the scheme by way of reactivity time-scale term, (s^{-1}) which characterises the time scale of a chemical reaction for ozone in the sea water. It is calculated as the product of the sea-surface concentration of the ozone reactant () and a second order rate constant () for the reaction in question

$$A_{ozone} = \sum C_i k_i.$$

Instead of using the three-resistance approach of Wesely [18], this parameterisation combines atmospheric and quasilaminar resistance into one resistance term which is calculated using the tropical ocean global atmosphere coupled ocean-atmosphere response experiment (TOGA-COARE) gas transfer model [27, 28]. The bulk of the parameterisation serves the computation of the turbulence-dependent surface resistance with an added chemical enhancement term. The scheme has been modified to allow for variation of iodide reaction rates with temperature [29], variation of ozone diffusivity in water due to temperature [30] and salinity [31], and variation of ozone solubility with temperature [32].

Sensitivity Analysis

Turbulent/Nonturbulent Scheme

The Fairall et al. [13] scheme computes surface resistance by first solving the conservation equation for the case of negligible turbulence. From this solution, the turbulent solution is calculated by considering a turbulent eddy diffusivity term obtained using surface-layer similarity scaling [27]. From the scheme, deposition velocities are effectively calculated for two cases: negligible turbulence and nonnegligible turbulence. Figure 1 depicts deposition velocities obtained from the turbulent and nonturbulent schemes as a function of reactivity for various wind speeds, displaying how the turbulent and nonturbulent schemes converge for very high water-side reactivity values. This convergence occurs due to destruction of ozone in the surface layer occurring so rapidly that ocean turbulence has no enhancing effect on deposition velocity. The magnitude of the effect of ocean surface turbulence on deposition velocity depends on the reactivity term. For high-wind speeds (conditions of significant ocean surface turbulence), the deposition velocity is less dependent on reactivity than in the nonturbulent case (lower wind speeds). This illustrates the dominance of turbulent forcing over the reactive sink at high-wind speed.

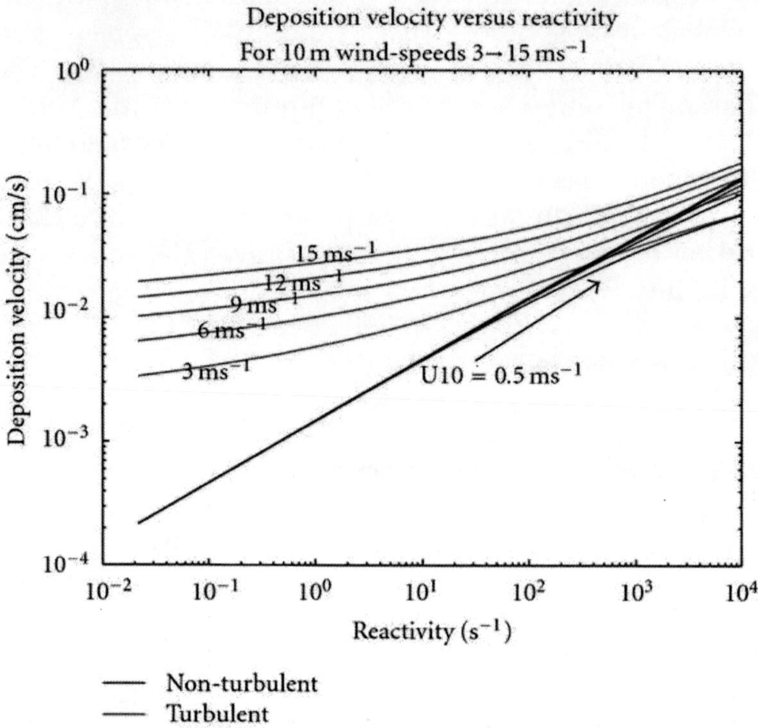

Figure 1: Comparison of deposition velocity as a function of reactivity for the nonturbulent and the turbulent Fairall et al. [13] scheme for differing 10 m wind-speeds (U10). Nonturbulent deposition velocity is obtained by solution of the basic ozone conservation equation, whereas the turbulent deposition velocity is obtained by solution of the conservation equation including a turbulent eddy diffusivity term that is obtained from surface-layer similarity scaling [27]. Note how both schemes converge for high reactivity values. This occurs because in incidences of very high oceanic reactivity, ozone is destroyed so rapidly in the ocean surface layer that ocean turbulence has no enhancing effect on deposition velocity.

Sensitivity of Deposition Velocity to Reactivity Term

The deposition velocities predicted by the Fairall et al. [13] scheme following a simulation in REMOTE were plotted against their corresponding wind speeds for various reactivity values, in order to look at the sensitivity of deposition velocity to the reactivity time scale factor (), as shown in Figure 2. Deposition velocities computed by the Wesely [18] scheme were also plotted in Figure 2, for comparative purposes. A significant increase in deposition velocity from the Fairall et al. [13] scheme is observed in Figure 2 for wind speeds over 4 m s^{-1} for reactivities exceeding 1000 s^{-1}. At lower reactivities, the Wesely [18] scheme predicts higher deposition velocities, especially at low-wind speeds.

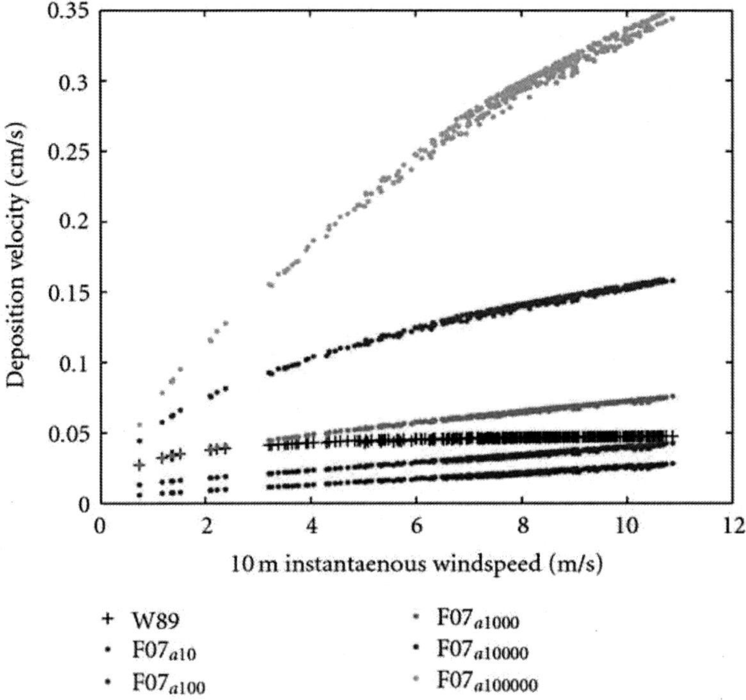

Figure 2: Deposition velocities computed from the Fairall et al. [13] (F07) and Wesely [18] (W89) schemes running within REMOTE. In the figure legend above, the subscript following the F07 label refers to the reactivity time-scale factor in operation for that scheme.

The reactivity value needed in the Fairall et al. [13] parameterisation to match deposition velocities computed by the Wesely [18] scheme depends on the wind speed value. For low-wind speeds (under 4 m s^{-1}), reactivity of 1000 s^{-1} yields similar deposition velocities to those predicted by Wesely [18]. For high-wind speeds (over 10 m s^{-1}), water-side reactivity of 100 s^{-1} would yield deposition velocities in the same range as those predicted by Wesely [18]. Therefore, for deposition velocities exceeding those predicted by the nonchemical, nonturbulent Wesely [18] scheme, reactivity in the ocean should exceed 1000 s^{-1}. At low reactivities, the Fairall et al. [13] parameterisation predicts a linear relationship between deposition velocity and wind speed, due to the dominance of the turbulent driven deposition over the chemical sink, as discussed in the previous section.

In their sensitivity analysis of the parameterisation, Fairall et al. [13] found that using the chemically enhanced turbulent deposition scheme, typical observed ozone deposition velocities of 0.05 cm s^{-1} would require a reactivity rate of 1000 s^{-1}. This can be observed from Figure 2. Gallagher et al. [17] observed ozone deposition velocities to the ocean as high as 0.1 cm s^{-1} in the North Sea. Deposition velocities of this magnitude would require a reactivity rate of the order of 10^4 s^{-1} using the Fairall et al. [13] scheme at low to moderate wind speeds.

Sensitivity of Deposition Velocity to Sea-Surface Temperature (SST)

As mentioned in Section 2.1, the Fairall et al. [13] parameterisation has been constrained to allow for variation of ozone diffusivity in water due to temperature [30] and salinity [33] and variation of ozone solubility with temperature [34]. Using box model simulations, the effect of SST on the deposition velocity was investigated for typical North Eastern Atlantic conditions (wind speed of 8 m s^{-1} and oceanic reactivity of 500 s^{-1}). See Figure 3.

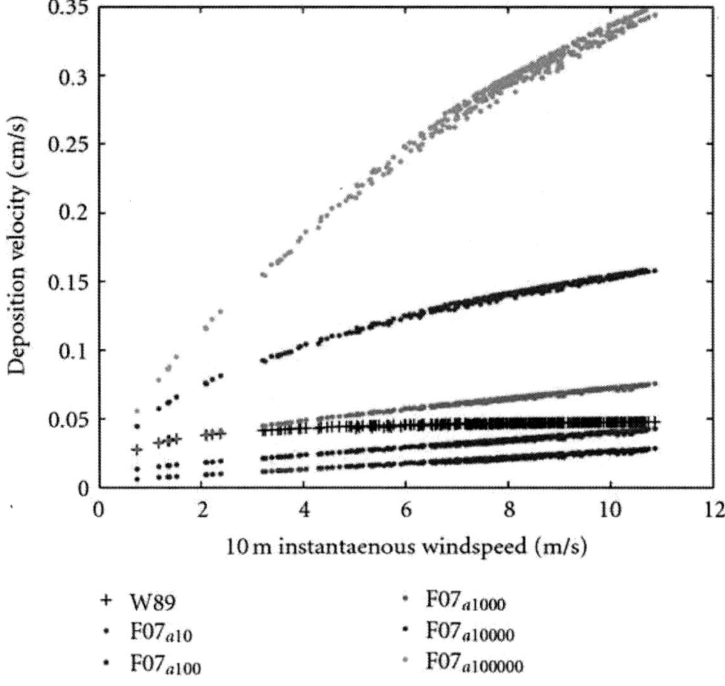

Figure 3: Box-model-derived relationship between deposition velocity and sea-surface temperature, using a typical Atlantic wind speed of 8 m s^{-1} and reactivity of 500 s^{-1}. The solid line depicts a relationship between deposition velocity and SST when only temperature dependence of ozone solubility in seawater is considered based on the theory of Kosak-Channing and Helz [34]. The filled circles represent a relationship between deposition velocity and SST when temperature dependence of both ozone solubility and molecular diffusivity in seawater is considered. Variation of ozone diffusivity in seawater is based on the theory of Johnson and Davis [30] and Jahne et al. [31]. Linear correlations between ozone deposition velocity and SST are given in the legend.

Results from box-model simulations confirm that inclusion of temperature dependence of molecular diffusivity renders the simulated deposition velocities less sensitive to SST variations. From Figure 3, the slope of the relationship between ozone deposition velocity and SST is 50% steeper when the effect of SST on molecular diffusivity of ozone is not considered; in this case, the computed deposition velocities are likely to be oversensitive to SST. As discussed in the introduction to this paper, the effect of including variability of ozone diffusivity with SST is thought to compensate for the limiting effect of low-ozone solubility on simulated ozone deposition velocities, compared to the results of the global model study of Ganzeveld et al. [5].

Chemical Scaling of the Reactivity Term

Accounting for chemical reactions in the Fairall et al. [13] parameterisation requires scaling of the chemical reactivity term , as defined in (1). In their exploration of impact of chemical reactions on ozone deposition, Chang et al. [20] identified iodide as the most likely chemical compound residing in the sea surface to drive ozone deposition compared to other substances (DMS, alkenes). DMS was recognised as having potential to enhance ozone deposition, but only at extreme oceanic concentrations. At mean oceanic concentrations of DMS, deposition velocity due to molecular gas transfer and chemical reactions of ozone with DMS was an order of magnitude lower than the deposition velocity due to molecular diffusion and chemical reactions of ozone and iodide and deposition velocity due to molecular gas transfer. Reactions of ozone and alkenes were a further two orders of magnitude lower again. For this reason, reactions of ozone with DMS and alkenes were not considered in this study, their chemically enhancing effects on ozone deposition being overshadowed by iodide reactions. Chang et al. [20] did not investigate the effect of chlorophyll on ozone deposition which Clifford et al. [23] found to have a significant enhancment effect of the same order of magnitude as for iodide reactions. In this study, analysis of chemical enhancement to ozone deposition is limited to iodide and chlorophyll reactions.

Iodide Reactions

To scale , the kinetics between ozone and it's various reactants needed to be determined. The enhancing effect of iodide reactions on ozone deposition has been long documented [20, 21]. From the kinetics of the ozone-iodide reaction derived by Magi et al. [29], relationships between ozone-iodide rate constant and water salinity were derived for different water temperatures. From this, a relationship between the second order reaction rate and saline water temperature T was deduced for water of ionic strength of seawater (0.7 M) resulting in the linear relationship of (2), where refers to the reaction rate of ozone in salt water of ionic strength 0.7 M and refers to water temperature in degrees Kelvin

$$k_{salt}(10^9\,s^{-1}) = -40.85 + 0.15\,T\,(K).$$

From the work of Magi et al. [29], the effect of salinity enhances ozone-iodide reaction rates by as much as 23% compared to those for pure water, and so it was considered important to account for the salinity of seawater in parameterising ozone-iodide reactions.

Oceanic iodide concentrations in the North Atlantic vary between 0 and 150 nM [35]. For this study, constant iodide concentrations were varied between 50 nM and 200 nM to evaluate the sensitivity of simulated ozone levels and resulting ozone fluxes to the effect of changing iodide concentrations, in contrast to the global modelling study of Ganzeveld et al. [5] who infer oceanic concentrations using an anticorrelation between nitrate and iodide developed by Campos et al. [36]. Furthermore, the global model study did not include the effect of salinity of seawater on ozone-iodide reaction kinetics, as is done in this study.

Organic Enhancement of Ozone Deposition

It has been long postulated that organic reactions in the sea surface layer can have a significant enhancing effect on ozone deposition [22]. Clifford et al. [23] suggest that the reaction between ozone and chlorophyll can increase ozone deposition velocity by up to a factor of 3 for wind speeds up to 20 m s^{-1}compared to deposition velocities computed solely on ozone-iodide reactions. Considering the ubiquity of chlorophyll in the ocean, the organic enhancement of ozone deposition could have a significant effect on ozone concentrations in the marine boundary layer. This organic enhancement of ozone deposition was mechanistically incorporated into the Fairall et al. [13] deposition scheme using satellite chlorophyll data from MODIS (moderate resolution imaging spectroradiometer) [37]. The upper and lower limits of oceanic chlorophyll concentrations were set to 3 mg m^{-3} and 0 mg m^{-3}, respectively, and oceanic iodide concentration was set to a typical value of 100 nM. Figure 4 shows monthly averaged chlorophyll concentrations for June 2003, as detected by MODIS. Although chlorophyll values as high as 30 mg m^{-3} in coastal regions are visible in Figure 4, the upper limit of chlorophyll concentration was set to 3 mg m^{-3}because open ocean chlorophyll concentrations in the North East Atlantic do not exceed this value. Also, when the resolution of the model is taken into consideration, chlorophyll values averaged over a single grid cell never exceed 3 mg m^{-3}, even in coastal areas. Therefore, 3 mg m^{-3} is a sensible upper limit for open ocean chlorophyll concentrations for this study.

Figure 4: Monthly averaged chlorophyll-a concentration (mg m^{-3}), as detected by MODIS for June, 2003. Taken from the NASA OceanColor website (http:// oceancolor.gsfc.nasa.gov/.)

Ozone deposition velocity was computed as before using the Magi et al. [29] iodide chemistry and increased according to a linear chlorophyll-dependent enhancement factor: deposition velocity increased by a factor of 3 at the upper limit of chlorophyll concentration values. Deposition velocity was left unchanged at the lower limit of chlorophyll concentration where deposition velocity was based only on iodide chemistry.

The technique employed in this study differs from that of Ganzeveld et al. [5] who applied linear dependence of reactivity term on chlorophyll concentrations whereas in this study, the deposition velocity increases linearly with chlorophyll and the organic reactions are not represented in . This organically enhanced version of the Fairall et al. [13] deposition scheme is herein referred to as the mechanistic ozone deposition scheme.

This scaling is a crude approach to the quantification of organic reactions in the computation of ozone deposition velocity and should be interpreted as a first-order representation of organic chemistry within the Fairall et al. [13] scheme, oceanic chlorophyll concentration being used as a proxy for biological activity. Further work is needed to parameterise the role of organic reactions in ozone deposition.

Ozone-Deposition Driven Upward Iodine Flux

The major source of iodine to the atmosphere is most likely due to emission of organoiodine compounds from the ocean [38]. Martino et al. [8] have recently shown that a proportion of organic iodide found in the marine atmosphere can be formed from volatile organoiodine compounds (VOIs) which are produced via ozone-iodide reactions in the sea surface, thereby uncovering a potentially significant iodine source to the marine atmosphere which is kick started by the deposition of ozone to the sea surface. The VOI species identified by Martino et al. [8] resulting from ozone deposition are CH_2I_2, CH_2ICl, and CHI_3. Based on the findings of Martino et al. [8] and the work of Garland et al. [21], the extreme upper limit of iodine vapours released to the atmosphere can be estimated from the ozone flux of ozone to the ocean, assuming all of the new VOIs are emitted from the ocean surface, without any being transported to the ocean mixed layer or being destroyed by photolysis at the sea-air interface. This newly defined halogen source in the marine boundary layer could have significant biogeochemical consequences, in terms of both marine boundary layer (MBL) ozone depletion and new particle formation.

SIMULATION RESULTS AND DISCUSSION

Monthly REMOTE simulations were carried out for the month of June, 2003. The model domain spans Northern Europe and is depicted in Figure 11. Resolution of the model is set to 0.5 deg, giving an average grid cell size of 50 km 50 km. Meteorological and chemical initial and boundary conditions are taken from ECMWF (European Centre for Medium-Range Weather Forecasts) and used to initialise all grid points at the start of each simulation. Lateral boundaries are enforced at six hourly intervals and emission scenarios were supplied from the EMEP site. In 2003, monthly average chlorophyll concentrations in the North East. Atlantic were at an annual peak in the month of June [37].

Ozone Concentration Fields

Figure 5 shows surface level mean monthly ozone concentrations simulated using different deposition schemes within REMOTE. There is little or no difference between simulated surface level ozone concentrations using the nonchemical Wesely [18] scheme and the Fairall et al. [13] scheme for 50 nM or 200 nM iodide concentrations, consistent with findings of the global model study of Ganzeveld et al. [5]. One can conclude from this that ozone concentrations predicted using the Fairall et al. [13] scheme within REMOTE are insensitive to iodide reactions at realistic Atlantic iodide concentrations—even at an upper limit of oceanic iodide concentration of 200 nM. Therefore, although variation in iodide concentration and reaction rate causes variation in ozone deposition velocity, these deposition velocities are not sufficiently significant to overcome other compensating effects in the climate model and substantially decrease simulated ozone concentrations. However, a notable decrease in ozone concentrations is observed around the North East Atlantic when the mechanistic ozone deposition scheme of Fairall et al. [13] is employed. Simulated ozone depletion in this region is due to the relatively high chlorophyll concentrations. In regions of low oceanic chlorophyll concentration (e.g., the Mediterranean), the mechanistic ozone deposition scheme models similar ozone concentrations as computed using the Wesely [18] scheme.

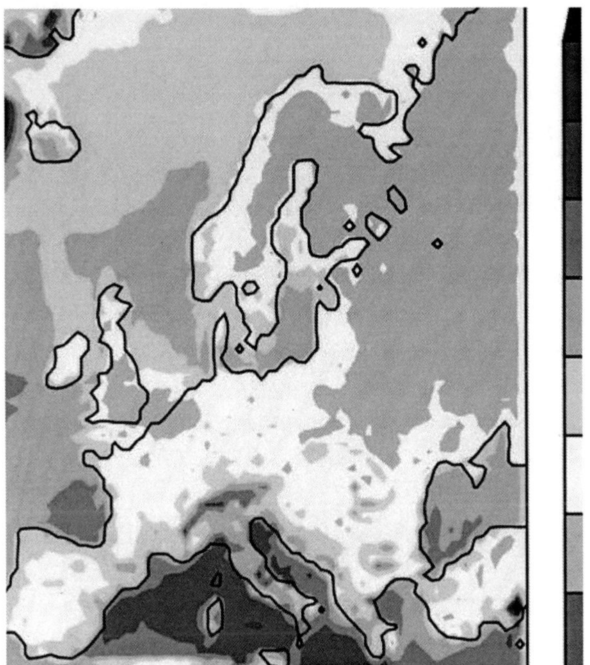

(a)

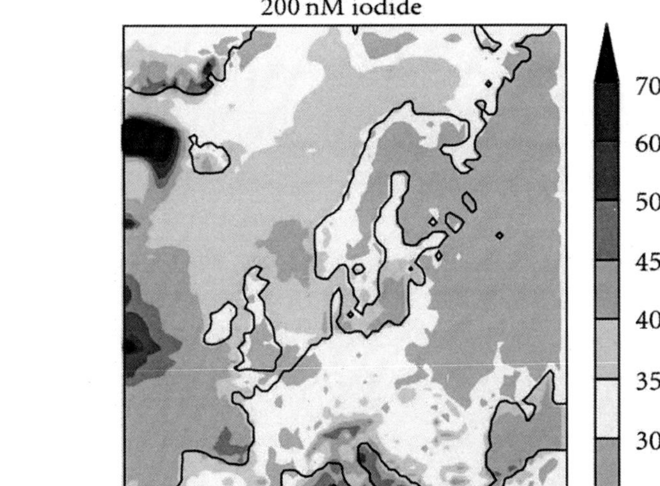

(b)

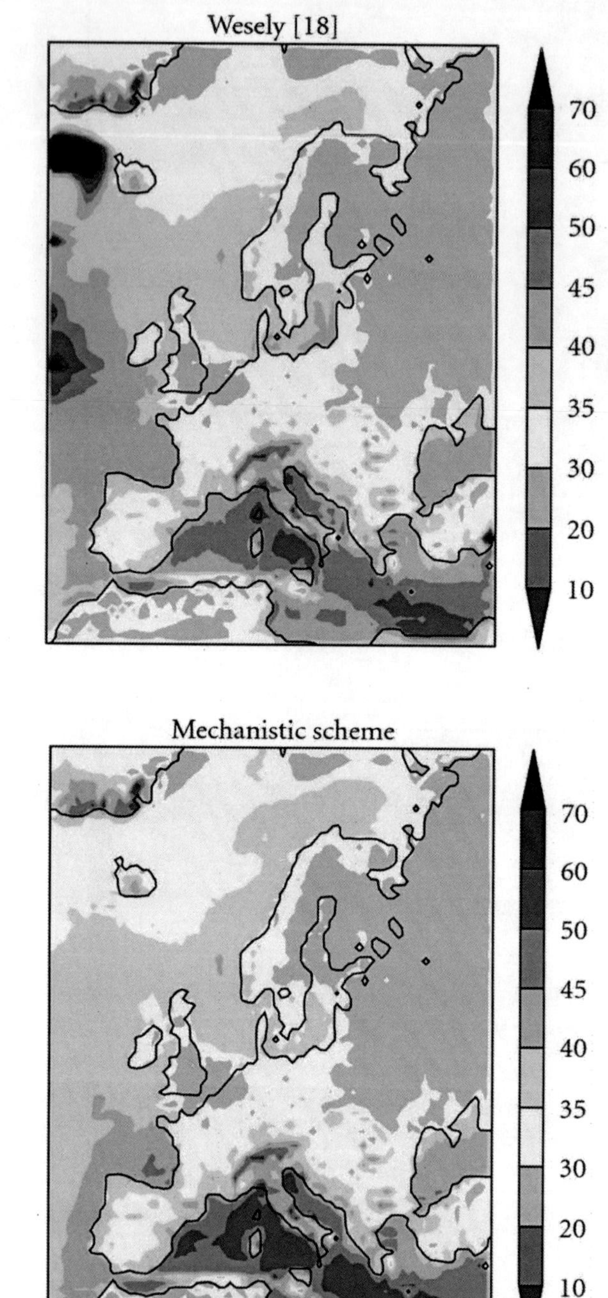

Figure 5: Monthly average O_3 concentrations in ppb as predicted by REMOTE

for June 2003 for various dry deposition schemes. 50 nM I plot and 200 nM I plot show simulated ozone concentrations (ppb) using the iodide-only chemical Fairall et al. [13] parameterisation with oceanic iodide concentrations set to 50 nM and 200 nM, respectively. The Wesely [18] plot shows the ozone concentrations simulated by the nonchemical scheme and the Mechanistic Scheme plot depicts the O_3 concentrations simulated using the Fairall et al. [13] scheme with inclusion of a first-approach chlorophyll-based organic chemistry.

Figure 6 depicts ozone concentrations in the North East. Atlantic region as predicted by REMOTE using three permutations of the Fairall et al. [13] scheme and using the nonchemical Wesely [18] scheme. The Fairall et al. [13] scheme including 100 nM iodide chemistry predicts ozone concentrations similar to those simulated using the Wesely [18] scheme. The impact of chlorophyll on simulated ozone levels in this region can be seen by comparing the mean monthly ozone levels simulated using the Fairall et al. [13] scheme scaled to include iodide chemistry alone (100 nM I scheme), and the mechanistic version scaled to include both iodide chemistry and organic reactions. Ozone levels simulated by the mechanistic dry deposition scheme are as much as 15 ppb lower than ozone levels simulated by both the Wesely [18] scheme and the 100 nM I scheme. The Fairall et al. [13] 100 nM I scheme and the mechanistic scheme differ only by organic enhancement of deposition velocity and so the lower ozone concentrations simulated by the mechanistic schemes are due to inclusion of organic chemistry alone.

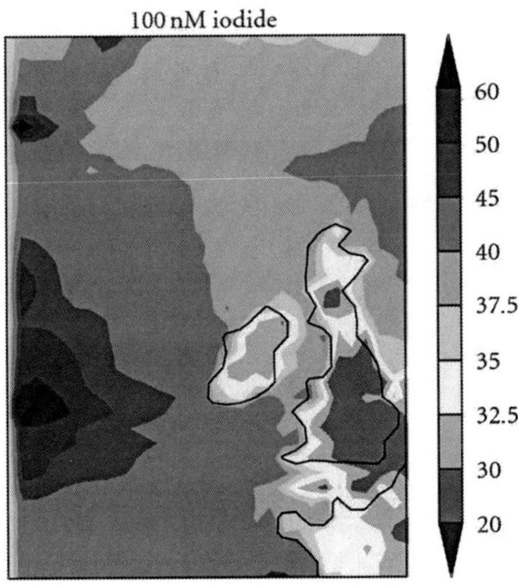

(a)

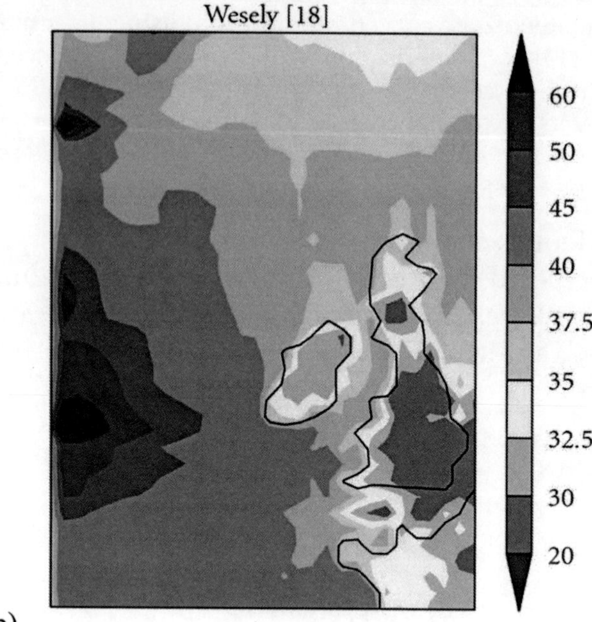
Wesely [18]
60
50
45
40
37.5
35
32.5
30
20

(b)

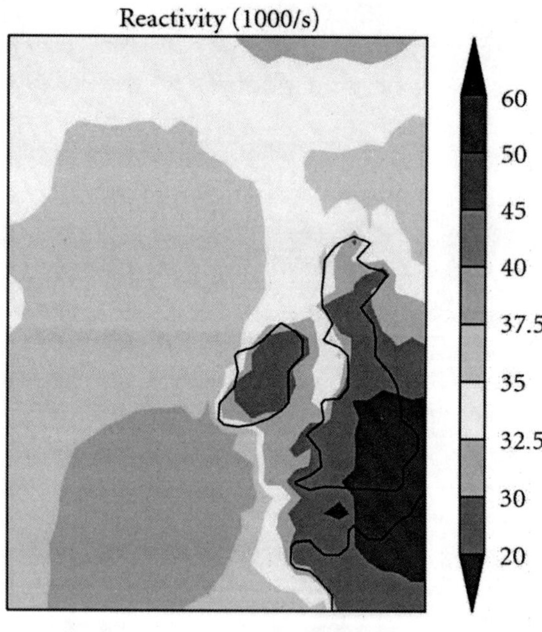
Reactivity (1000/s)
60
50
45
40
37.5
35
32.5
30
20

(c)

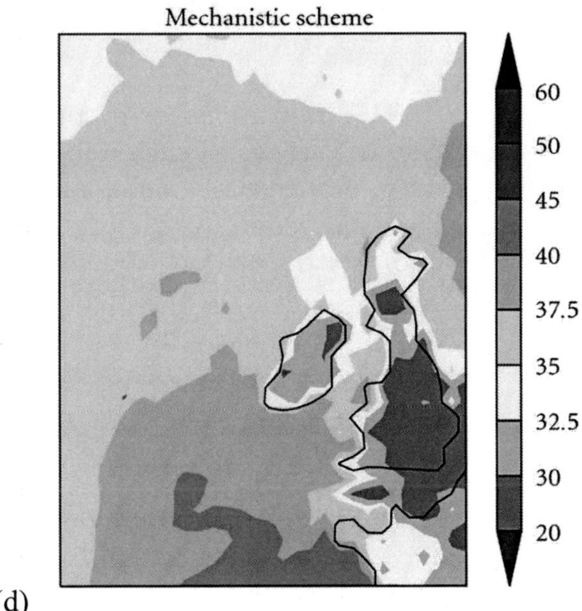

Figure 6: Monthly average O_3 concentrations in ppb in the North East Atlantic region as predicted by REMOTE for June 2003 for various dry deposition schemes. 100 nM I plot depicts simulated ozone concentrations (ppb) using the iodide-only chemical parameterisation of Fairall et al. [13] with oceanic iodide concentrations set to 100 nM. The Wesely [18] plot shows the ozone concentrations simulated by the nonchemical scheme. Reactivity 1000/s plot shows ozone concentrations simulated using the Fairall et al. [13] scheme with reactivity term set to constant 1000 s^{-1} and mechanistic scheme plot depicts the ozone concentrations simulated using the Fairall et al. [13] scheme with inclusion of first-approach chlorophyll-based organic chemistry.

Figure 7 compares ozone concentrations measured at the Mace Head atmospheric research station measurement site off the west coast of Ireland to modelled ground-level ozone concentrations in this region. Ozone levels were measured at Mace Head using a continuous ozone analyser by UV photometry. See Tripathi et al. [39] for full description of this data and measurements. All simulated ozone concentrations agree quite well with observations, but the Mace Head ozone concentrations simulated by the mechanistic scheme correlate better with the observed ozone concentrations than other schemes. The root mean square (RMS) deviation between in situ ozone measurements at Mace Head and modelled ozone concentration for this region was 7.7 ppb when using the mechanistic scheme whereas the next closest correlations were obtained using the Fairall et al. [13] scheme with reactivity set to 1000 s^{-1}, and the Wesely [18] scheme which gave RMS deviations of 9.1 ppb and 11.6 ppb, respectively. This shows that on average, deviations between simulated ozone concentrations and in situ measurements are less for results obtained using the mechanistic ozone dry deposition scheme than those generated using other dry deposition schemes. However, firm conclusions on accuracy of dry deposition parameterisation cannot be drawn here, as simulated ozone-mixing ratio is dependent on other model processes including atmospheric chemistry, transport, and boundary conditions. In addition, the in situ measurements represent a point measurement of ozone-mixing ratio, whereas the model output represents average ozone mixing ratio over a grid cell which spans 50 km 50 km. In this analysis, attempts have been made to account for this by comparing modelled ozone concentrations with in situ ozone measurements averaged over 1.75 hours, the time taken for an air mass to traverse the length of one grid cell (50 km), assuming an average easterly wind speed of 8 m s^{-1}. However, localised effects influencing ozone concentrations at Mace Head will not be represented in model simulations due to the effect of averaging over the grid cell area and so simulated ozone concentrations will not have the same variation as ozone measurements at Mace Head. This would explain large variations between simulated and measured ozone concentrations for example, measurements between Julian Day 166 and 171 are much lower than all model results.

From Figure 7, it can be seen that the 100 nM I Fairall et al. [13] scheme and the Wesely [18] scheme predict higher ozone concentrations than observations or other model set ups during periods of low-wind speeds (e.g., between days 165 and 171). The reactivity of the 100 nM I scheme is not large enough to account for ozone water surface transfer at low-wind speed and the Wesely [18] scheme does not account for chemical ozone transfer at all. Therefore, using either of these schemes, REMOTE over predicts ozone concentrations at periods of low-wind speed. Simulations using the Fairall et al. [13] scheme with constant reactivity set to 1000 s^{-1} predict very similar ozone concentrations around Mace Head to the simulations using the mechanistic scheme (reactivity due to 100 nM Iodide and organic enhancement).

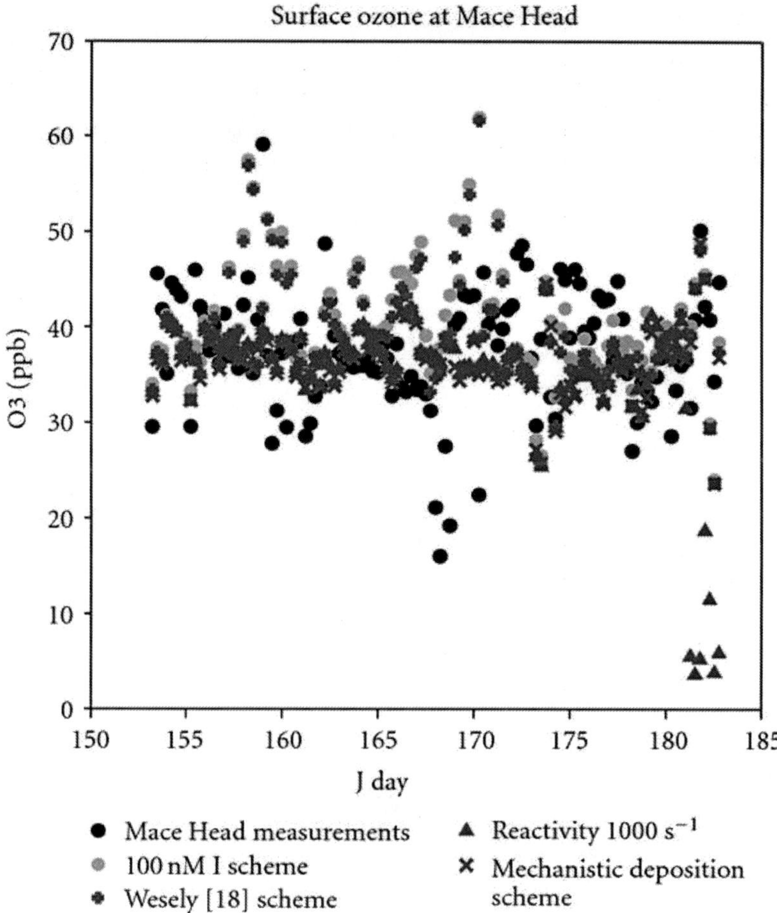

Figure 7: Ozone concentrations (ppb) simulated by REMOTE and measured ozone concentrations (ppb) at Mace Head for June 2003.

Therefore, around the western Irish coast, it can be deduced that effect of organic chemistry causes oceanic reactivity of the order of $1000 s^{-1}$. Figure 8 shows monthly average values of deposition velocities computed for the various deposition schemes. In the mechanistic scheme, organic reactions are parameterised by increasing deposition velocity according to oceanic chlorophyll concentration, and so organic reactions are not represented in the reactivity term, of the Fairall et al. [13] parameterisation. However, by comparing average deposition velocities depicted in Figure 8 with relationship between reactivity and deposition velocity shown in Figures 1 and 2, it can be deduced that the deposition velocities obtained around the Irish coast using the mechanistic scheme (0.6–0.1 cm s^{-1}) would require reactivity values exceeding 1000 s^{-1}, assuming moderate windspeeds. Variation of in the Wesely [18] scheme and the Reactivity 1000 s^{-1} scheme occurs due to turbulent effects alone.

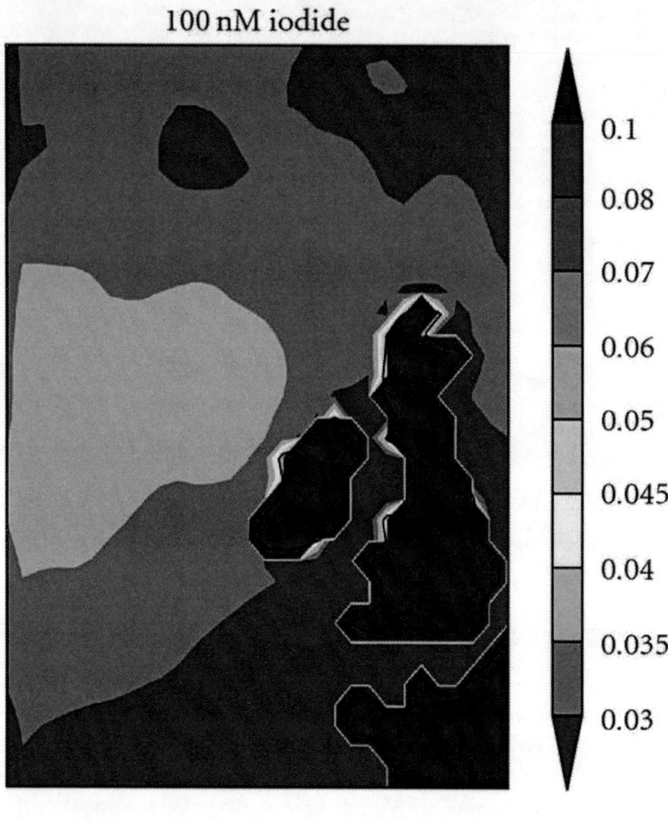

(a)

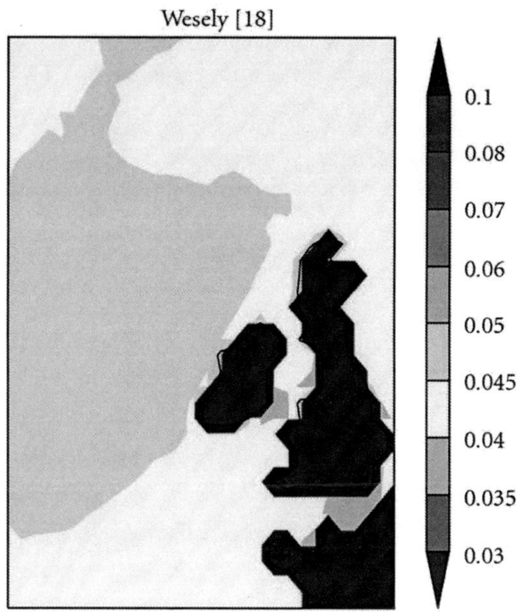
Wesely [18]
0.1
0.08
0.07
0.06
0.05
0.045
0.04
0.035
0.03

(b)

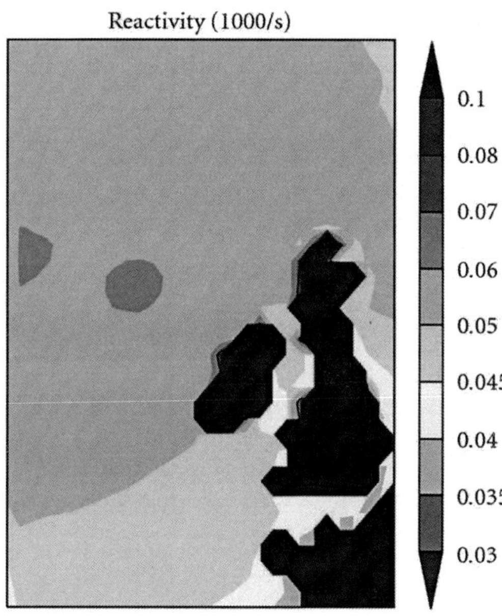
Reactivity (1000/s)
0.1
0.08
0.07
0.06
0.05
0.045
0.04
0.035
0.03

(c)

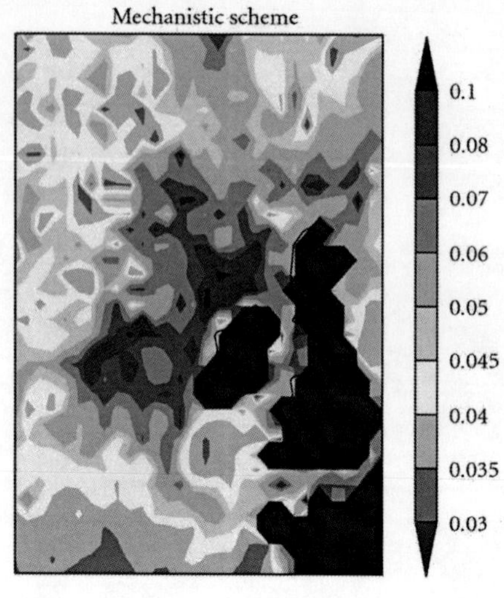

(d)

Figure 8: Monthly mean deposition velocity in the North East Atlantic region as predicted by REMOTE for June 2003 for various dry deposition schemes. 100 nM I plot depicts deposition velocity using the Fairall et al. [13] parameterisation with oceanic iodide concentrations set to 100 nM. The Wesely [18] plot shows deposition velocity simulated by the nonchemical scheme. Reactivity 1000/s plot shows deposition velcity simulated using the Fairall et al. [13] scheme with reactivity term set to constant 1000 s^{-1} and mechanistic scheme plot depicts the deposition velocity simulated using the Fairall et al. [13] scheme with inclusion of first-approach chlorophyll-based organic chemistry.

Ozone-Loss Rate

The loss rate of ozone from the mixing volume to the ocean due to dry deposition was calculated in order to assess the impact of varying ozone deposition on the rate of atmospheric ozone loss to the ocean as a result of dry deposition and the consequential effect on ambient ozone concentration. The loss rate is given bywhere is the dry deposition flux of ozone to the ocean (the product of deposition velocity and ambient ozone concentration) and L is the boundary layer height. Deposition velocities and ozone concentrations were extracted from REMOTE and boundary layer height taken as a constant 800 m. Figure 9 shows ozone-loss rates computed for the various dry deposition schemes.

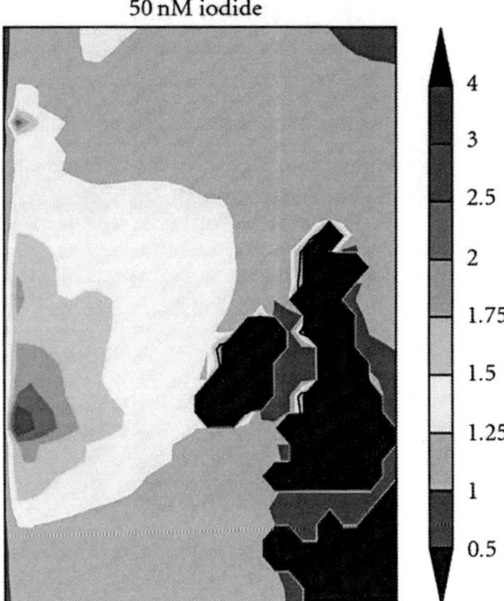
50 nM iodide
4
3
2.5
2
1.75
1.5
1.25
1
0.5

(a)

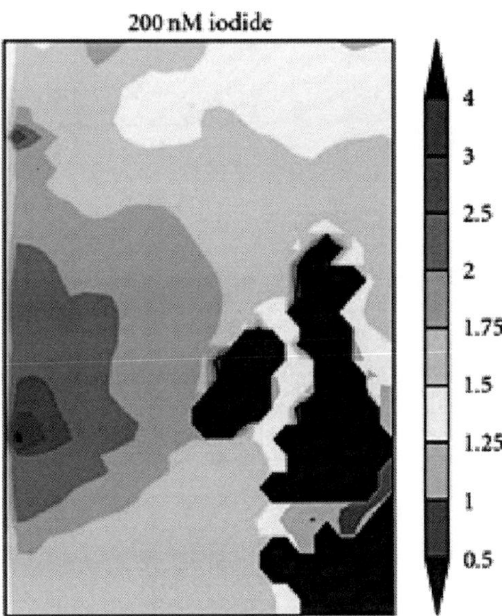
200 nM iodide
4
3
2.5
2
1.75
1.5
1.25
1
0.5

(b)

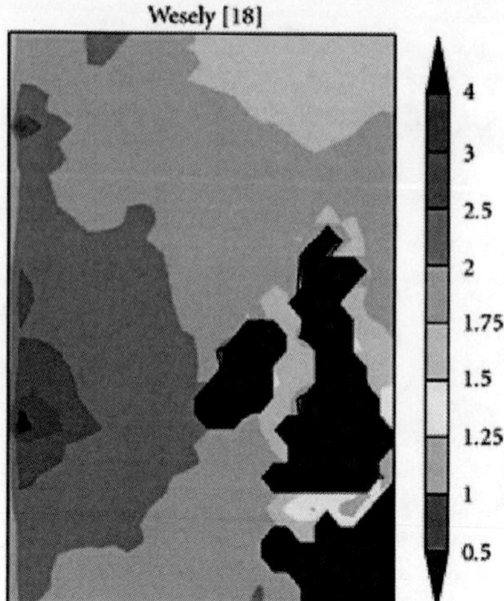

(c)

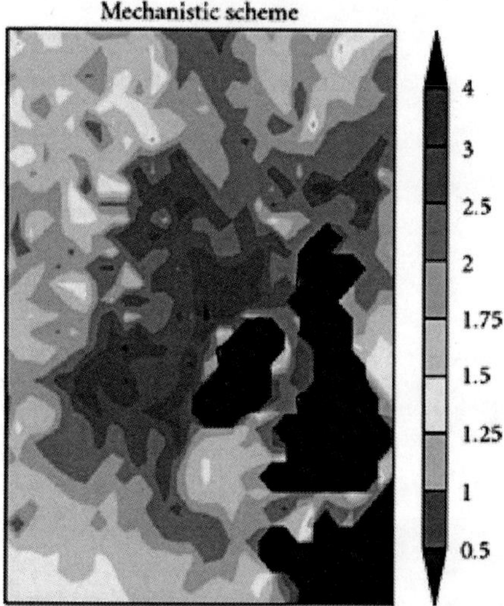

(d)

Figure 9: Monthly mean O_3 loss rates (ppb per day) as predicted by REMOTE June 2003 for various dry deposition schemes. Simulated loss rates using the Fairall et al. [13] parameterisation with oceanic iodide concentrations set to 50 nM and 200 nM, respectively, are shown. The Wesely [18] plot shows the ozone-loss rates simulated by the nonchemical Wesely [18] scheme and the mechanistic scheme shows the O_3 loss rate simulated using the Fairall et al. [13] scheme with inclusion of first-approach chlorophyll-based organic chemistry.

A higher deposition velocity does not necessarily denote a higher ozone-loss rate, as the loss rate is proportional to both ozone concentration and deposition velocity.

The loss rate of ozone to the ocean varies with oceanic iodide concentration as seen from the first two plots of Figure 9, using the iodide-only constrained scheme of Fairall et al. [13]. This occurs due to the different deposition velocities generated from the dry deposition schemes based on reactivity. However, it can be seen that the variation of the iodide concentration has no significant effect on ground level ozone concentration and so the variation in ozone-loss rate due to differences in iodide concentration is too subtle to have a bearing on ambient ozone concentrations, indicating the dominance of other processes over the dry deposition sink in regulation of ambient boundary layer ozone concentrations within REMOTE for lower dry depositional loss rates.

The Wesely [18] scheme predicts a greater ozone-loss rate than the mechanistic ozone deposition scheme. This occurs due to the depletion of ambient ozone concentrations because of enhanced dry deposition in the presence of marine biological activity.

Upward Iodine Flux

20% of ozone deposited to the sea surface reacts with iodide [21]. 1% of iodide oxidised by ozone reacts with organic matter to form the VOIs (CH_2I_2, CH_2ICl, or CHI_3) that are released from the sea surface [8]. The upper limit of potential iodine flux to the marine atmosphere as a result of ozone deposition is predicted using these relationships. Resulting VOI fluxes are shown in Figure 10. The upward VOI flux is derived directly from the downward ozone flux, and so the same factors influence both the downward ozone flux and the upward iodine flux. The upward VOI flux also varies with oceanic iodide concentration. It is stressed that this potential VOI flux to the atmosphere represents an extreme upper limit, and is to be interpreted as such.

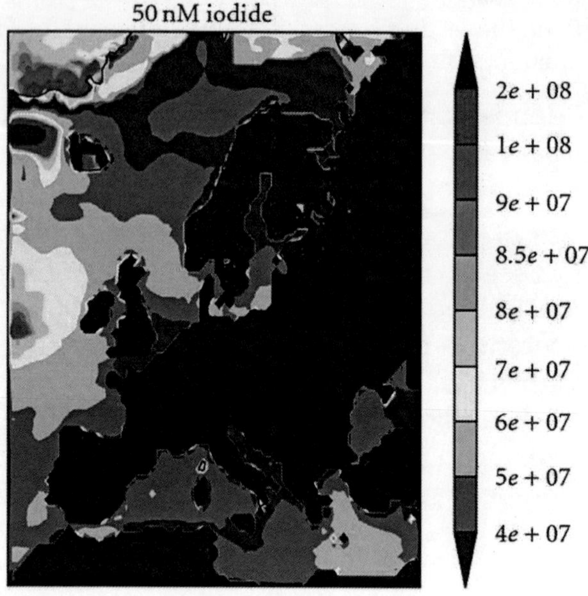
50 nM iodide
2e + 08
1e + 08
9e + 07
8.5e + 07
8e + 07
7e + 07
6e + 07
5e + 07
4e + 07

(a)

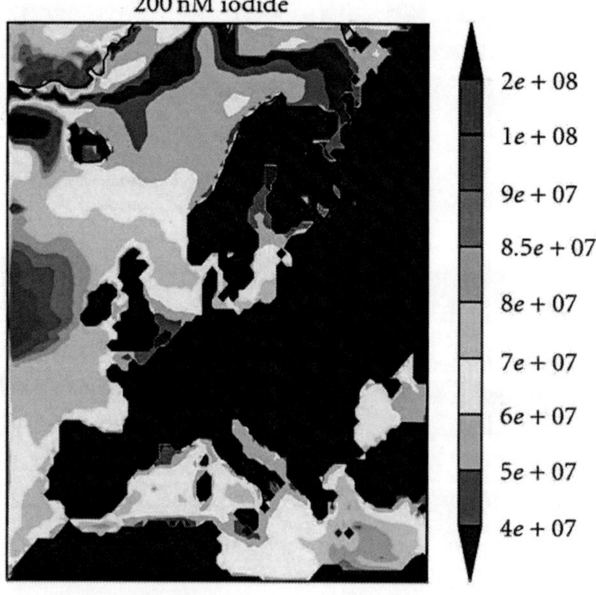
200 nM iodide
2e + 08
1e + 08
9e + 07
8.5e + 07
8e + 07
7e + 07
6e + 07
5e + 07
4e + 07

(b)

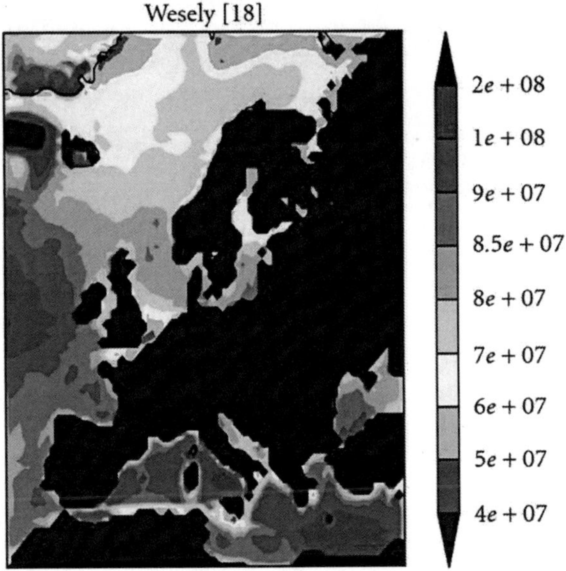

(c)

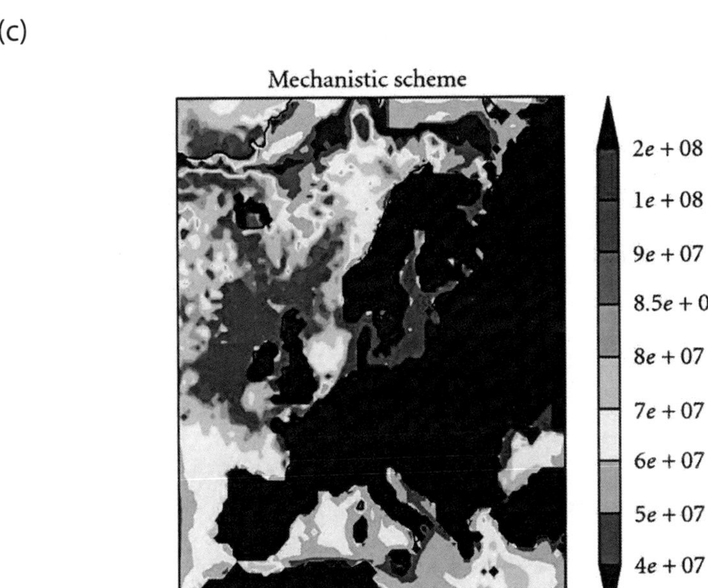

(d)

Figure 10: Monthly average upward iodine flux in molecules per $cm^{-2} s^{-1}$, as predicted by REMOTE for June 2003 shown for the European region. Simulated flux using the Fairall et al. [13] parameterisation with oceanic iodide concentrations set to 50 nM and 200 nM, respectively, are shown. The Wesely [18] plot shows the flux simulated by the nonchemical Wesely [18] scheme and the mechanistic scheme shows the flux simulated using the Fairall et al. [13] parameterisation using a first-approach organic chemistry scaling.

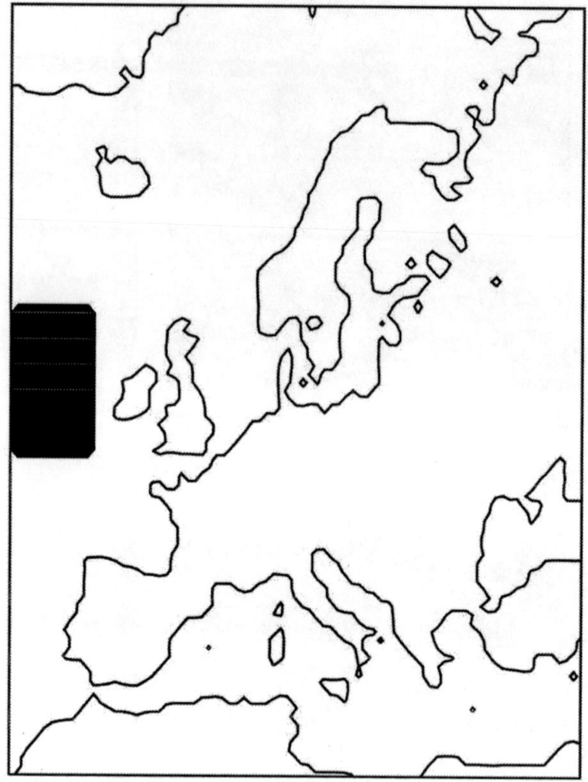

Figure 11: Location of boxed region in the North Atlantic off the west coast of Ireland in which the area-averaged analysis was performed.

The upward VOI flux predicted using the Wesely [18] scheme exceeds that predicted by the Fairall et al. [13] scheme constrained by iodide reactions. The mechanistic ozone dry deposition scheme predicts VOI flux exceeding that predicted using Wesely [18] in some regions (north east of Britain) while the flux is less than that predicted using Wesely [18] in other areas (off the southern Irish coast).

This ozone deposition-driven flux of iodine from the ocean is likely to have significant biogeochemical consequences. Marine aerosols and cloud condensation nuclei (CCN) can be formed from iodine vapours in coastal environments [11, 40]. The addition of new aerosols to the coastal atmosphere could have a significant impact on the global solar radiation budget due to their role in scattering of incoming solar radiation, and therefore accurate prediction of the VOI flux is imperative in predicting future climatic scenarios. To fully assess the atmospheric implications of this extra iodine source in the MBL, the upward organoiodine flux and resulting chemical reactions and particle formation would have to be included in the chemical scheme of REMOTE.

AREA-AVERAGED STUDY OF SIMULATION RESULTS

A quantitative comparison of simulation results using the various dry deposition schemes was performed by taking an area average of various parameters over a boxed region in the North Atlantic, off the Irish coast—a region of relatively high biological activity. The region is displayed in Figure 11. In this area, average, maximum, and minimum deposition velocities from the various schemes were compared, as were the average simulated ozone concentrations, average ozone-loss rates, average upward VOI flux and derived average increase in the daily rate of the VOI-mixing ratio. Results are tabulated in Table 1.

Table 1: Results from area average analysis carried out over a boxed region in the North Atlantic.

Scheme	Average deposition velocity (cm/s)	Maximum deposition velocity (cm/s)	Minimum deposition velocity (cm/s)	Average [O3] in ppb	Average O3 loss rate (ppb per day)	Average upward VOI flux (molecules $cm^{-2} s^{-1}$)	Average mixing ratio VOI increase (ppt per day)
Reactivity of 50nM Iodide	0.0305	0.0509	0.00123	47.8647	1.57081	7.11E+07	3.14E+00
Reactivity of 100nM Iodide	0.0355	0.0573	0.00127	47.3876	1.82221	8.26E+07	3.64E+00
Reactivity of 150nM Iodide	0.0392	0.0625	0.00129	47.3911	2.0009	9.07E+07	4.00E+00
Reactivity of 200nM Iodide	0.0411	0.0661	0.0013	47.2334	2.1445	9.72E+07	4.29E+00
Reactivity $1000s^{-1}$	0.0573	0.0848	0.00134	36.6113	2.38844	1.03E+08	4.53E+00
Wesely [18] scheme	0.0468	0.0495	0.0019	46.6433	2.34064	1.06E+08	4.68E+00
Mechanistic scheme (Reactivity of 100nM Iodide and Organic reactions)	0.0547	0.161	0.00173	37.2185	2.20902	9.98E+07	4.40E+00
Percentage difference between Wesely [18] and Mechanistic Scheme	-16.171	-230.38	8.86149	20.5466	5.87979	5.87979	5.87979
Percentage difference between Wesely [18] and 100nM I⁻ scheme	23.1573	-17.915	33.0837	-1.5888	22.1495	22.1495	22.1493

Deposition Velocity

Deposition velocities simulated using the Fairall et al. [13]

parameterisation vary with oceanic iodide concentration, but as we have seen earlier, these variations are too subtle to have a significant effect on ambient ozone concentrations. The mechanistic scheme computes deposition velocities in this region exceeding those computed using the [18] scheme by 18%. Neither the Wesely [18] scheme nor the Fairall et al. [13] scheme constrained to iodide chemistry can account for deposition velocities as high as the observations of Gallagher et al. [17]. Surface resistance is set to a constant 2000 m^{-1} s in the Wesely [18] scheme, and so maximum deposition velocity computed by this scheme deviates very little from the average deposition velocity; using this nonchemical scheme, deposition velocities cannot match the upper limit of observations under any conditions.

Additional chemical reactions must be included in the dry deposition scheme in order to realise deposition velocities as high as observations. Only the mechanistic ozone deposition scheme predicts deposition velocities higher than observations of Gallagher et al. [17]. Therefore, only by integration of organic chemistry into the dry deposition scheme can deposition velocities as high as observations be realised.

Ozone Concentration

Simulated average ozone concentrations in this area are relatively insensitive to oceanic iodide concentration. Inclusion of first-order organic chemistry parameterisation based on Clifford et al. [23] into the dry deposition parameterisation causes a decrease in simulated ozone concentration in the area of 20.5% or 9.5 ppb compared to concentrations simulated using the Wesely [18] scheme. This increase in ozone deposition occurs due to the enhancement of ozone deposition velocity due to inclusion of organic reactions in the mechanistic ozone deposition scheme, as discussed above. The Wesely [18] scheme predicts an average ozone concentration in this area 1.6% lower than that predicted using the Fairall et al. [13] parameterisation constrained by an oceanic iodide concentration of 100 nM.

Ozone-Loss Rate

Average ozone-loss rate computed by Wesely [18] exceeds that computed by the mechanistic ozone deposition scheme by nearly 6% in this region. The lower loss rate is due to the lower ambient ozone concentrations in the boundary layer due to organic enhancement of ozone deposition computed by the mechanistic ozone deposition scheme. The less reactive scheme constrained only by iodide reactions also computes ozone-loss rate less than that of the Wesely [18] scheme due to the higher deposition velocity computed by the Wesely [18] scheme. The relatively small changes in loss rates are shown not to have a large effect on ambient ozone concentrations in REMOTE, but accurate estimation of ozone dry deposition flux is necessary to evaluate ozone-deposition driven upward VOI flux.

Potential Upward VOI Flux

This flux is estimated linearly from the downward ozone flux. The simulated potential VOI flux emitted from the ocean in this area is of the order of 10^8 molecules $cm^{-2}\,s^{-1}$. Martino et al. [8] compute a typical downward depositional iodine flux of 3 10^7 atoms $cm^{-2}\,s^{-1}$. The potential upward VOI flux exceeds the depositional iodine flux, thus constituting a significant additional source of atmospheric iodine in marine environments, especially in regions of high biological activity when organic enhancement of ozone deposition is factored into the dry deposition parameterisation.

Increase of VOI-Mixing Ratio

Recent measurements of iodocarbon fluxes taken in the North East. Atlantic [41] have found open ocean sea-air fluxes of the ozone-deposition derived VOIs (CH_3I, CH_2I_2, and CH_2ICl) as high as 9.6 10^7molecules cm^{-2} day^{-1}, which would lead to a mixing ratio increase of 4.24 ppt per day. The average flux value observed in this region would lead to a VOI increase of 1.2 ppt per day. In the more biologically active shelf and coastal regions, measured VOI sea-air fluxes have been measured that would result in mixing ratio increases as high as 9.3 ppt per day and 18.1 ppt per day, respectively. The measurements were taken in the biologically active regions off the west coast of

Ireland where organic enhancement of ozone deposition is likely to occur.

The average simulated increase of potential VOI-mixing ratio in this area is of the order of 4 ppt per day and the range of ozone-driven iodocarbon flux in this region is 2.5–500 M molec cm^{-2} s^{-1} which would correspond to a mixing-layer enhancement of VOI concentrations of 0.1–22.0 ppt per day. This enhancement to the mixing ratio was calculated using ideal gas theory and assuming a boundary layer height of 800 m and should be interpreted as the extreme upper limit of potential VOI flux occurring as a result of ozone deposition – in reality, a proportion of this flux would be mixed downwards and a further proportion will be photolysed at the sea surface [8]. Taking this into consideration, simulated VOI-mixing ratio enhancement is well within range of the recently observed iodocarbon fluxes [41], especially considering large iodocarbon fluxes observed in shelf and coastal regions. Considering the molar mixing ratio of total organic iodine ranges between 3–4 pptv in the boundary layer [42–44], even a small mixing ratio enhancement of these proportions would enhance atmospheric iodine concentrations considerably.

The extra iodine compounds released to the atmosphere could further add to ozone depletion in the marine boundary layer, thus forming a catalytic cycle of ozone destruction. Read et al. [45] investigated halogen-mediated ozone destruction over the tropical Atlantic Ocean and found that based on typical organic iodine molar mixing ratios of 3–4 ppt in the MBL [44], resulting IO concentrations of 1 ppt would cause ozone loss of 1.24 ppb per day due to atmospheric reactions with IO. Crudely assuming a linear relationship between organic iodine concentrations and resulting ozone loss, additional influx of 22 ppt organic iodine per day into the mixing layer due to the ozone-deposition driven VOI flux would result in a further ozone loss of 7.8 ppb per day. Even though this case represents the upper limit VOI flux to MBL, it is evident that the halogen-mediated ozone destruction cycle is likely to have significant consequences for ozone concentrations in the marine boundary layer, depleting ozone mixing ratios to the order of a ppb per day. To further investigate this feedback mechanism, REMOTE would need to be adapted to include halogen chemistry and the upward VOI flux. This is scheduled for further work.

Conclusions

Consistent with results of global model study conducted by Ganzeveld et al. [5], our results show that ozone concentrations predicted by the Fairall et al. [13] parameterisation within REMOTE are insensitive to realistic variations of oceanic iodide concentrations. Furthermore, the ozone concentrations predicted by this turbulent and chemically enhanced deposition scheme do not exceed those predicted by the highly parameterised Wesely [18] scheme when the parameterisation is constrained by iodide chemistry alone for typical oceanic iodide concentrations. In order for the new scheme to simulate deposition rates as high as field observations, oceanic reactivity must be in the order of $1000\,s^{-1}$. The source of this extra reactivity is most likely due to ozone reacting with organic matter and chlorophyll in the sea surface [22, 23].

Deposition velocities as high as the observations of Gallagher et al. [17] were realised in regions of high biological activity using the mechanistic ozone deposition scheme, in which a crude first-approach scaling of organic ozone reactions was applied to the Fairall et al. [13] parameterisation. Use of this mechanistic ozone deposition scheme including organic reactions yielded boundary layer ozone concentrations much lower than those predicted using the highly parameterised, constant surface-resistance deposition scheme of Wesely [18]. This result is in contrast with findings of Ganzeveld et al. [5] who found the use of mechanistic approach to evaluation of surface resistance in fact reduced total dry deposition flux of ozone to the ocean compared to simulations ran using the Wesely [18] scheme with constant surface-resistance, even though the global model study incorporated additional chemical reactions of ozone with DMS, C_2H_4, and C_3H_6, which were deemed negligible in this study. However, variations between the global model results and results from this are most likely due to variation between the temporal and spatial resolution of the two model studies: the focus of this study is limited to a biologically active region for a time of year in which biological activity is at a maximum and so it would be expected that biochemical effects would be significant in resulting simulations. In contrast, the Ganzeveld et al. [5] study was performed on a global scale for a yearly period and simulated annual mean mixing ratios were analysed which would not reflect short-term seasonal effects. As stated in the introduction, the Wesely [18] scheme serves well in simulating ozone deposition for large-scale simulations, but the mechanistic Fairall et al. [13] parameterisation serves best for simulating ozone deposition velocity to the ocean for particular regional or seasonal effects, as in the case of high biological activity due to a phytoplankton bloom. In addition, Ganzeveld et al. [5] used variable inferred oceanic iodide concentration fields as opposed to constant oceanic iodide fields utilised in this study. Also, the global model study did not allow for the enhancing effect of high sea surface temperature (SST) on diffusivity of ozone in water and so ozone transfer in regions of high SST was limited by solubility and the enhancing effect of seawater salinity on ozone-iodide reaction kinetics was not considered in the global model study.

Based on the findings of this study, it is postulated that high ozone fluxes can be theoretically explained only by consideration of reactions of ozone with ocean dwelling organic matter.

Simulated ozone concentrations agree closely with in situ measurements at Mace Head. On average the mechanistic ozone deposition scheme displayed least deviation from measurements. This indicates a closer correlation between actual ozone concentrations and simulated ozone concentrations using the mechanistic scheme (RMS deviation of 7.7 ppb) than using the iodide-only scheme or the nonchemical Wesely [18] scheme (RMS deviation of 11.6 ppb).

Ozone dry deposition flux depends on both ozone concentration and deposition. Variations in ozone flux do not necessarily have a significant effect on ambient ozone concentration due to dominance of other model processes over the dry deposition processes at low deposition velocities. To the best of authors' knowledge, this is the first study to use the advanced mechanistic ozone deposition scheme of Fairall et al. [13] to quantitatively evaluate the newly discovered ozone deposition-driven upward flux of iodine from the ocean outlined by Martino et al. [8], which is likely to have significant biogeochemical consequences. Ambient ozone concentrations are insensitive to oceanic iodide concentrations, but ozone dry deposition flux (and resulting upward iodine flux) varies with oceanic iodide concentration and so iodide reactions must be explicit within dry deposition models to adequately simulate biogeochemical consequences of dry depositional ozone flux.

Acknowledgments

The authors gratefully acknowledge the Environmental Protection Agency (EPA) for its support (EPA Project 2006-AQ-MS-50: Ozone levels, changes and trends over Ireland—an Integrated Analysis). Chris Fairall of NOAA is also gratefully acknowledged for making available the parameterisation used in this study. The authors would also like to thank anonymous reviewers for their comments and reviews that served to significantly improve the quality of the paper.

REFERENCES

1. IPCC, 2001.
2. IPCC, "Climate Change 2007: the Physical Science Basis. Contribution of Working Group I to the Fourth Assessment Report of the Intergovernmental Panel on Climate Change," 2007.
3. K. A. Hunter and P. S. Liss, "Organic sea surface films," in Marine Organic Chemistry: Evolution, Composition, Interactions and Chemistry of Organic Matter in Seawater, Elsevier, Amsterdam, The Netherlands, 1981.
4. US-EPA, "Ozone: good up high, bad nearby," edited by U.S. Environmental Protection Agency, 2003, http://www.epa.gov/oar/oaqps/gooduphigh/.
5. L. Ganzeveld, D. Helmig, C. W. Fairall, J. Hare, and A. Pozzer, "Atmosphere-ocean exchange: a global modeling study of biogeochemical, atmospheric, and waterside turbulence dependencies,"Global Biogeochemical Cycles, vol. 23, Article ID GB4021, 16 pages, 2009. View at Publisher · View at Google Scholar
6. L. Ganzeveld and J. Lelieveld, "Dry deposition parameterization in a chemistry general circulation model and its influence on the distribution of reactive trace gases," Journal of Geophysical Research, vol. 100, no. D10, pp. 20999–21012, 1995.
7. D. A. Hauglustaine, C. Granier, G. P. Brasseur, and G. Megie, "The importance of atmospheric chemistry in the calculation of radiative forcing on the climate system," Journal of Geophysical Research, vol. 99, no. 1, pp. 1173–1186, 1994. View at Publisher · View at Google Scholar · View at Scopus
8. M. Martino, G. P. Mills, J. Woeltjen, and P. S. Liss, "A new source of volatile organoiodine compounds in surface seawater," Geophysical

Research Letters, vol. 36, no. 1, Article ID L01609, 5 pages, 2009. View at Publisher · View at Google Scholar · View at Scopus

9. D. I. Reeser, C. George, and D. J. Donaldson, "Photooxidation of halides by chlorophyll at the air-salt water interface," Journal of Physical Chemistry A, vol. 113, no. 30, pp. 8591–8595, 2009. View at Publisher · View at Google Scholar · View at Scopus
10. D. Davis, J. Crawford, S. Liu et al., "Potential impact of iodine on tropospheric levels of ozone and other critical oxidants," Journal of Geophysical Research D, vol. 101, no. 1, pp. 2135–2147, 1996. View at Scopus
11. C. D. O'Dowd, J. L. Jimenez, R. Bahreini et al., "Marine aerosol formation from biogenic iodine emissions," Nature, vol. 417, no. 6889, pp. 632–636, 2002. View at Publisher · View at Google Scholar · View at Scopus
12. A. S. Mahajan, J. M. C. Plane, H. Oetjen, et al., "Measurement and modelling of reactive halogen species over the tropical Atlantic Ocean," Atmospheric Chemistry and Physics Discussions, vol. 9, no. 6, pp. 24281–24316, 2009.
13. C. W. Fairall, D. Helmig, L. Ganzeveld, and J. Hare, "Water-side turbulence enhancement of ozone deposition to the ocean," Atmospheric Chemistry and Physics, vol. 7, no. 2, pp. 443–451, 2007. View at Scopus
14. J. H. Seinfeld and S. N. Pandis, Atmospheric Chemistry and Physics, John Wiley & Sons, New York, NY, USA, 1997.
15. M. Z. Jacobson, Fundamentals of Atmospheric Modeling, Cambridge University Press, Cambridge, Mass, USA, 2005.
16. M. L. Wesely and B. B. Hicks, "A review of the current status of knowledge on dry deposition,"Atmospheric Environment, vol. 34, no. 12–14, pp. 2261–2282, 2000. View at Publisher · View at Google Scholar · View at Scopus
17. M. W. Gallagher, K. M. Beswick, G. McFiggans, H. Coe, and T. W. Choularton, "Ozone dry deposition velocities for coastal waters," Water, Air and Soil Pollution: Focus, vol. 1, no. 5-6, pp. 233–242, 2001.
18. M. L. Wesely, "Parameterization of surface resistances to gaseous dry deposition in regional-scale numerical models," Atmospheric Environment, vol. 23, no. 6, pp. 1293–1304, 1989. View at Scopus
19. D. H. Lenschow, R. Pearson Jr., and B. B. Stankov, "Measurements of ozone vertical flux to ocean and forest," Journal of Geophysical Research, vol. 87, no. 11, pp. 8833–8837, 1982. View at Scopus
20. W. Chang, B. G. Heikes, and M. Lee, "Ozone deposition to the sea

surface: chemical enhancement and wind speed dependence," Atmospheric Environment, vol. 38, no. 7, pp. 1053–1059, 2004. View at Publisher · View at Google Scholar · View at Scopus

21. J. A. Garland, A. W. Etzerman, and S. A. Penkett, "The Mechanism for dry deposition of ozone to seawater surfaces," Journal of Geophysical Research, vol. 85, no. C12, pp. 7488–7492, 1980.
22. S. Schwartz, "Factors governing dry deposition of gases to surface water," in Precipitation Scavenging and Atmosphere-Surface Exchange, pp. 789–801, Hemisphere Publishing Corporation, Washington, DC, USA, 1992.
23. D. Clifford, D. J. Donaldson, M. Brigante, B. D'Anna, and C. George, "Reactive uptake of ozone by chlorophyll at aqueous surfaces," Environmental Science and Technology, vol. 42, no. 4, pp. 1138–1143, 2008. View at Publisher · View at Google Scholar · View at Scopus
24. B. Langmann, "Numerical modelling of regional scale transport and photochemistry directly together with meteorological processes," Atmospheric Environment, vol. 34, no. 21, pp. 3585–3598, 2000. View at Scopus
25. B. Langmann, "REMOTE-Regional Model with Tracer Extension," 2005.
26. C. W. Fairall, E. F. Bradley, J. S. Godfrey, G. A. Wick, J. B. Edson, and G. S. Young, "Cool-skin and warm-layer effects on sea surface temperature," Journal of Geophysical Research C, vol. 101, no. 1, pp. 1295–1308, 1996. View at Scopus
27. C. W. Fairall, J. E. Hare, J. B. Edson, and W. McGillis, "Parameterization and micrometeorological measurement of air-sea gas transfer," Boundary-Layer Meteorology, vol. 96, no. 1-2, pp. 63–105, 2000. View at Scopus
28. J. E. Hare, C. W. Fairall, W. R. McGillis, J. B. Edson, B. Ward, and R. Wanninkhof, "Evaluation of the National Oceanic and Atmospheric Administration/ Coupled-Ocean Atmospheric Response Experiment (NOAA/COARE) air-sea gas transfer parameterization using GasEx data," Journal of Geophysical Research C, vol. 109, no. 8, Article ID C08S11, 11 pages, 2004. View at Publisher · View at Google Scholar · View at Scopus
29. L. Magi, F. Schweitzer, C. Pallares, S. Cherif, P. Mirabel, and C. George, "Investigation of the uptake rate of ozone and methyl hydroperoxide by water surfaces," Journal of Physical Chemistry A, vol. 101, no. 27, pp. 4943–4949, 1997. View at Scopus
30. P. N. Johnson and R. A. Davis, "Diffusivity of ozone in water," Journal

of Chemical and Engineering Data, vol. 41, no. 6, pp. 1485–1487, 1996. View at Scopus

31. B. Jahne, G. Heinz, and W. Deitrich, "Measurement of the diffusion coefficients of sparingly soluble gases in water," Journal of Geophysical Research, vol. 92, no. C10, pp. 10767–10776, 1987.
32. L. F. Kosak-Channing and G. R. Helz, "Solubility of ozone in aqueous solutions of 0–0.6 M ionic strength at 5–30C∘," Environmental Science and Technology, vol. 17, no. 3, pp. 145–149, 1983. View at Scopus
33. B. Jahne, T. Wais, and M. Barabas, "A new optical bubble measuring device: a simple model for bubble contribution to gas exchange," in Gas Transfer at Water Surfaces, pp. 237–246, D. Reidel, Norwell, Mass, USA, 1984.
34. L. F. Kosak-Channing and G. R. Helz, "Solubility of ozone in aqueous solutions of 0–0.6 M ionic strength at 5–30C∘," Environmental Science and Technology, vol. 17, no. 3, pp. 145–149, 1983. View at Scopus
35. V. W. Truesdale, A. J. Bale, and E. M. S. Woodward, "The meridional distribution of dissolved iodine in near-surface waters of the Atlantic Ocean," Progress in Oceanography, vol. 45, no. 3-4, pp. 387–400, 2000. View at Publisher · View at Google Scholar · View at Scopus
36. M. L. A. M. Campos, R. Sanders, and T. Jickells, "The dissolved iodate and iodide distribution in the South Atlantic from the Weddell Sea to Brazil," Marine Chemistry, vol. 65, no. 3-4, pp. 167–175, 1999. View at Publisher · View at Google Scholar · View at Scopus
37. MODIS, "Oceancolor Web," http://oceancolor.gsfc.nasa.gov/.
38. G. McFiggans, J. M.C. Plane, B. J. Allan, L. J. Carpenter, H. Coe, and C. O'Dowd, "A modeling study of iodine chemistry in the marine boundary layer," Journal of Geophysical Research D, vol. 105, no. D11, pp. 14371–14385, 2000. View at Scopus
39. O. P. Tripathi, S. G. Jennings, C. D. O'Dowd, et al., "Statistical Analysis of Eight Surface Ozone Measurement Series for various sites in Ireland," Journal of Geophysical Reseach. In press.
40. G. McFiggans, H. Coe, R. Burgess et al., "Direct evidence for coastal iodine particles from Laminaria macroalgae—linkage to emissions of molecular iodine," Atmospheric Chemistry and Physics, vol. 4, no. 3, pp. 701–713, 2004. View at Scopus
41. C. E. Jones, K. E. Hornsby, R. Sommariva, et al., "Quantifying the contribution of marine organic gases to atmospheric iodine," Geophys. Res. Lett, in press. View at Publisher · View at Google Scholar
42. K. A. Rahn, R. D. Borys, and R. A. Duce, "Tropospheric halogen gases:

inorganic and organic components," Science, vol. 192, no. 4239, pp. 549–550, 1976. View at Scopus

43. S. Yoshida and Y. Muramatsu, "Determination of organic, inorganic and particulate iodine in the coastal atmosphere of Japan," Journal of Radioanalytical and Nuclear Chemistry, vol. 196, no. 2, pp. 295–302, 1995. View at Publisher · View at Google Scholar · View at Scopus
44. R. Vogt, R. Sander, R. Von Glasow, and P. J. Crutzen, "Iodine chemistry and its role in halogen activation and ozone loss in the marine boundary layer: a model study," Journal of Atmospheric Chemistry, vol. 32, no. 3, pp. 375–395, 1999. View at Publisher · View at Google Scholar · View at Scopus
45. K. A. Read, A. S. Mahajan, L. J. Carpenter et al., "Extensive halogen-mediated ozone destruction over the tropical Atlantic Ocean," Nature, vol. 453, no. 7199, pp. 1232–1235, 2008. View at Publisher · View at Google Scholar · View at Scopus

Citations

CHAPTER 1

Awatef Mohamed El-Maghraby, "Green Chemistry: New Synthesis of Substituted Chromenes and Benzochromenes via Three-Component Reaction Utilizing Rochelle Salt as Novel Green Catalyst,"Organic Chemistry International, vol. 2014, Article ID 715091, 6 pages, 2014. doi:10.1155/2014/715091

CHAPTER 2

Alfredo Ricci, "Asymmetric Organocatalysis at the Service of Medicinal Chemistry," ISRN Organic Chemistry, vol. 2014, Article ID 531695, 29 pages, 2014. doi:10.1155/2014/531695

CHAPTER 3

Yunus Bekdemir and Kürşat Efil, "Microwave Assisted Solvent-Free Synthesis of Some Imine Derivatives," Organic Chemistry International, vol. 2014, Article ID 816487, 5 pages, 2014. doi:10.1155/2014/816487.

CHAPTER 4

V. Natarajan, J. Sundar, P. Selvarajan, M. Arivanandhan, K. Sankaranarayanand, S. Natarajan and Y. Hayakawa, "Crystal Growth, Thermal, Mechanical and Optical Properties of a New Organic Nonlinear Optical Material: Ethyl P-Dimethylamino Benzoate (EDMAB)," Journal of Minerals and Materials Characterization and Engineering, Vol. 10 No. 1, 2011, pp. 1-11.

CHAPTER 5

Hiroaki Onoda, Takeshi Sakumura, Synthesis and Pigmental Properties of Nickel Phosphates by the Substitution with Tetravalent Cerium Cation, DOI: 10.4236/msa.2011.211211

CHAPTER 6

R. Rajendran, T. Freeda, U. Kalasekar and R. Peruma, "Synthesis, Crystal Growth and Characterization of Organic NLO Material: M-Nitroacetanilide," Advances in Materials Physics and Chemistry, Vol. 1 No. 2, 2011, pp. 39-43. doi: 10.4236/ampc.2011.12007.

CHAPTER 7

Janja Makarević, Milan Jokić, Leo Frkanec, Vesna Čaplar,Nataša Šijaković Vujičić and Mladen Žinić,, Oxalyl retro-peptide gelators. Synthesis, gelation properties and stereochemical effects, doi:10.3762/bjoc.6.106.

CHAPTER 8

Guijun Wang, Hao Yang, Sherwin Cheuk and Sherman Coleman, Synthesis and self-assembly of 1-deoxyglucose derivatives as low molecular weight organogelators, doi:10.3762/bjoc.7.31

CHAPTER 9

C. Anitha, C. D. Sheela, P. Tharmaraj, and R. Shanmugakala, "Studies on Synthesis and Spectral Characterization of Some Transition Metal Complexes of Azo-Azomethine Derivative of Diaminomaleonitrile," International Journal of Inorganic Chemistry, vol. 2013, Article ID 436275, 10 pages, 2013. doi:10.1155/2013/436275

CHAPTER 10

José C. F. Alves, "A Review on the Chemistry of Eremanthine: A Sesquiterpene Lactone with Relevant Biological Activity," Organic Chemistry International, vol. 2011, Article ID 170196, 35 pages, 2011. doi:10.1155/2011/170196

CHAPTER 11

L. Coleman, S. Varghese, O. P. Tripathi, S. G. Jennings, and C. D. O'Dowd, "Regional-Scale Ozone Deposition to North-East Atlantic Waters,"

INDEX

V

X